Jeff Siegel
A Grammar of Nama

Pacific Linguistics

Managing editor
Alexander Adelaar

Editorial board members
I. Wayan Arka
Danielle Barth
David Bradley
Don Daniels
Bethwyn Evans
Nicholas Evans
David Nash
Bruno Olsson
Bill Palmer
Andrew Pawley
Malcolm Ross
Dineke Schokkin
Jane Simpson

Volume 668

Jeff Siegel

A Grammar of Nama

A Papuan Language of Southern New Guinea

DE GRUYTER
MOUTON

ISBN 978-3-11-107661-4
e-ISBN (PDF) 978-3-11-107701-7
e-ISBN (EPUB) 978-3-11-107711-6
ISSN 1448-8310

Library of Congress Control Number: 2023934442

Bibliographic information published by the Deutsche Nationalbibliothek
The Deutsche Nationalbibliothek lists this publication in the Deutsche Nationalbibliografie;
detailed bibliographic data are available on the internet at http://dnb.dnb.de.

© 2023 Walter de Gruyter GmbH, Berlin/Boston
Cover image: Jeff Siegel
Typesetting: Integra Software Services Pvt. Ltd.
Printing and binding: CPI books GmbH, Leck

www.degruyter.com

In memory of Majam (Tony) Emoia
12 February 1963 – 18 March 2022

Acknowledgements

My biggest thanks go to Murry Dawi, who first connected me with the Nama-speaking community and started teaching me his language when he was visiting Australia in the 1990s. Thanks also to the many Nama speakers who worked with me when I visited PNG. First and foremost is Tony Emoia, who left this world far too early in March 2022. The other main consultants were all members of the Dawi family: Murry, Yoshie, Francis, Mawai, and Les. Thanks also go to other consultants who worked with me for shorter times, including Daniel Dawi, Eli Dawi, Joy Dawi, Ezekiel Erick, Daink Kuro, Kemsy Kuro, Gaita Mawari, Stanley Mawari, Zephaniah Mawi, Harry Murry, Guma Subam and Warapa Samadari.

On my visits to Morehead station, wonderful hospitality was given to me by Yoshie Dawi, Joy Dawi and her then husband Kwari Kapa, and Murry Dawi and his wife Naka Pawar. In Daraia village I was also housed and amply fed by Mawai Dawi and Francis Dawi and their families.

Support for this research came from an Australian Research Council Discovery Grant: "The Languages of Southern New Guinea" (DP110100307). Many thanks to Nick Evans for inviting me to take part in the project and for his enthusiastic support for its duration. Thanks also to other members of the project – especially Christian Döhler, Eri Kasihma, Julia Miller, Matthew Carroll and Dineke Schokkin – for their help and friendship.

I'd like to thank Kurang Wekum and Sali Subam for assisting me when transiting through Port Moresby.

Much gratitude to my research assistant and eagle-eyed proof-reader in Armidale, Vicki Knox, who unknowingly has become an expert in the Nama language. Also, I am indebted to Christian Döhler and Nick Evans for their detailed and insightful comments on the first draft of this grammar, and to another anonymous reviewer.

And finally, my heartfelt gratitude to my son, Ben Siegel, and my wife, Diana Eades, who supported and encouraged me during this project. Ben, a Nama speaker himself, helped me fill in many gaps back in Armidale. And Diana put up with me during the Covid-19 outbreak when I was working on this grammar like a manic PhD student.

https://doi.org/10.1515/9783111077017-202

Contents

Acknowledgements —— VII

List of tables —— XV

List of figures —— XVII

Abbreviations of grammatical terms —— XIX

1   **Introduction** —— **1**
1.1   The Nama language —— **1**
1.2   Geographical and cultural context —— **4**
1.2.1   Landscape —— **4**
1.2.2   The Nama-speaking people —— **7**
1.2.3   The importance of yams —— **12**
1.2.4   Modern history and development —— **17**
1.2.5   Language use —— **20**
1.3   Data and fieldwork —— **21**
1.3.1   Background —— **21**
1.3.2   Fieldwork —— **23**
1.3.3   Sources of data —— **25**
1.3.4   Presentation of examples from the data —— **28**
1.4   Grammatical overview and organisation of this grammar —— **28**
1.4.1   Morphological processes —— **29**
1.4.2   Syntax —— **32**
1.4.3   Organisation —— **32**

2   **Phonology** —— **34**
2.1   Consonants —— **34**
2.1.1   Stops —— **35**
2.1.2   Fricatives and affricates —— **35**
2.1.3   Nasals —— **36**
2.1.4   Liquids and glides —— **38**
2.1.5   The velarised bilabials —— **38**
2.1.6   Geminate consonants —— **39**
2.2   Vowels —— **39**
2.2.1   Monophthongs —— **39**
2.2.2   The status of /ə/ —— **41**
2.2.3   Diphthongs —— **42**

X —— Contents

2.3	Orthography —— **44**	
2.4	Phonotactics —— **45**	
2.4.1	Syllable structure —— **45**	
2.4.2	Consonant clusters within the syllable —— **45**	
2.4.3	Stress —— **46**	
2.4.4	Vowel syncope and epenthesis —— **47**	

3	**Nominal morphology —— 51**	
3.1	Morphophonemic processes —— **51**	
3.2	Number marking suffixes —— **51**	
3.2.1	Nonsingular marking *-af, -f, -of* —— **51**	
3.2.2	Singular marking *-o* —— **53**	
3.3	Major case-marking suffixes —— **54**	
3.3.1	Ergative (ERG) *-am, -m* —— **54**	
3.3.2	Dative (DAT) *-e* —— **57**	
3.3.3	Instrumental (INST) *-e* —— **60**	
3.3.4	Genitive (GEN) *-ne* —— **60**	
3.3.5	Comitative (COM) *-afè, -fè* —— **63**	
3.3.6	Privative (PRIV) *-ofnar, -afnar* —— **65**	
3.3.7	Locative —— **67**	
3.3.8	Ablative (ABL) *-ta* —— **71**	
3.3.9	Perlative (PERL) *-mè* —— **73**	
3.3.10	Allative (ALL) *-t* —— **74**	
3.3.11	Purposive (PURP) *-ot* —— **77**	
3.3.12	Associative (ASS) *-faf* —— **79**	
3.3.13	Originative (ORIG) *-mèn* —— **82**	
3.3.14	Semblative (SEMB) *-nit* —— **85**	
3.3.15	Summary —— **85**	
3.4	Peripheral case-marking suffixes —— **85**	
3.4.1	Attributive —— **85**	
3.4.2	Temporal —— **87**	
3.5	Exclusive and restrictive suffixes —— **88**	
3.5.1	Exclusive (EXCL) *-yo* —— **89**	
3.5.2	Restrictive (RSTR) *-ro* —— **93**	
3.6	Affixes in word formation —— **97**	
3.6.1	Case suffixes used in word formation —— **97**	
3.6.2	Derivational nominal affixes —— **98**	
3.7	Possessive prefixing —— **100**	
3.8	Reduplication —— **102**	

4	**Verbal morphology** —— **105**	
4.1	Nominalising suffix (nom) -*gh* —— **105**	
4.2	Core argument indexing affixes —— **107**	
4.2.1	A- and S_A-indexing suffixes —— **107**	
4.2.2	P- and S_P-indexing prefixes —— **107**	
4.3	Morphological subclasses —— **112**	
4.4	Deictic prefixes —— **112**	
4.5	Other verbal prefixes —— **115**	
4.5.1	Reflexive/reciprocal prefix —— **115**	
4.5.2	Applicative prefix —— **116**	
4.5.3	Autobenefactive prefix —— **120**	
4.5.4	Ordering of prefixes —— **121**	
4.6	Perfectivity marking —— **121**	
4.6.1	Imperfective vs perfective aspect —— **121**	
4.6.2	Morphophonemic changes —— **123**	
4.7	Imperfective tenses and aspects —— **127**	
4.7.1	Imperfective current and recent tenses —— **128**	
4.7.2	Imperfective remote tense —— **133**	
4.8	Perfective tenses and aspects —— **135**	
4.8.1	Inceptive tenses —— **135**	
4.8.2	Remote punctual tense —— **141**	
4.8.3	Remote inceptive tense —— **142**	
4.9	Inceptive imperfective aspect —— **144**	
4.10	Summary of tense/aspect marking —— **146**	
4.11	P-Aligned intransitive verbs —— **147**	
4.11.1	Differences between P-aligned intransitives and the other verb types —— **148**	
4.11.2	The copula and related forms —— **151**	
4.12	Valency reduction: A-aligned intransitive —— **156**	
4.12.1	Antipassive —— **156**	
4.12.2	Anticausative —— **157**	
4.12.3	Reflexive/reciprocal —— **159**	
4.12.4	Autocausative —— **160**	
4.12.5	Multiple functions —— **161**	
4.12.6	Discussion —— **162**	
4.13	Valency reduction: P-aligned intransitive —— **164**	
4.13.1	Statives —— **164**	
4.13.2	Transitive verbs with both P-aligned and A-aligned forms —— **166**	
4.13.3	Distinguishing P-aligned intransitives from transitives —— **168**	

XII —— Contents

5 **Additional word classes and phrase structure** —— **170**
5.1 Nominal phrase: Subclasses of nominals —— **170**
5.1.1 Nouns —— **170**
5.1.2 Pronouns —— **178**
5.1.3 Demonstrative —— **179**
5.1.4 Quantifiers —— **181**
5.1.5 Temporal nominals —— **187**
5.2 Nominal phrase structure —— **190**
5.3 Verb phrase: Other word classes —— **190**
5.3.1 Tense marker —— **190**
5.3.2 Aspect markers —— **192**
5.3.3 Modals (modality markers) —— **196**
5.3.4 Combinations of TAM markers —— **202**
5.3.5 Quasi-modals —— **203**
5.3.6 Adverbs (ADV) —— **210**
5.4 Verb phrase clitics —— **215**
5.4.1 Proximal clitics (PROX) —— **215**
5.4.2 Mirative proclitic (MIR) —— **217**
5.5 Verb phrase structure —— **218**

6 **Simple sentences** —— **220**
6.1 Simple sentence structure —— **220**
6.1.1 P-focussed constructions —— **221**
6.1.2 Phrasal coordination —— **224**
6.1.3 Multi-verb construction —— **225**
6.2 Declarative sentences —— **226**
6.3 Negative sentences —— **226**
6.4 Interrogative sentences —— **227**
6.4.1 Yes/no questions —— **227**
6.4.2 Question word questions —— **229**
6.5 Imperative sentences —— **234**
6.5.1 Immediate imperatives —— **234**
6.5.2 Future imperatives —— **241**
6.5.3 Negative imperatives —— **244**
6.5.4 Imperatives with deictic prefixes —— **245**
6.5.5 Special imperative forms —— **246**
6.5.6 Benefactive imperatives —— **247**
6.5.7 Third person imperatives —— **250**
6.6 Minor sentences —— **251**
6.7 Exclamative sentences —— **257**

Contents — **XIII**

6.8	Discourse markers in simple sentences — **258**	
6.9	Morphological discord as a stylistic device — **262**	

7   **Compound and complex sentences — 265**
7.1   Compound sentences — **265**
7.2   Complex sentences with adverbial clauses — **267**
7.2.1   Adverbial clauses of time — **267**
7.2.2   Adverbial clauses of place — **272**
7.2.3   Concessive adverbial clauses — **273**
7.2.4   Adverbial clauses of purpose — **275**
7.2.5   Conditional adverbial clauses — **277**
7.2.6   Adverbial clauses of reason — **279**
7.3   Complex sentences with relative clauses — **280**
7.4   Complex sentences with complement clauses — **287**
7.4.1   Complement clauses introduced by *nde* 'that' — **287**
7.4.2   Complement clauses without a complementiser — **290**
7.4.3   Complement clauses with an interrogative/relative pronoun or adverb — **291**
7.5   Complex sentences with focus marking clauses — **295**

8   **Typological implications — 299**
8.1   Relative clauses — **299**
8.2   Verbs — **300**
8.2.1   Tense — **300**
8.2.2   Perfectivity — **301**
8.2.3   Dual versus nondual — **302**
8.2.4   Morphological discord — **304**

Appendix A: Dialectal differences — 309

Appendix B: Recordings — 311

Appendix C: List of grammatical morphemes — 313

Appendix D: Sample text — 319

References — 343

Index — 347

List of tables

Table 1.1	Numbers used in yam counting	16
Table 1.2	Numbers used in the yam counting ceremony	16
Table 1.3	Dates of fieldwork and locations	24
Table 1.4	Major case marking suffixes in Nama	29
Table 2.1	Consonant phonemes in Nama	34
Table 2.2	Distribution of stop phonemes in Nama	35
Table 2.3	Minimal/subminimal pairs for stops in Nama	36
Table 2.4	Distribution of fricative and affricate phonemes in Nama	36
Table 2.5	Minimal/subminimal pairs for fricatives and affricates in Nama	37
Table 2.6	Distribution of nasal phonemes in Nama	37
Table 2.7	Minimal/subminimal pairs for nasals in Nama	37
Table 2.8	Distribution of liquids and glides in Nama	38
Table 2.9	Minimal/subminimal pairs for liquids and glides in Nama	38
Table 2.10	Vowel phonemes in Nama	40
Table 2.11	Distribution of vowels in Nama	40
Table 2.12	Minimal pairs for vowels in Nama	41
Table 2.13	Groups of words showing vowel differences in Nama	41
Table 2.14	Minimal pairs with and without initial and final /ə/	42
Table 2.15	Diphthongs in Nama	43
Table 2.16	Distribution of diphthongs in Nama	43
Table 2.17	Minimal/subminimal pairs for diphthongs in Nama	43
Table 2.18	Syllable types in Nama	45
Table 2.19	Consonant clusters as onsets in Nama lexical items	46
Table 2.20	Consonant clusters with proclitic markers on verbs	46
Table 2.21	Examples of stress patterns in Nama	47
Table 3.1	Nonsingular nominals with -*af*	52
Table 3.2	Nonsingular nominals with -*f*	52
Table 3.3	Plural vs paucal	52
Table 3.4	Ergative and absolutive pronouns in Nama	57
Table 3.5	Dative pronouns in Nama	58
Table 3.6	Number-marked genitives	62
Table 3.7	Possessive pronouns in Nama	63
Table 3.8	Singular/nonsingular distinctions in comitative case	64
Table 3.9	Comitative pronouns in Nama	65
Table 3.10	Ablative suffix with number-marked nominals	72
Table 3.11	Locational/positional nominals marked by -*ta*	72
Table 3.12	Paucal/plural distinctions with allative -*t*	75
Table 3.13	Number-marked associatives	82
Table 3.14	Summary of major case-marking suffixes in Nama	86
Table 3.15	Adverbs formed by the use of case markers	97
Table 3.16	Conjunctive adverbs formed by the use of case markers	98
Table 3.17	Free and possessed bound forms of relationship nominal	101
Table 3.18	Relationship nominals that have no free form	101
Table 4.1	A/S_A-indexing suffixes for imperfective current and recent tenses	107

https://doi.org/10.1515/9783111077017-204

XVI —— List of tables

Table 4.2	P/S$_P$-indexing and ØP prefixes —— **108**	
Table 4.3	P-indexing prefixes plus the applicative prefix —— **116**	
Table 4.4	Prefix ordering in Nama —— **121**	
Table 4.5	Perfectivity markers in Nama —— **122**	
Table 4.6	Suffixes for the imperfective current and recent tenses —— **128**	
Table 4.7	Verb endings with remote delimited suffixes —— **133**	
Table 4.8	Verb endings with remote durative suffixes —— **134**	
Table 4.9	Current inceptive verb endings —— **136**	
Table 4.10	Verb endings with perfective remote punctual suffixes —— **141**	
Table 4.11	Verb endings with remote inceptive suffixes —— **143**	
Table 4.12	Verb endings with inceptive imperfective suffixes —— **144**	
Table 4.13	Prefix sets for tense/aspect categories —— **147**	
Table 4.14	Remote tense marking (nondual) —— **150**	
Table 4.15	Forms of the copula in Nama —— **154**	
Table 4.16	Some transitives with corresponding antipassive intransitives —— **158**	
Table 4.17	Some transitives with corresponding anticausative intransitives —— **159**	
Table 4.18	Some "deponents" in Nama —— **163**	
Table 4.19	Some transitive and corresponding P-aligned intransitive verbs —— **166**	
Table 4.20	Some verbs with corresponding forms in all three verb classes —— **167**	
Table 5.1	TAM markers and quasi-modals in Nama —— **191**	
Table 6.1	Immediate imperative forms for \tar/'dig' —— **242**	
Table 6.2	Immediate and future imperative forms for \fenè/ 'feed' —— **243**	
Table A.1	Dialectal differences in Nama —— **309**	
Table B.1	Recordings of narratives —— **311**	
Table B.2	Daraia clips —— **312**	

List of figures

Figure 1.1	Location of the Nama language —— 1
Figure 1.2	Morehead District centre and Ngaraita village —— 2
Figure 1.3	Locations of the Yam family languages —— 3
Figure 1.4	Morehead River near Morehead station —— 4
Figure 1.5	Savanna —— 5
Figure 1.6	House in Mata village —— 6
Figure 1.7	An old woman (the late Kafuk Kuti) carrying a load —— 8
Figure 1.8	Earth ovens —— 9
Figure 1.9	Woven blind —— 10
Figure 1.10	Examples of material culture —— 11
Figure 1.11	Singsing —— 12
Figure 1.12	Namesake ceremony —— 13
Figure 1.13	Identifying types of yams —— 14
Figure 1.14	Yam garden —— 15
Figure 1.15	Yam storage house —— 15
Figure 1.16	Morehead Highway —— 18
Figure 1.17	Newly constructed Mata-Daraia Road (2011) —— 20
Figure 1.18	Murry Dawi and his grandson (2012) —— 22
Figure 1.19	Consultants (working on fish names) —— 26
Figure 2.1	Nama vowel space —— 39
Figure 8.1	Four types of grammatical distinctions based on number (Croft 2004) —— 303
Figure 8.2	Dual/nondual grammatical distinction —— 303

https://doi.org/10.1515/9783111077017-205

Abbreviations of grammatical terms

1	first person	LOC	locative
2	second person	MIR	mirative
3	third person	MOD	modal
A	A argument	ND	nondual
ABIL	abilitative	NEG	negative
ABL	ablative	NOM	nominaliser
ABS	absolutive	NSG	nonsingular;
ADV	adverb	OBL	obligative
ALL	allative	ORIG	originative
AND	andative	P	P argument
APP	applicative	PA	P-aligned;
ASS	associative	PERL	perlative
ATT	attemptive	PERM	permissive
ATTR	attributive	PFV	perfective
AUTO	autobenefactive	PL	plural
CHAR	characterised	PM	predicate marker
COM	comitative	POSB	possibility
CONT	continuative	POT	potential
COP	copula	PRF	perfect
CUR	current tense	PRIV	privative
DA	dual argument	PROL	prolonged action
DAT	dative	PROX	proximate
DEM	demonstrative	PUNC	punctual
DLT	delimited	PURP	purposive
DU	dual	QUES	question marker
DUB	dubitative	REAL	realis
DUR	durative	REC	recent
EMPH	emphatic	REFL	reflexive/reciprocal
ERG	ergative	REM	remote tense
EVID	evidential	RSTR	restrictive
EXCL	exclusive	S_A	S argument of A-aligned intransitive
FUT	future	S_P	S argument of P-aligned intransitive
GEN	genitive	SEMB	semblative
IMP	imperative	SG	singular
INC	inceptive	TA	tense aspect
INCH	inchoate	TAM	tense aspect modality
IPFV	imperfective	TEMP	temporal
INT	intentional	TR	transitiviser
INTJ	interjection	VAL	validative
INTR	intransitive	VEN	venitive
INV	investigative	ØP	prefix indicating absence of a P argument
IRR	irrealis		

https://doi.org/10.1515/9783111077017-206

1 Introduction

1.1 The Nama language

Nama (pronounced ['nəmə]) is a Papuan (non-Austronesian) language spoken near the district centre of Morehead in the South Fly District of the Western Province of Papua New Guinea (PNG). This is in the Morehead Rural LLG (Local Level Government area), approximately 100 km east of the border with Indonesia and 80 km north of the Torres Strait. In the international standard for representing names of languages (ISO 639-3) its code is nmx.

Nama has approximately 1200 speakers living in three villages: Daraia (8.61637°S 141.733576°E), Mata (8.674546°S 141.743133°E) and Ngaraita (8.599511°S 141.714869°E).[1] The location of where Nama is spoken is shown in Figure 1.1. The Morehead District centre and Ngaraita village are shown in Figure 1.2.

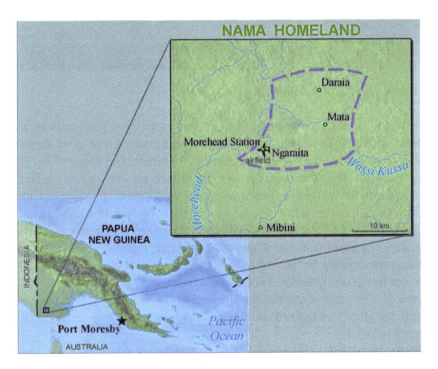

Figure 1.1: Location of the Nama language.
(https://joshuaproject.net/people_groups/19557/pp)

---

[1] Note that for Nama speakers, the first consonant of Daraia is actually [ɣ] rather than [d].

https://doi.org/10.1515/9783111077017-001

Figure 1.2: Morehead District centre and Ngaraita village.[2]

Nama has two dialects. One is spoken in Ngaraita, which is just northeast of Morehead station (as the administrative centre is referred to). The other dialect is spoken in both Mata, 12 km to the northeast of Ngaraita, and Daraia, 7 km to the north of Mata. The differences between the dialects are only in a few lexical items (see Appendix A). However, people living in Ngaraita are in close contact with speakers of neighbouring languages, especially Namat (spoken in Mibini, 16 km to the south), who now live near the administrative centre. As a result, Ngaraita people often mix words from Namat and other languages into their speech. Therefore, Nama speakers say the "purest" form of their language is spoken in Mata and Daraia, and this dialect is the focus of this grammatical description.

Nama is a member of the Nambu subgroup of the Yam family of languages (formerly the Morehead-Upper Maro family). The location of these languages is shown in Figure 1.3.

The first published references to the Nama language are in the works of anthropologists who focussed on Nama speakers' closest neighbours. Williams (1936)

---

2 The airstrip, school (blue buildings) and administrative centre are upper left and Ngaraita village spread out on the right.

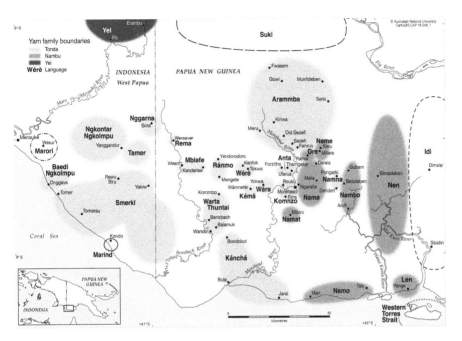

Figure 1.3: Locations of the Yam family languages.

studied the Keraki people to the east (who speak Nambo), and Ayres (1983) the Farem people to the west (Komnzo speakers). Both of these works remain as the most detailed ethnographic descriptions of people of the Morehead District.

I began the first linguistic study of Nama in the 1990s (see section 1.3.1 below), and a few examples from the language appeared in Crowley, Lynch, Siegel and Piau (1995). A trial orthography was developed for Nama at a workshop run by SIL International (Price 2000). This orthography was then used in teaching initial literacy in Nama to children in elementary schools in Daraia, Mata and Ngaraita. (This lasted until 2014, when the national government returned to an English-only policy.)

My more intensive research on the language started in 2011 and recent publications describe the verbal system (Siegel 2014, 2017) and relative clauses (Siegel 2019). Bible translators from SIL have been coming to Daraia for ten years, and in consultation with the community have produced a translation of the gospel of Mark (Youngdu Church 2018). An areal survey (Evans et al. 2017) provides information about Nama's relationship with other languages of southern New Guinea. Carroll (2020) provides a survey of the morphology of the Yam languages. Other Yam family languages that have been studied in detail are Komnzo (Döhler 2018), Ngkolmpu (Carroll 2016), Nambo (Kashima 2020) and Nen (Evans 2009, 2012a, 2012b, 2014, 2015a, 2015b, 2017, 2019a, 2019b; Evans & Miller 2016).

This rest of this chapter begins by presenting the geographical and cultural context in which the Nama language is spoken. Then fieldwork, data collection and sources for this grammar are described. The chapter ends with an overview of the grammar of the language, and an outline of the following chapters.

1.2 Geographical and cultural context

1.2.1 Landscape

The Morehead District is in the centre of the southern New Guinea region located southwest of the Fly River, the second largest river in PNG. The dominating feature of the landscape of the area where Nama is spoken is the Morehead River (Figure 1.4), which flows south to Ngaraita, then west past Rouku (where Komnzo is spoken) and south again for approximately 80 km into the Torres Strait.

Figure 1.4: Morehead River near Morehead station.

Morehead District, and the Trans-Fly region in general, is geographically very different from the rest of Papua New Guinea. First, it is flat, mostly below 50 metres above sea level. Even on the clearest day no mountains can be seen in the distance.

Second, it is characterised by savanna grasslands, with areas of eucalyptus, albizzia and melaleuca woodland (see Figure 1.5). There are also more sparsely vegetated areas dotted with red or brown termite mounds. With acacia and banksia also abundant, the similarity to northern Australia is striking. Other ecological zones are also found: thick patches of monsoonal rain forest and swampy wetlands.

Figure 1.5: Savanna.

Many kinds of fauna are also reminiscent of northern Australia, including wallabies, pademelons, echidnas, bandicoots, cassowaries, large goannas, crocodiles and taipan snakes. There are also wild pigs and introduced deer that have crossed the border from Indonesia. Birdlife is abundant and rivers, creeks and swamps are full of fish and shellfish.

Also like northern Australia, Morehead has a monsoonal climate, with a distinct wet season from approximately January to April and a dry season from June to November. Large areas along the Morehead River are under water in the wet season, and most of the small creeks have little if any water at the end of the dry season.

Another feature of the landscape is the fenced gardens people establish to grow food (section 1.2.2). A system of shifting agriculture is in place, where a plot of land is cleared for a garden, and after the harvest left fallow for at least seven years. Because of the low population density, people have often been clearing new areas of land rather than going back to old garden plots. However, people have now become aware of the shrinking areas of natural vegetation, and the resulting loss of nearby wildlife, and young people have started to promote conservation of uncultivated areas.

People live permanently in villages on higher ground that does not become flooded in the wet season. Individual families have a compound that includes their house and an outside kitchen (see Figure 1.6). The compounds are usually situated along tracks or roads, and boundaries are lined with decorative plants such as crotons and heliconia. In addition to villages, there are many "garden places", such as Sarghar near Daraia, where people have houses and stay temporarily while working in nearby gardens.

Figure 1.6: House in Mata village.

1.2 Geographical and cultural context — 7

1.2.2 The Nama-speaking people

Nama speakers are divided into three groups, which they call "tribes" in English and *mèrèn* in Nama (also meaning 'family'). These are associated with ownership in particular locations: *Tèndáf* (Mata), *Walal* (Daraia) and *Fesua* (Ngaraita). However, some younger people from Daraia have started referring to their tribe as *Sènkomèngo*.[3]

Like other groups in the Morehead District, all Nama speakers belong to one of three sections, which they call "clans" in English and *tèfèn* in Nama (also meaning 'base or stump of a tree' or 'origin'). The sections are Mbangu, Maiawè and Sángárè. They are often described as ordered, with Mbangu first or at the front, Maiawè in the middle. and Sángárè last or at the rear. The sections are also described using a tree analogy: Mbangu are *sèrèmb ár* ('root people'), Maiawè *fègh ár* 'trunk people' and Sángárè are *njámbám ár* 'treetops people'. Each section is said to have descended from a different ancestor: Mbangu from Aikumi, Maiawè from Tèmngokor and Sángárè from Ndimbèn.

Customarily, a person is not allowed to marry anyone from their own tribe or their own clan (i.e. section), or from a neighbouring tribe or clan. In Nama, the term *njarar* refers to a permissible marriage partner. For a Mbangu person, a *njarar* can be from the Sángárè clan or from part of the Maiawè clan. Similarly, for a Sángárè person, a *njarar* can be from most of the Mbangu clan or from a different part of the Maiawè clan. However, this prohibition reflects an ideal rather than actual practice, especially in recent decades.

For Nama speakers, as throughout the Morehead District, the ideal system of marriage is what people call "sister exchange" in English. When a man wants to marry a woman from another tribe and clan, he has to provide a sister, or another close female relative of the same generation, for that woman's brother to marry. So two couples get married at the same time. (This does not always happen, but it is still frequent, and the older generation describe it as a pillar of traditional culture.) As in neighbouring languages, Nama has special kin terms used for those involved in an exchange marriage. The term *-mbrè* (with a possessive prefix), usually translated as 'exchange man', refers to the brother (or close male relative) of the woman in the exchange – e.g. *yámbrè* 'his exchange man'. The term *-blalè* 'exchange girl or woman' refers to the sister (or close female relative) of the man – e.g. *yámblalè* 'her exchange girl'. So, both men in an exchange marriage call each other *tambrè* 'my exchange man', and both women call each other *tamblalè* 'my exchange woman'.

---

3 In Nama orthography, <è> represents /ə/, <á>/æ/, <f> /ɸ/ and <gh> /ɣ/. (See section 2.2.)

The relationship term *mwitè* is used for the children of both couples, who are special cousins.

In daily life, there is little physical separation of the genders in Nama-speaking society, and men and women socialise and eat together. Both men and women work in the gardens and go fishing, and men play a large part in raising and looking after children after they are able to walk. However, other roles in society are gender-specific. Women do all the cooking and are generally responsible for fetching water. Like other areas of PNG, women carry heavy loads, such as baskets of produce, using straps over their forehead (see Figure 1.7). Men build the houses and generally do the hunting, although women may be involved in communal hunting by making noise and chasing the game toward the men. However, in terms of power, the men are clearly in charge.

Figure 1.7: An old woman (the late Kafuk Kuti) carrying a load.

Nama speakers are almost entirely self-sufficient with regard to food. The main vegetable staples are root crops which are grown in their gardens. In order of importance, these are yams, cassava, taro and sweet potatoes. Also in the diet are coconuts, cooking bananas, sago and sugar cane. (For those who have money and access to a trade store, rice may also be part of the diet, as well as instant noodles, and flour is sometimes used to make scones or doughnuts.) Local leafy green vegetables, such as *aibika*, are sometimes cultivated. Seasonal fruits include mangoes, pineapples and soursop. Fish is the most common protein in the menu, and over 60 different types of fish have been identified. Sources of meat, which are normally acquired by hunting, include wallabies, deer, cassowaries, bandicoots, pademelons, wild pigs and some kinds of pythons and goannas. Baby pigs and cassowaries are occasionally captured, brought back to the village and raised in pens until they are big enough to be eaten for a special occasion.

Fishing is still done with traditional nets and fish traps, as well as with store-bought hooks and fishing line. Spears with wire prongs are also used. Hunters generally use homemade spears and bows and arrows. Bows are made of bamboo (the same word *námb* is used for both), and arrows and spears are made from bamboo or reeds, often with tips made from scrap metal. Knives and machetes are now store-bought. Dogs are frequently used for hunting. Animals are also trapped using traditional spring and deadfall traps.

Cooking is done either on an open wood fire, using store-bought pots and pans, or in an earth oven. Because of the lack of any rock or stones in the region, chunks of termite mounds are used in earth ovens. (The word *karèf* means both 'termite mound' and 'earth oven'.) Unlike in other areas, earth ovens are not dug into the ground; rather layers of termite mound chunks are heated with fire and then covered by sheets of tree bark (see Figure 1.8).

Figure 1.8: Earth ovens.

Root vegetables are boiled in water, sometimes with the addition of coconut cream. For special occasions, cassava is grated, mixed with coconut cream, wrapped in banana leaves and cooked in an earth oven. This is called *kwèrmè*.

Houses are still built entirely out of locally acquired materials, except for nails. Only some government buildings and teacher's quarters at Morehead District School are constructed with manufactured materials and have corrugated iron roofs. Traditional materials used for houses include planks and poles from a variety of trees, flooring made from a kind of pandanus, blinds and walls made from bark, bamboo or woven spines of sago fronds, and roofing made from tree bark or sago fronds. Various traditional patterns are used for the woven blinds (see Figure 1.9).

Figure 1.9: Woven blind.

Other items of material culture, in addition to the blinds and hunting weapons already mentioned, are dugout canoes, a variety of types of baskets, clay pots, bamboo flutes and panpipes, wooden drums with animal- or lizard-skin heads (called *kundu* in PNG), headdresses made out of cassowary feathers used in traditional singing and dancing (called *singsing* in PNG) and grass skirts, used for traditional occasions. All other clothing is modern and store-bought. (See Figure 1.10.)

With regard to cultural practices, all Nama speakers that I have met are Christians of various denominations, including the Evangelical Church of Papua, Seventh Day Adventists, the United Church and CLC (Christian Life Church). Weekly church services and large church gatherings of people from different villages are

1.2 Geographical and cultural context — 11

Figure 1.10: Examples of material culture.
(clockwise from top left: kundu drum, cassowary feather headdress, grass skirt, panpipes, fish trap. baskets)

very popular. Nevertheless, many customs and beliefs from the past remain. One of these is the widespread belief in sorcery, and that the actions of a sorcerer can be the cause of a death.

Because of Christian missionaries, singsings (traditional singing and dancing) were taboo for many years. However, in the 1990s there was a resurgence of these practices and traditional singsings are now performed on PNG Independence Day (16 September) and at other special occasions (see Figure 1.11).

Figure 1.11: Singsing.

One custom that has continued without interruption is the *tagase awerègh* 'namesake holding ceremony'. This occurs with a mother, her child who has reached the age of one year and the child's namesake. The mother pretends to be handing the child over to the namesake but then pulls him/her back when the namesake reaches out to hold him/her. This happens twice. But the third time the mother does hand over the child. The namesake lifts the child high up and says *Fèm yabun so namote!* 'May you grow big!', and then holds the child down to the ground and then cuddles the child (see Figure 1.12).

1.2.3 The importance of yams

As in the rest of the Morehead District, yams play a very important part in Nama-speaking society. As already mentioned, yams are the main staple, and they are also used in ceremonial exchanges, for example between the two families involved in an exchange marriage. A person's standing in the community is often determined by the quantity and size of the yams they can grow.

Figure 1.12: Namesake ceremony.

Nama speakers have identified 54 different named varieties of yams. Figure 1.13 shows one session in which people laid out different varieties and told me their names, colour inside (white or purple) and what category they belong to. The general term for yams is *wem*. But there is a special category called *mai*, usually long rather than round and often purple inside. These yams are considered the tastiest and are eaten on special occasions. Of the 54 named varieties, 13 are *mai*. Yams are also categorised into three different types according to how they are planted. For one type, mounds are built up and one big yam or 3 or 4 small ones are planted. The other two types are just planted in a hole in the ground. For one of them, leaves are put in the hole before planting; for the other, no leaves are used.

Planting is done from as early as October to as late as January. Stakes are used for the yam vines to climb up as they grow (see Figure 1.14). Yams are harvested from June to August.

Yam are also valued over other root crops because they can be stored and used through the rainy season when other food is in short supply. This is done in a *wembègh mèngo* 'storage house' which is especially used for storing yams (see Figure 1.15).

As in other Yam family languages, Nama has a complex base-6 (or senary) system for the ceremonial counting of yams (see Donohue 2008; Hammarström 2009; Evans 2009; Carroll 2016; Döhler 2018). This system is surprising, considering that the language has only three separate morphemes for numbers normally used for everyday counting: *ámbiro* 'one', *sembyo* 'two' and *nambyo* 'three'. Combinations of these are used for higher numbers: *sembyo sembyo* 'four', *sembyo nambyo* 'five'. (In everyday counting,

Figure 1.13: Identifying types of yams.

1.2 Geographical and cultural context — 15

Figure 1.14: Yam garden.

Figure 1.15: Yam storage house.

16 —— 1 Introduction

the term *ngangè brubru* can also be used, referring to the fist and the number 5.) Nowadays, borrowings from English are most commonly used for the numbers 5 and above.

In the base-6 yam counting system, however, there are words for six, six to the power of 2, six to the power of 3, etc, going up to six to the power of 5 (6^5), which is 7776 in the decimal system. These are shown in Table 1.1.

Table 1.1: Numbers used in yam counting.

Numbers	Base-6	Value		
si	6		=	6
fètè	(= 6 *si*)	= 6^2	=	36
tarumbè	(= 6 *fètè*)	= 6^3	=	216
ndamno	(= 6 *tarumbè*)	= 6^4	=	1296
wárámákè	(= 6 *ndamno*)	= 6^5	=	7776

The number of *si* (groups of six) is indicated by the usual numbers (*ámbiro, sembyo*, etc.), but when ceremonial counting is taking place, the different groups of six have special names that are called out. These are shown in Table 1.2.

Table 1.2: Numbers used in the yam counting ceremony.

Numbers	Value	Ceremonial number	Value	
ámbiro si	= 1 x 6	*ñambi*	=	6
sembyo si	= 2 x 6	*yèndè*	=	12
nambyo si	= 3 x 6	*yèro*	=	18
sembyo sembyo si	= 4 x 6	*ásár*	=	24
nambyo sembyo si or *witma brubru si*	= 5 x 6	*tambèno*	=	30
nambyo nambyo si or *witma brubru ámbiro si*	= 6 x 6	*nimbo*	=	36

I have not been able to witness a yam counting ceremony, but from what I have been told it proceeds as follows. Two men each carry three yams from a heap and place them on the ground. The word *ñambi* is called out, indicating the first group of 6. Then the two men carry three more yams each and put them on top of the others; *yèndè* is called out, indicating the second group of 6 or the total of 12. Six more yams are piled on by the two men again and *yèro* is called out, indicating the third group of six or a total of 18. This continues with the following called out for the next groups of six: *ásár* (fourth group, total 24), *tambèno* (fifth group, total 30) and *nimbo* (sixth group, total 36). The heap now contains one *fètè* (group of 36), and this is marked either by putting a yam next to the pile or by tearing a segment off a coconut frond counter. The whole procedure is repeated, but the groups of 6 yams in the second *fètè* are piled onto the first. This is done until there are six *fètè*, which make up one *tarumbè* (group of 216). This is the basic storage heap. The whole process begins again, with a new heap of yams. Six *tarumbè* (i.e. six separate heaps of 216 yams) then make up a *ndanmo*. According to Williams (1936: 227), in the past this was the minimum amount of yams a "conscientious gardener" needed to store.[4]

1.2.4 Modern history and development

The Australian colonial administration established the first government station (i.e. administrative centre) in the area in 1951, at a place to the west of the current village of Rouku (where the Komnzo language is spoken). A school was set up, run by the London Missionary Society (Döhler 2018: 23). The station was moved to its present location in 1959, where the current airstrip was constructed and government school established. Oil exploration in the area took place in the 1950s by an Australian company, and in the late 1980s by a New Zealand company. The companies constructed the first roads in the district. Although the periods of exploration lasted only a few years, they resulted in many Nama speakers having their first prolonged contact with outsiders and their first paid employment.

Papua New Guinea gained its independence in 1975, but since that time conditions for Nama speakers have generally got worse rather than better, and many people look back nostalgically to the days of the Australian administration. In a study of the Morehead District in the 1990s, Budai Tapari[5] (1995) found there was "a decline in prosperity for the majority of the rural people" (p.17). He blamed this on two things: (1) the failure of both the provincial and national governments to create new infrastructure and opportunities for people to get into the cash economy and (2) the poor work ethic and financial management skills of rural civil servants. He concluded (p.17): "On the whole, roads, health services and government buildings have deteriorated rather than improved in most parts of the District."

In the early 1990s, when I first went to Morehead, it had a police station, a post office, a bank branch (Westpac), and three commercial flights a week from Port Moresby. There was a trade store (Tonda Trading) where people could buy supplies such as rice, flour and sugar at reasonable prices (as these things came by barge from Port Moresby up the Morehead River). When I went back in 2009,

---

4 A video recording of a yam counting ceremony in Rouku village can be seen at https://vimeo.com/21058525.
5 Budai Tawari was from the Bensbach area, southwest of Morehead station and attended Morehead Primary School. He had a PhD in geography from University of Waikato in New Zealand. Sadly, he died an early death in 2003.

none of these existed. Houses and offices built by the Australian government were still being used, but their walls were crumbling and their water tanks had rusted away. One or two small "canteens" sold goods at inflated prices. And the roads had become bush tracks (see Figure 1.16).

Figure 1.16: Morehead Highway.

To get to a bank, trade store or hospital, people now have to go to Daru, the provincial capital, which is on an island in the Torres Strait. This involves first getting to Arufi (35 km away) by walking, bicycle or, if one is lucky, by one of the two or three 4-wheel drive vehicles in the district. From there, one has to hire an aluminium dinghy and travel down the Wassi Kussa River for two hours and then on the open sea of the Torres Strait for another three to four hours. This costs from one thousand to two thousand kina (Australian $500–$1000). (An alternative for shopping is to ride a bicycle to Sota, 100 kms to the west across the Indonesian border.)

Promised economic initiatives, such as farming to produce ti-tree oil, never got off the ground. I have not seen a house in any of the villages or at Morehead station that has running water, or even one with functioning gutters and a tank to collect rain water. The only modern amenity supplied by the government is electricity from a diesel generator, but only from sunset until about 10 pm, and only for houses at the station and in Ngaraita village and nearby settlement areas. With the

Ok Tedi copper and gold mine, the Western Province should be one of the richest in PNG, but it has remained the poorest.

In 2007, a Nama speaker, Sali Subam, was elected to national Parliament as the member from the South Fly Open electorate. He had the best intentions for development of the Morehead District, including building a high school and improving the roads.[6] Early in his tenure, he managed to get new classrooms built at the Morehead Community School, using a Chinese contractor.[7] But the project, which was supposed to include the rebuilding of teachers' quarters, was not completed because of lack of funds. The national government had committed funding for this project and others, such as the high school and road improvement, and sent the money to the provincial government. However, Sali Subam had political enemies in the provincial finance department, and they reportedly refused to release these funds for his projects.

In another example, Subam hired men from Nama-speaking villages and elsewhere to build a new road between Mata and Daraia. This was done with just picks and spades, without any modern equipment, and was completed in 2011 (see Figure 1.17). All that was left was to build small log bridges over several creeks. But funding for the workers was not released, and since they did not get paid, they refused to build the bridges. The road was never used and is now covered with thick forest.

Those projects that have been successful were not funded by the national government. One was the construction of an airstrip near Daraia in 2018. According to people from the village, the SIL Bible translator there told them to build the airstrip for him. (This was reportedly to remove the need to hire a vehicle to travel the 19 km from the Morehead airstrip to Daraia.) Like the Mata-Daraia road, the airstrip was constructed entirely by hand, without any modern equipment, and the people were not paid for their labour. (There are no reports yet of the airstrip ever having been used.)

The most beneficial project was funded by the Papua New Guinea Sustainable Development Program (a not-for-profit limited liability company, incorporated in Singapore). This was the construction in 2011 of a telecommunication tower at Morehead station, one of 48 constructed throughout the province. This has enabled people to use mobile phones, and as there was previously only one land-line telephone in the district, it revolutionised communication. Today, most people have a mobile phone and many are on social media such as Facebook. But reception is

---

6 Landowners were paid K26,000 in the early 1980s for land to build a high school near Morehead (Tapari 1995: 9), but at that time it had still not been built.
7 Although the classrooms are modern structures with corrugated iron roofs, they still do not have gutters or tanks to collect rain water.

Figure 1.17: Newly constructed Mata-Daraia Road (2011).

limited to around Morehead station and Ngaraita, except for a few "hot spots" in or nearby the other villages. People in these villages usually have to go to the district centre to charge their phones, although solar panels are now being used by some.

1.2.5 Language use

The Nama language is spoken by all generations of Walal, Tèndáf and Fesua people living in Daraia and Mata and in or nearby Ngaraita. Intergenerational transmission has not been broken, and no other language has usurped any important traditional communication contexts. However, English and Tok Pisin (the expanded pidgin/creole language of wider communication) are often used in new domains involving education and religion.

At the same time, multilingualism is dominant. Nama speakers, like others in the Morehead District, are most commonly receptive bilinguals. People understand other languages, especially those from nearby groups, but in interactions they speak their own language rather than code-switch. This is partially the result of the fact that women generally marry into a different linguistic group and learn that group's language. Some women switch completely to their husband's language, while others continue to use their first language with their children. In such cases however, the children eventually use only their father's language, even when talking to their mothers.

In interactions with speakers from unknown languages, a lingua franca is used. As the Western Province is in the former colony of Papua, the lingua franca in the Morehead District was Police Motu (also known as Hiri Motu), a pidginised form of the Motu language spoken near Port Moresby. A few loanwords from Motu have come into the Nama language, such as *dibura* 'prison, jail', *nandi* 'stone' and *tuwari* 'comb'.

Police Motu is known today by only a few old people as it was eventually replaced by PNG English (Smith 1978, 1988). Like other indigenised varieties of English, PNG English is characterised lexically by loanwords from indigenous languages (e.g. *kaukau* 'sweet potato'), semantic shift (e.g. *rascal* 'criminal, thug') and grammatical shift (e.g. *he was magic-ing her* 'he was doing sorcery on her'). Morphosyntactically, it does not always have distinctions between mass and count nouns (e.g. *informations*) or between past and nonpast tense (e.g. *yesterday I talk to my sister*). It also has some grammatical constructions that may be confusing to speakers of other varieties of English, such as *use to* marking habitual aspect (e.g. *I use to play football* 'I (regularly) play football').

People from the Western Province have prided themselves in speaking English rather than Tok Pisin, which is used as the lingua franca in other parts of the country. However, in recent years, many people from the Morehead District have lived for a time in Tok Pisin-speaking areas, and speakers of Tok Pisin have moved to Morehead. As a result, knowledge and use of Tok Pisin has been increasing.

1.3 Data and fieldwork

1.3.1 Background

My connection to Papua New Guinea began in 1976, when I took up a teaching position at the PNG University of Technology in Lae. I worked there for six years before going to the Australian National University in 1982 to do my PhD. I joined the University of New England in 1988 and returned to PNG to do research in the early 1990s. My work there at that time was in applied linguistics: examining the effect of using Tok Pisin in formal education in the Ambunti district of the East Sepik Province (Siegel 1997). I also went to PNG in 1995 as part of a research project for AusAid on vernacular education in the South Pacific (Siegel 1996).

My link with Nama speakers and interest in the language precedes the formal study represented in this grammar. Nama is the first language of my former partner,

Yoshie Dawi, who I met in Lae. It is also the first language of our son, Ben, who lived with his mother in the Morehead District until he came to Australia at the age of eight. As he went back to Morehead each summer holiday until he finished high school, and has made many visits since then, he has maintained fluency in the language. And since 2011, when the Morehead telecommunication tower was built, he has been in regular phone contact with his Nama-speaking relatives.

I began linguistic work on Nama in 1993, when Ben's uncle and my good friend, Murry Dawi visited us in Armidale for three months. (See Figure 1.18 for a more recent photo.)

Figure 1.18: Murry Dawi and his grandson (2012).

At that time, I recorded a short story from Murry and collected a few words and sentences, which enabled me to determine the basic structure of the language. During another visit from Murry in 1999, we made more recordings and collected more data in an attempt to analyse the verb morphology. But because of work pressures and frustration over the difficulty of the language, we did not get very far. In 2009, Ben wanted me and my wife, Diana Eades, who had been his Australian stepmother for more than 15 years, to visit Morehead. (I had not been back there since 1994.) In that trip, I did a bit more work on Nama when we were stranded in Arufi, waiting for road transport to Morehead, after having travelled five and a half hours from Daru in a crowded dinghy (see section 1.2.5 above).

In 2010, Nick Evans heard about my connection with Nama and invited me to join the Languages of Southern New Guinea project that he was applying for funding for from the Australian Research Council. The application was successful, and I began fieldwork in 2011.

I should make it clear that I am not a descriptive linguist or a typologist, and that working on Nama has been a retirement project. In my previous life, I was a contact linguist, studying pidgins, creoles and new dialects, with a side applied interest in the use of these varieties in formal education. Doing language documentation has involved a steep learning curve for me and this grammar is a result of my first effort, as I'm sure will be evident from its many shortcomings.

1.3.2 Fieldwork

From 2011 to 2018, I made nine trips to Papua New Guinea to work on the Nama language. Seven of these were to the Morehead District. There I spent time at either Morehead station, staying at teachers quarters, or with my friend Murry Dawi near Ngaraita village, During all but one of these visits, I also spent time in Daraia village, and in one visit, in Mata village as well. Getting to Morehead required going through Daru, the provincial capital. Many Nama speakers live there, and while waiting for transport to Morehead, I was also able to work on the language. Two visits were only as far as Port Moresby. One was planned, as I knew some of my consultants would be there and I didn't have the time to go all the way to Morehead. Another was unplanned, as my flight from Port Moresby to Daru was cancelled at the last minute, which made all ongoing travel arrangements impossible. Luckily, there were several Nama speakers in Port Moresby who I could work with.

I have already described the difficulty and expense of travelling between Daru and Morehead, and my initial travel was by dinghy and road. Fortunately, however, MAF (Mission Aviation Fellowship) started to carry non-church-affiliated passengers between Daru and Morehead at a reasonable cost. Flights were only once a week, and not always reliable, but they made travel much easier. For one fieldtrip, however, the Morehead airstrip was closed because of lack of maintenance, and I flew into Gubam, three hours by car to Morehead on a rough bush track.[8] Dates of fieldwork and locations are shown in Table 1.3.

---

8 In order to catch the early morning flight, the return trip was before sunrise. Since the car had no headlights (or brakes), young men sitting on the bonnet held battery torches to light the way.

24 — 1 Introduction

Table 1.3: Dates of fieldwork and locations.

Dates	Location(s)
14 September – 4 October, 2011	Morehead, Daraia
19 – 26 April, 2012	Port Moresby
16 September – 5 October, 2012	Morehead, Daraia
8 – 16 April, 2013	Morehead
29 August – 22 September, 2013	Morehead, Daraia, Mata
23 – 30 April, 2014	Port Moresby
14 – 24 July, 2015	Morehead, Daraia
5 – 13 January 2017	Morehead, Daraia
6 – 18 August, 2018	Morehead, Daraia

Subtracting travel time, the combined duration of my field work was approximately 16 weeks. Most days of these 16 weeks involved very intense work. My hopes that Nama speakers would want to participate in the documentation of their language were met with a level of enthusiasm beyond my expectations. In each trip I stayed with one or more of my consultants. They arranged for a full day's work on the language, except for when we travelled between Morehead station and the villages. I normally worked with them and others in their houses, although there were times when we walked around to document names of plants and items of material culture . Work usually started right after breakfast (around 8 am) and continued with only a few breaks, including lunch, until around 4.30 pm. Then I had some time to myself before bath time – immersing in a creek when there was water (not in the Morehead River because of crocodiles), or a bucket bath in a structure near water wells). This was followed by dinner, and then work began again, often until past 10 pm. A few times I was ready for bed but people said "let's just work for another hour". As an example, during the short field trip in July 2015, I worked with consultants for 7 days for an average of 9 hours a day, and paid 12 different consultants for a total of 297 hours of work.

The exception was the trip in 2017 when my wife and I went to Morehead with my son, Ben, and his wife, Brooke, for their PNG wedding. They had a big Australian wedding in December 2015, but they also wanted a big customary wedding back in Morehead for all the family there. No such wedding had occurred for decades among Nama speakers, but Ben convinced people to revive the traditions – including the associated singing and dancing, decoration of the bride and groom and the actual ceremony. Since Ben is a Mbangu man, Brooke had to be adopted by the Sángárè line (see section 1.2.2). Another couple – a Sángárè man and one of Ben's female cousins – agreed to get (re)married at the same time.

One additional short period of work on the language was done when Ben's mother, Yoshie, my friend Murry and his wife Naka, and Tony Emoia (my main consultant) visited Australia from 26 September to 8 October, 2016. (Their plans to attend Ben and Brooke's Australian wedding the year before were frustrated by a long delay in the issuing of their Australian visas.)

1.3.3 Sources of data

The data for this study and the examples given come from three sources: (1) question and elicitation sessions with consultants, (2) audio and video recordings of narratives, conversations and commentaries, and (3) written text.

Consultants

Because of family connections, it was not difficult to find consultants and start work straight away. My expert consultant was Tony Emoia, who sadly passed away in March 2022. He was a natural linguist who quickly understood the gist of my questions and the structure of his language as it emerged. The other main consultants were Murry Dawi, Yoshie Dawi and Francis Dawi. Francis provided most of the 350 Nama names I collected for trees, plants and grasses in the environment. All these consultants are Mbangu and from Daraia, except for Tony who was from Mata. Two other people also spent a lot of time helping in the project: Mawai Dawi and Les Dawi. Many other consultants worked with me for shorter times, including Daniel Dawi, Eli Dawi, Joy Dawi, Ezekiel Erick, Daink Kuro, Kemsy Kuro, Gaita Mawari, Stanley Mawari, Zephaniah Mawi, Harry Murry, Guma Subam and Warapa Samadari. Figure 1.19 is a photo of some of the consultants at work.

Question and elicitation sessions were normally with two or more consultants, and although there was some coming and going, people made sure I always had someone to work with. As most of the consultants had lived for a time outside of the Morehead District in other areas of PNG, they know Tok Pisin, which I am also fluent in, and the sessions were conducted mainly in that language, as well as sometimes in English.

The information from the sessions was written down mainly in notebooks, which eventually totalled 10 in number. Examples in the grammar from the notebooks are labelled "NBn:", where "n" is the notebook number, followed by the page number. Some examples also come from my own loose-leaf notes and lists of questions and from verb paradigms written out by Tony or Murry. These are marked "VN:" (verb notes) followed by the page number or "RC-Qs" (relative clause questions). There are also a few examples from personal communication (PC), usually by phone with either Tony Emoia (TE), Murry Dawi (MD) or Ben Siegel (BS). (I also had small pocket notebooks which I used outside the elicitation sessions to write down lexical items, overheard snippets of language and other information.)

Figure 1.19: Consultants (working on fish names).
(left to right: Yoshie Dawi, Eli Dawi, Francis Dawi, Tony Emoia, Joy Dawi, Gaita Mawari)

Consultants also helped with transcribing and translating audio/video recordings, as described in the following subsection.

Recordings

Audio/video recordings were made for this study with the following devices: either a Zoom Q3HD audio/video recorder or a combination of a Zoom H4n audio recorder and a Canon Legria HF M52 video camera (except for the earlier recordings made in 1993 and 1999). Recordings were of four types.

The first type consists of 29 narratives. These were mainly traditional stories and personal experiences. Some were the result of questions asked to elicit particular grammatical structures – for example, "How do you make sago?" and "What would you do if someone gave you one million kina [the PNG currency]?". The narratives ranged in length from less than a minute to nearly 18 minutes. They are listed in Table B.1 in Appendix B.

The computer software ELAN was used to transcribe and translate the narratives (see ELAN 2021 in the references), using a Macbook Air laptop. I created ELAN documents with the recordings, segmented them, and in the initial stages, transcribed them and roughly translated them directly into ELAN with the help of consultants. I also transferred some of the early recordings to CDs and these were transcribed in notebooks by Tony Emoia between my visits, using a Discman player. Later, I taught Francis Dawi to use ELAN to transcribe recordings that I had already segmented. He wrote the transcriptions in notebooks, and I later entered them into the ELAN files, and then entered rough translations with the help of other consultants.

Examples in the grammar from these narratives are indicated with the file name, which consists of the speaker's initials, the last two digits of the year of recording, and the number of the recording, if more than one that year. The recordings and the ELAN files are archived in PARADISEC (Pacific and Regional Archive for Digital Sources in Endangered Cultures) https://www.paradisec.org.au.

The second type of recording is a series of 37 very short audio/video clips, recorded with the Q3HD audio/video recorder in Daraia village on 28 September 2011. These clips record Murry Dawi going around the village in the early morning and having short conversations with people as they go about their daily business – doing things like washing pots and making mats and baskets. Others record people demonstrating traditional items and practices, such as a jew's harp and initiation. These recordings are identified as Daraia Clips (DC); those that were used and their topics are listed in Table B2 in Appendix B. Again, these clips were transcribed and translated with the help of consultants, using ELAN. The file names, here "DC" plus the number of the recording, are given for examples from these recordings.

The third type of recording is of Tony Emoia and Yoshie Dawi doing the "Picture Task" for eliciting language (Carroll et al 2009; San Roque et al 2012). In this task, the participants are given 16 pictures relating to a story about "family problems". These show a man getting drunk, hitting his wife, getting arrested, going to jail and eventually returning home. Participants are given these pictures out of order, and asked to describe each one. Next they are asked to put them in order to form a coherent story, and then tell this story. This was recorded with the Zoom H4n audio recorder and the Canon Legria HF M52 video camera at Morehead station on 19 September 2013. The recording lasts 43 minutes. As with the other recordings, it was transcribed and translated with the help of consultants, using ELAN. Examples from this recording used in the grammar are indicated by "PT:" followed by the line number of the transcription in ELAN.

The fourth type of recording is of Ben and Brooke's traditional wedding (mentioned in section 1.3.2). It consists of videos taken from 7 – 9 January 2017 of the welcome of guests at Morehead airstrip and at Murry Dawi's house, the engagement ceremony, preparations for the wedding, the wedding ceremony and celebrations afterwards. The videos were recorded by myself and others using my Canon video camera, and by Ben and others using his iPhone. Smart phone videos from other people were also included. These were edited and compiled by Patsy

Asch into a single recording lasting nearly 58 minutes. The sections of the recording with audible dialogue and commentary by various masters of ceremony were transcribed and translated as above with ELAN. Examples from this recording used in the grammar are indicated by "B&B:" followed by the line number of the transcription in ELAN.

Altogether there are approximately three hours of recordings, with the transcriptions containing more than 16,800 words, most of them morphologically complex. While not a large corpus, it provided enough material for grammatical analysis and for naturalistic examples illustrating nearly all of the grammatical features obtained by elicitation. (The grammar points out the few instances where an example of a feature described by consultants is not found in the recordings.)

Written text
The written text used for data in this study is the Nama translation of the Gospel of Mark (Youngdu Church 2018). This consists of 12,750 words. Examples from this source are indicated by "Mark" followed by the chapter and verse.

1.3.4 Presentation of examples from the data

With the exception of the following section, the numbered examples in this grammatical description were provided by consultants or taken from recordings. The first line of the example, given in italics, represents the example as spoken in Nama, using the phonemic orthography adopted for the language (see chapter 2). Examples that are clearly complete stand alone sentences begin with a upper case letter. Those that are part of a longer stretch of speech begin with a lower case letter. The second line indicates the underlying morphology of the example. The third line gives morpheme by morpheme glosses in English. Grammatical labels are given in small capitals. When it is necessary for a single morpheme to be glossed by more than one word or grammatical label, these are separated by a full stop (period), with some exceptions, such as with the label for an argument, A, P or S. The fourth line is the translation, using the same capitalisation conventions as the first line.

1.4 Grammatical overview and organisation of this grammar

The Nama language has very complex morphology. This is described mainly in chapters 3 and 4. (Grammatical morphemes in Nama are listed in Appendix C.) Here I give a brief overview of Nama morphosyntax in order to make examples in

1.4 Grammatical overview and organisation of this grammar —— **29**

these chapters more understandable, as well as to give the reader some idea of the structure of the language. This is followed by an outline of the grammatical description in the following chapters.

1.4.1 Morphological processes

The predominant productive morphological process in Nama is affixation. Affixes are of two distinct types and characterise the two core word classes of the language: nominal affixes occur only with nominals (which include pronouns and nominalised verbs) and verbal affixes only with verbs.

Productive nominal affixes are almost entirely suffixes. There are 19 case suffixes, 15 major (shown in Table 1.4) and 4 peripheral. In addition, there are 2 suffixes indicating exclusivity and restrictiveness, and 2 suffixes, limited in usage, indicating number. Prefixation to show possession also occurs for a subset of nominals (see section 3.7). Reduplication (complete) with several functions also exists with nominals but not with verbs (section 3.8). Nominal morphology is described in detail in chapter 3.

Table 1.4: Major case marking suffixes in Nama.

case	suffix	example
ergative	-am	*kambanam* 'the snake (agent)'
dative	-e	*Bene* 'for Ben'
instrucmental	-e	*wambate* 'with a spade'
genitive	-ne	*ausane* 'the old woman's'
comitative	-afè	*àghafè* 'with the dog'
privative	-ofnar	*njamkeofnar* 'without food'
locative 1 (inessive)	-n	*mèngon* 'in the house'
locative 2 (addessive/supressive)	-an	*merekinan* 'on the plate'
ablative	-ta	*mèngota* 'from the house'
perlative	-mè	*endmè* 'along the road'
allative	-t	*mèngot* 'to the house'
purposive	-ot	*susiot* 'for fishing'
associative	-faf	*ndaufaf* 'place where the garden is'
originative	-mèn	*wèrimèn* 'because of drunkenness'
sembaltive	-nit	*árnit* 'like a person'

Verbal affixation is more complex (see chapter 4). First, all verb stems have a prefix or suffix or both that index the person and number of core verbal arguments. Second, in most contexts verbs also have suffixes indicating aspect (perfective/ imperfective, inceptive, punctual, delimited, durative) and tense (current, recent,

30 —— 1 Introduction

remote).[9] Note that current tense covers progressive, habitual or iterative events that are occurring at the time of speaking or that have occurred earlier in the day. (See section 4.7.1)

For transitive verbs, which have two core arguments, a verb-final suffix indexes the A argument, most often the semantic agent. If the A argument is overtly specified, it is marked by an ergative suffix. A prefix on the verb stem indexes the P argument, most often the semantic patient. The encoded number distinction is singular (SG) versus nonsingular (NSG). (Dual versus plural is indicated by other means.) In some cases the particular set of the prefix used also indicates a particular tense. Some examples follow, constructed for clear illustration, and with simplified glosses (in current tense with imperfective [IPFV] aspect).

(1) a. *Ghakram*   *mèrès*   *yèfrangote.*
         ghakèr-am   mèrès   y-frango-ta-e
         boy-ERG   girl   3SGP-leave-IPFV-2|3SGA
         'The boy is leaving the girl.'

     b. *Ghakram*   *mèrès*   *yèfrangotat.*
         ghakèr-am   mèrès   y-frango-ta-t
         boy-ERG   girl   3SGP-leave-IPFV-3NSGA
         'The boys are leaving the girl.'

     c. *Ghakram*   *mèrès*   *efrangote.*
         ghakèr-am   mèrès   e-frango-ta-e
         boy-ERG   girl   3NSGP-leave-IPFV-2|3SGA
         'The boy is leaving the girls.'

In (1a) and (1c), the suffix -*e*, following the verb stem *frango* 'leave' and the imperfective (IPFV) suffix -*ta*, indicates that the A argument (*ghakèr* 'boy') is singular. (A morphophonemic change results in the combined suffix -*te*.) In (1b) the suffix -*t* indicates the A argument is nonsingular. In (1a) and (1b), the prefix *y*- indicates that the P argument (*mèrès* 'girl') is singular and in (1c) the prefix *e*- indicates that it is nonsingular. The combination of imperfective marker and choice of particular prefix set and suffix gives the reading of the current tense (see section 4.7.1).

Intransitive verbs in Nama, which have only one core argument (the S argument), divide into two classes according to the way this single argument is indexed. This is in terms of alignment with one or the other of the two core arguments of

---

9 Note several changes from the terminology used in Siegel (2015). Here, current (CUR) replaces immediate, instead of being a cover term for immediate and proximate, and recent (REC) replaces proximate. (The term proximate (PROX) is used for two verbal clitics [see section 5.4.1].)

transitive verbs, A and P. For A-aligned intransitives, the S argument is indexed by the same sets of suffixes as those for A arguments of transitive verbs. These are indicated by S_A in glosses, as shown in example (2) with the verb stem *ásáfo* 'work'. The prefix *n-* indicates that absence of a P argument:

(2) a. *Ghakèr   násáfote.*
     ghakèr   n-ásáfo-ta-**e**
     boy   ØP-work-IPFV-**2|3SGS$_A$**
     'The boy is working.'

  b. *Ghakèr   násáfotat.*
     ghakèr   n-ásáfo-ta-**t**
     boy   ØP-work-IPFV-**3NSGS$_A$**
     'The boys are working.'

For P-aligned intransitives, the single S argument is indexed by the same sets of prefixes as those for P arguments of transitive verbs. These are indicated by S_P in glosses, as shown in (3) with the verb stem *mor* 'stay, live':

(3) a. *Mèrès   mèngotuan   yèmor.*
     mèrès   mèngotu-an   **y**-mor
     girl   village-LOC   **3SGS$_P$**-stay
     'The girl stays in the village.'

  b. *Mèrès   mèngotuan   emor.*
     mèrès   mèngotu-an   **e**-mor
     girl   village-LOC   **3NSGS$_P$**-stay
     'The girls stay in the village.'

Thus Nama has a "split S" system (Dixon 1979) or a system of "split intransitivity" (Merlan 1985), more recently referred to as "semantic alignment" (Donohue & Wichmann 2008).

Other verbal affixes include a nominalising suffix (section 4.1) and two verbal deictic prefixes (section 4.4), one indicating coming towards or nearness to the speaker (venitive) and one going away from or farness from the speaker (andative). There is also a reflexive/reciprocal prefix, an applicative prefix and an autobenefactive prefix (section 4.5).

32 — 1 Introduction

1.4.2 Syntax

As shown by the examples above, the basic word order in Nama is SOV, or more accurately APV or SV. However, this word order is not absolute, and AVP is frequent. Overt specification of the A and/or P argument is often absent when it is clear from the context, but the person and number of an existing argument is always indexed on the verb.

Also as shown in example (1) above, the specified A argument of a transitive verb is marked by an ergative suffix and the P argument is unmarked. The S argument of A-aligned intransitive verbs is also unmarked. If pronouns are used, the A argument of a transitive verb is represented by an ergative pronoun such as *yèmo* '3SG.ERG' and the P argument is represented by an absolutive pronoun such as *fá* '3.ABS'. (There is no number distinction for absolutive pronouns.) The S argument of A-aligned intransitive verbs is also represented by an absolutive pronoun. See example (4):

(4) a. *Yèmo*   *fá*   *yèfrangote.*
         yèmo   fá   y-frango-ta-**e**
         3SG.ERG   3.ABS   **3SGP-leave-IPFV-2|3SGA**
         'She is leaving him.'
     b. *Fá*   *násáfote.*
         fá   n-ásáfo-ta-**e**
         3.ABS   **ØP-work-IPFV-2|3SGS$_A$**
         'He is working.'

However, as we have also seen, the S argument of A-aligned intransitives is indexed by the same suffixes as those for indexing the A argument of transitive verbs. Thus, Nama is characterised by having split ergativity: an ergative–absolutive pattern with regard to case marking, but a nominative–accusative pattern with regard to agreement.

1.4.3 Organisation

The grammatical description of Nama in the remaining chapters is organised as follows. Chapter 2 gives a brief overview of the phonology of the language, and also discusses orthography and phonotactics. As already mentioned, chapter 3 describes nominal morphology, and chapter 4 verbal morphology. Other word classes and subclasses are introduced in chapter 5, as well as phrase structure. Subclasses of nominals in a nominal phrase include nouns, pronouns, quantifiers, temporal nominals and a demonstrative. The verb phrase can contain other word classes. These include adverbs and markers for tense, aspect and modality. Chapter 6 describes the structure of simple sentences (comprising one clause) and depicts five types – declarative, negative, interrogative, imperative and exclamative, as well as minor sentences. Three other word classes are introduced – coordinating conjunctions, discourse particles and interjections – as well as two subclasses: interrogative pronouns and interrogative adverbs. Chapter 7 deals with compound sentences and three types of complex sentences: those with adverbial clauses (introducing the subordinating conjunction word class), those with relative clauses, those with clausal arguments and those with focus marking clauses. Chapter 8 presents the typological implications of some of the features described for Nama.

2 Phonology

This chapter provides a survey of the consonants and vowels of Nama, and a description of the writing system. It also discusses phonotactics, and the thorny issue of vowel syncope versus epenthesis.

2.1 Consonants

Nama has 26 consonant phonemes. These are shown in Table 2.1, with their orthographic representations in angled brackets. Marginal phonemes are in round brackets (parentheses).

Table 2.1: Consonant phonemes in Nama.

	Bilabial	Velarised bilabial	Alveolar	Post-alveolar	Palatal	Velar	Labialised velar
Stop	(p) b		t d			k g	k^w (g^w)
	<p> <b>		<t> <d>			<k> <g>	<kw> <gw>
Prenasalised stop	^mb	$^mb^w$	nd			^ng	$^ng^w$
	<mb>	<mbw>	<nd>			<ng>	<ngw>
Affricate				dʒ			
				<j>			
Prenasalised affricate				$^ndʒ$			
				<nj>			
Nasal	m	m^w	n		ɲ	ŋ	
	<m>	<mw>	<n>		<ny>	<ñ>	
Trill/tap			r				
			<r>				
Fricative	ɸ	$ɸ^w$	s			ɣ	
	<f>	<fw>	<s>			<gh>	
Lateral			l				
			<l>				
Approximant					j		w
					<y>		<w>

https://doi.org/10.1515/9783111077017-002

2.1.1 Stops

Stops (or plosives) in Nama have four main points of articulation: bilabial, alveolar, velar and labialised velar. Each of these points of articulation has voiceless, voiced and prenasalised representations. However, two of the stops are marginal: /p/ which occurs only in loanwords (in 6 out of 4000+ dictionary entries) and /gʷ/ (only 3 entries). There is also a velarised prenasalised bilabial stop /ᵐbʷ/.

The alveolar and velar voiceless stops are slightly aspirated – i.e. [tʰ] and [kʰ]. The point of articulation for alveolar stops varies from dental to alveolar: thus /t/: [t̪ʰ] ~ [tʰ], /d/: [d̪] ~ [d], and /nd/: [ⁿd̪] ~ [ⁿd]. The stop component of a prenasalised stop can be devoiced word-finally.

All stops occur in initial, medial and final position in words except for /p/, /b/, /ᵐbʷ/ and /gʷ/ which to not occur word-finally (see Table 2.2).

Table 2.2: Distribution of stop phonemes in Nama.

Stops	Initial	Medial	Final
(p)	/pawə/ *pawè* 'electricity'	/talapia/ *talapia* 'fish type'	—
b	/besi/ *besi* 'arrow type'	/yabun/ *yabun* 'big'	—
ᵐb	/ᵐbi/ *mbi* 'sago'	/kaᵐban/ *kamban* 'snake'	/suᵐb/ *súmb* 'deep pool'
ᵐbʷ	/ᵐbʷito/ *mbwito* 'rat'	/ɣəᵐbʷe/ *ghèmbwe* 'tree type'	—
t	/tuk/ *tuk* 'top'	/latu/ *latu* 'flute'	/sot/ *sot* 'bone'
d	/diær/ *diár* 'arrow type'	/ɸader/ *fader* 'shoulder'	/səd/ *sèd* 'tree type'
ⁿd	/ⁿdau/ *ndau* 'garden'	/təⁿdo/ *tèndo* 'side'	/eⁿd/ *end* 'road'
k	/kal/ *kal* 'ankle'	/ɣakər/ *ghakèr* 'boy'	/ɸak/ *fak* 'ashes'
g	/gastol/ *gastol* 'fish type'	/dəgin/ *dègin* 'fish type'	/wəg/ *wèg* 'spoiled food'
kʷ	/kʷam/ *kwam* 'smoke'	/wəkʷər/ *wèkwèr* 'liquid'	/akʷ/ *akw* 'morning'
(gʷ)	/gʷərad/ *gwèrad* 'honey type'	/gʷərgʷər/ *gwèrgwèr* 'tree type'	—
ⁿg	/ⁿgar/ *ngar* 'skin, bark'	/məⁿgo/ *mèngo* 'house'	/næⁿg/ *náng* 'grass skirt'
ⁿgʷ	/ⁿgʷaiɲ/ *ngwainy* 'mosquito'	/ⁿdaⁿgʷəs/ *ndangwès* 'elbow'	/ⁿdʒaⁿgʷ/ *njèngw* 'lid'

Minimal and subminimal pairs for the stop phonemes are shown in Table 2.3.

2.1.2 Fricatives and affricates

Nama has four fricative phonemes: voiceless bilabial, velarised voiced bilabial, voiceless alveolar and voiced velar. The voiceless bilabial fricative /ɸ/ can become voiced intervocalically or word-finally. The voiced velar fricative /ɣ/ can also be realised as an approximate [ɰ] in some contexts. There are two affricates, both voiced postalveolar, but one prenasalised.

36 —— 2 Phonology

Table 2.3: Minimal/subminimal pairs for stops in Nama.

Stops	Examples of minimal/subminimal pairs	
b – ᵐb	/bøke/ *bóke* 'throat'	/ᵐbøk/ *mbók* 'water rat'
ᵐb – ᵐbʷ	/ᵐbi/ *mbi* 'sago'	/ᵐbʷito/ *mbwito* 'rat'
t – d	/təutəu/ *tèutèu* 'fast movement'	/dəudəu/ *dèudèu* 'confusion'
d – ⁿd	/ɸader/ *fader* 'shoulder'	/ɸaⁿdər/ *fandèr* 'thorn'
k – g	/kiri/ *kiri* 'tree type'	/giri/ *giri* 'small knife'
g – ⁿg	/giri/ *giri* 'small knife'	/ⁿgir/ *ngir* 'sticky liquid'
k – kʷ	/kən/ *kèn* 'yam type'	/kʷən/ *kwèn* 'sugarcane'
kʷ – gʷ	/kʷərkʷər/ *kwèrkwèr* 'bushfire'	/gʷərgʷər/ *gwèrgwèr* 'tree type'
ⁿg – ⁿgʷ	/ⁿgaiɲ/ *ngainy* 'tree type'	/ⁿgʷaiɲ/ *ngwainy* 'mosquito'

Two of the fricatives and affricates, /fʷ/ and /dʒ/ do not occur word-finally. The others occur in all positions (see Table 2.4).

Table 2.4: Distribution of fricative and affricate phonemes in Nama.

Fricatives/ Affricates	Initial	Medial	Final
ɸ	/ɸan/ *fan* 'savannah'	/aɸə/ *afè* 'father'	/saɸ/ *saf* 'clear place'
ɸʷ	/ɸʷiᵐb/ *fwimb* 'arrow'	/koɸʷe/ *kofwe* 'back of neck'	—
s	/sæɸ/ *sáf* 'face'	/besi/ *besi* 'spear type'	/jus/ *yús* 'grass'
ɣ	/ɣakər/ *ghakèr* 'boy'	/æɣə/ *ághè* 'dog'	/kaɣ/ *kagh* 'ankle'
dʒ	/dʒuar/ *juar* 'tree type'	/mudʒajær/ *mujayár* 'fish type'	—
ⁿdʒ	/ⁿdʒaɸar/ *njafar* 'sky'	/næⁿdʒi/ *nánji* 'banana'	/ⁿgeⁿdʒ/ *ngenj* 'leech'

Minimal and subminimal pairs for the fricative and affricate phonemes are shown in Table 2.5.

2.1.3 Nasals

Nama has four nasal consonants with the same points of articulation as the stops and an additional palatal one. Like the corresponding stop, the alveolar nasal can also have dental articulation.

2.1 Consonants — **37**

Table 2.5: Minimal/subminimal pairs for fricatives and affricates in Nama.

Fricatives/ Affricates	Minimal/subminimal pairs	
ɸ – ɸʷ	/kaɸe/ *kafe* 'cockatoo'	/kaɸʷe/ *kafwe* 'branch'
ɸ – s	/ɸəɸ/ *fèf* 'breadfruit'	/ɸəs/ *fès* 'fire'
k – ɣ	/kəm/ *kèm* 'front of neck'	/ɣəm/ *ghèm* 'river bend'
ɣ – g	/ɣərarə/ *ghèrarè* 'moon'	/gərar/ *gèrar* 'cassowary type'
ⁿd – ⁿdʒ	/məⁿd/ *mend* 'vomit'	/məⁿdʒ/ *mènj* 'vine'
dʒ – ⁿdʒ	/dʒəro/ *jèro* 'leaves used for dancing'	/ⁿdʒər/ *njèr* 'dregs'

Two of the nasals are relatively rare. The velarised bilabial /mʷ/ occurs in only 28 dictionary entries and the velar nasal /ŋ/ in 20.[10] In the vast majority of these, the phoneme occurs word-initially; neither occurs word-finally, except /ŋ/ in one loanword. The distribution with examples is shown in Table 2.6.

Table 2.6: Distribution of nasal phonemes in Nama.

Nasals	Initial	Medial	Final
m	/men/ *men* 'bird'	/ⁿdumal/ *ndumal* 'horrnet'	/kʷam/ *kwam* 'smoke'
mʷ	/mʷiɣə/ *mwigè* 'thought'	/mæmæmʷiɣə/ *mámámwighè* 'worry'	—
n	/nu/ *nu* 'water'	/mane/ *mane* 'tree type'	/min/ *min* 'nose'
ɲ	/ɲaɣ/ *nyagh* 'slaughter'	/niɲə/ *ninyè* 'witch'	/ⁿgʷaɪɲ/ *ngwainy* 'mosquito'
ŋ	/ŋærær/ *ñárár* 'bandicoot'	/tiŋa/ *tiña* 'parrot type'	/sliŋ/ *sliñ* 'slingshot'

Minimal and subminimal pairs for the nasal phonemes are shown in Table 2.7.

Table 2.7: Minimal/subminimal pairs for nasals in Nama.

Nasals	Minimal/subminimal pairs	
m – mʷ	/min/ *min* 'nose'	/mʷinə/ *mwinè* 'barren'
mʷ – ᵐbʷ	/mʷitə/ *mwitè* 'cousin'	/ᵐbʷito/ *mbwito* 'rat'
m – n	/sim/ *sim* 'snot'	/sin/ *sin* 'cooking pot'
n – ŋ	/tinə/ *Tinè* 'woman's name'	/tiŋa/ *tiña* 'parrot type'
ŋ – ⁿg	/ŋari/ *ñari* 'insect type'	/ⁿgar/ *ngar* 'skin'
n – ɲ	/nawəɣ/ *nawègh* 'block'	/ɲawəɣ/ *nyawègh* 'put inside'

---

10 There is evidence that initial velar nasals have been historically lost in most of the languages of the Nambu subbranch. It might be that contact with Tonda languages (especially neighbouring Komnzo) acounts for those remaining in Nama.

38 —— 2 Phonology

2.1.4 Liquids and glides

Nama has one rhotic phoneme, one lateral and two glides. The rhotic /r/ is a trill [r], which is realised as a tap [ɾ] in rapid speech and in consonant clusters. This and the lateral /l/ are both alveolar. The glides (or semivowels) are the palatal /j/ and the rounded labialised velar /w/. The liquids occur in all word positions; the glides, when acting as consonants, do not occur word-finally, as shown in Table 2.8. (There may be some argument that they occur word-finally following another vowel, rather than being part of a diphthong in this context. But see the discussion in section 2.2.3 below.)

Table 2.8: Distribution of liquids and glides in Nama.

Liquids/ Glides	Initial	Medial	Final
r	/raɸ/ *raf* 'tide'	/ⁿdʒəri/ *njèri* 'bubbles'	/mer/ *mer* 'good'
l	/latu/ *latu* 'flute'	/bulu/ *bulu* 'old man'	/ⁿdiᵐbal/ *ndimbal* 'big'
j	/jaʊ/ *yau* 'no, not'	/æʉjə/ *áuyè* 'cassowary'	—
w	/wem/ *wem* 'yam'	/nawin/ *nawin* 'fish type'	—

Minimal and subminimal pairs for the liquids and glides are shown in Table 2.9.

Table 2.9: Minimal/subminimal pairs for liquids and glides in Nama.

Liquids/ Glides	Minimal/subminimal pairs	
r – l	/waroɣ/ *warogh* 'cut through'	/waloɣ/ *walogh* 'set free'
r – l	/sor/ *sor* 'footprint'	/sol/ *sol* 'salt'
r – d	/sər/ *sèr* 'pus'	/səd/ *sèd* 'tree type'
j – ɲ	/yjoɣ/ *úyogh* 'rise up'	/ɣɲəɣ/ *únyègh* 'lie'
w – ɣ	/war/ *war* 'fish type'	/ɣar/ *ghar* 'sorcery type'
w – gʷ	/wərwər/ *wèrwèr* 'dried banana leaf'	/gʷərgʷər/ *gwèrgwèr* 'tree type'

2.1.5 The velarised bilabials

As indicated above, three of the bilabial consonants – /ᵐb/, /f/ and /m/ have velarised counterparts: /ᵐbʷ/, /fʷ/ and /mʷ/. They are also rounded, like the semivowel /w/, which accounts for the use of the /ʷ/ diacritic in representing them. As also indicated, these phonemes have relatively restricted distribution and two of them (/ᵐbʷ/ and /mʷ/ have only subminimal pairs.

2.1.6 Geminate consonants

Two consonant phonemes, /n/ and /r/, occur as geminates. Geminate /n/ is realised as a longer pronunciation of /n/. It occurs as the result of affixation – for example, /jənnəm/ *yènnèm* 'we come', which derives from the verb stem /nəm/ *nèm* plus the 1st person nonsingular S_P-indexing prefix /jən-/ *yèn-*. This contrasts with /jənəm/ *yènèm* 'he/she comes', in which the 3rd person singular prefix is /jə-/ *yè-*. Geminate /r/ is realised a longer trill than normal /r/, which is either a short trill or a tap. It occurs in a small number of lexical items, such as /ⁿdʒərri/ *njèrri* 'black flying fox', which contrasts with /ⁿdʒəri/ *njèri* 'bubbles'. Geminate /r/ also occurs as the result of affixation – for example, /ærro/ *árro* 'only the man', /ær/ *ár* 'man' plus the restrictive suffix /-ro/ *-ro*.

2.2 Vowels

2.2.1 Monophthongs

Nama has 10 monophthong vowel phonemes, shown in Figure 2.1. The inventory, orthographic symbols, description and variation are given in Table 2.10.

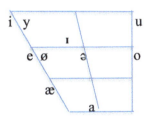

Figure 2.1: Nama vowel space.

With one exception, all vowels occur in word-initially, medially and finally (see Table 2.11). The exception is /ɪ/, which has limited distribution as a phoneme, occurring in contrast to other vowels in only 14 lexical items. In all these items, /ɪ/ is the nucleus of a stressed syllable. Phonetically, [ɪ] can occur as an allophone of /ə/, normally in unstressed syllables.[11]

---

[11] An unusual case is the frequently occurring word for 'tree' or 'plant'. Although it is normally pronounced as [wɪn] and not [wən], speakers consistently insist that it should be spelled *wèn* rather than *wìn*.

40 —— 2 Phonology

Table 2.10: Vowel phonemes in Nama.

Vowels	Orthographic symbol	Description	Variation
i	\<i\>	close front unrounded vowel	
y	\<ú\>	close front rounded vowel	can be slightly centralised
ɪ	\<ì\>	near-close near-front vowel	
u	\<u\>	close back vowel	
e	\<e\>	close-mid front unrounded vowel	can be more open (mid)
ø	\<ó\>	close-mid front rounded vowel	can be slightly centralised
ə	\<è\>	mid central vowel	can be raised and fronted
o	\<o\>	close-mid back vowel	can be more open (mid)
æ	\<á\>	near-open near-front vowel	
a	\<a\>	open central vowel	can be slightly closed and backed

Table 2.11: Distribution of vowels in Nama.

Vowels	Initial	Medial	Final
i	/ineɸi/ *inefi* 'yam type'	/min/ *min* 'nose'	/æki/ *áki* 'grandfather'
y	/yrər/ *úrèr* 'trap'	/ɸyt/ *fút* 'belly'	/nəsy/ *nèsú* 'tree type'
ɪ	—	/wɪt/ *wìt* 'thorn'	—
u	/uliɣ/ *uligh* 'get in'	/kuⁿdʒ/ *kunj* 'bee'	/anu/ *anu* 'washing'
e	/emo/ *emo* 'who'	/wem/ *wem* 'yam'	/jame/ *yame* 'mat'
ø	/øɸər/ *ófèr* 'tree type'	/ᵐbøk/ *mbók* 'water rat'	/ⁿdø/ *ndó* 'goanna type'
ə	/əkᵐboɣ/ *èkmbogh* 'stop'	/ɸəs/ *fès* 'fire'	/toᵐbə/ *tombè* 'long'
o	/oɸə/ *ofè* 'light (weight)'	/kone/ *kone* 'arm pit'	/kæⁿdʒo/ *kánjo* 'thanks'
æ	/æs/ *ás* 'coconut'	/wærær/ *wèrár* 'wallaby'	/ɸæ/ *fá* '3ʳᵈ pers. pronoun'
a	/amə/ *amè* 'mother'	/ɸan/ *fan* 'savannah'	/kaka/ *kaka* 'near'

Minimal pairs for the Nama vowels are shown in Table 2.12, and groups of words that differ only in vowels are given in Table 2.13.

It should be noted that the distinction between /æ/ and /a/ in multisyllabic words is not always clear. Speakers often disagree and are not always consistent about what they think the vowel is in certain words – for example, /rokær/ *rokár* vs /rokar/ *rokar* 'thing' and /mənæt/ *mènát* vs /mənat/ *mènat* 'so that'. The reasons for this are not understood, and further research is needed.

2.2 Vowels — **41**

Table 2.12: Minimal pairs for vowels in Nama.

Vowels	Minimal pairs	
i – ɪ	/wim/ *wim* 'smell'	/wɪm/ *wìm* 'stone'
i – y	/nini/ *nini* 'tusk scraper'	/nyny/ *núnú* 'pudding'
y – u	/syᵐb/ *súmb* 'pool in a creek'	/suᵐb/ *sumb* 'tree type'
ɪ – e	/ɸɪrəy/ *firègh* 'get dark'	/ɸerəy/ *ferègh* 'untie'
e – ø	/ⁿde/ *nde* 'like this/that'	/ⁿdø/ *ndó* 'goanna type'
ø – o	/søk/ *sók* 'fish type'	/sok/ *sok* 'sharp'
ø – y	/søt/ *sót* 'empty'	/syt/ *sút* 'wrapped'
e – i	/men/ *men* 'bird'	/min/ *min* 'nose'
e – æ	/es/ *es* 'mist'	/æs/ *ás* 'coconut'
o – u	/koⁿdʒ/ *konj* 'spell type'	/kuⁿdʒ/ *kunj* 'bee'
ə – ɪ	/ᵐbərᵐbər/ *mbèrmbèr* 'spirit'	/ᵐbɪrᵐbɪr/ *mbìrmbìr* 'lips'
ə – e	/wəm/ *wèm* 'I am'	/wem/ *wem* 'yam'
a – ə	/naᵐbət/ *nambèt* 'day after tomorrow'	/nəᵐbət/ *nèmbèt* 'penis'
æ – a	/ⁿgæmə/ *ngámè* 'tongs'	/ⁿgamə/ *ngamè* 'spell'

Table 2.13: Groups of words showing vowel differences in Nama.

Vowels	Examples showing vowel differences		
i	/siɸ/ *sif* 'sheep'	/sin/ *sin* 'cooking pot'	—
y	/syɸ/ *súf* 'soap'	/syn/ *sún* 'startled'	—
ɪ	/sɪɸ/ *sìf* 'hair'	—	—
u	—	—	/ju/ *yu* 'place'
e	—	/sen/ *sen* 'upper back'	/je/ *ye* 'crying'
ø	/søɸ/ *sóf* 'wave'	—	—
ə	/səɸ/ *sèf* 'bundle'	/sən/ *sèn* 'teeth'	—
o	—	/son/ *son* 'plant type'	/jo/ *yo* 'exclusive suffix'
æ	/sæɸ/ *sáf* 'face'	—	/jæ/ *yá* 'for him/her'
a	/saɸ/ *saf* 'clear place'	—	—

2.2.2 The status of /ə/

In the languages of the Yam family that have previously been documented, the mid
central schwa vowel [ə] is described as having limited distribution and occurring
predominantly as a result of epenthesis between consonants to create a syllable
nucleus where none exists underlyingly. In Ngkolmpu, its occurrence is entirely
predictable according to a series of phonotactic constraints, and it is therefore
analysed as being non-phonemic (Carroll 2016: 46). In Komnzo [ə] is analysed as a

42 —— 2 Phonology

marginal phoneme, as its occurrence is predictable except word-finally in a limited number of morphemes and lexemes (Döhler 2018: 57–8). In Nen, it similarly occurs almost entirely as a result of epenthesis except for "a couple of open mono-syllables", which lead to it being analysed as a phoneme (Evans & Miller 2016).

In Nama, however, the mid central vowel [ə] is clearly a phoneme. First of all, it does not have a restricted distribution, frequently occurring word-finally (in 396 entries in the dictionary) as well as initially (23 entries).[12] It occurs finally not only in words of one syllable – e.g. /tə/ *tè* 'already' – but also in multi-syllabic words – e.g. /baɸə/ *bafè* 'uncle', /kʷəɸitè/ *kwèfitè* 'black' and /maⁿdəɸarə/ *mandèfarè* 'shot right through'. In addition to minimal pairs with the vowel word-medially (see Tables 2.12 and 2.13), there are minimal pairs with the vowel word-finally – e.g. /jənə/ *yènè* 'this' vs /jəna/ *yèna* 'here' and /oɸə/ *ofè* 'light (in weight)' vs /oɸe/ *ofe* 'children's game' . There are also minimal pairs of words with and without initial and final /ə/, as shown in Table 2.14.

Table 2.14: Minimal pairs with and without initial and final /ə/.

With initial or final /ə/	Without initial or final /ə/
/əjəɣ/ *èyègh* 'become planted'	/jəɣ/ *yègh* 'plant (something)'
/əwiɣ/ *èwigh* 'argue'	/wiɣ/ *wigh* 'throw'
/taɸə/ *tafè* 'frond'	/taɸ/ *taf* 'yam type'
/ⁿdʒəⁿdʒə/ *njènjè* 'crayfish'	/ⁿdʒəⁿdʒ/ *njènj* 'bubbles'
/səɸə/ *sèfè* 'while'	/səɸ/ *sèf* 'bundle'

Rather than general epenthesis of [ə], Nama exhibits syncope, as described in section 2.4.4 below.

2.2.3 Diphthongs

Nama has seven diphthongs, all falling and closing (see Table 2.15).

Most diphthongs occur at the end of a syllable, and could be analysed as the combination of a vowel nucleus and a glide coda. However, some do occur as the syllable nucleus, therefore justifying the diphthong analysis. Examples are /mʷəɪɲ/ *mwèiny* 'sorcery type' and /ɸaɪs/ *fais* 'fist', the diphthong in the latter contrasting with monophthongs in /ɸəs/ *fès* 'fire' and /ɸus/ *fus* 'cat'. Table 2.16 shows the distribution

---

12 Note that all lexical items with initial /ə/ are intransitive verbs.

Table 2.15: Diphthongs in Nama.

Diphthong		Orthographic symbol
aɪ	[aɪ̯]	<ai>
æɪ	[æɪ̯]	<ái>
əɪ	[əɪ̯]	<èi>
aʊ	[aʊ̯]	<au>
æʉ	[æʉ̯]	<áu>
əʊ	[əʊ̯]	<èu>
ɔɪ	[ɔɪ̯]	<oi>

of diphthongs in Nama. Note that /ɔɪ/ is marginal in the language, occurring only five times in the dictionary.

Table 2.16: Distribution of diphthongs in Nama.

Diphthongs	Initial	Medial	Final
aɪ	/aɪkumi/ *Aikumi* 'founder of the Mbangu clan'	/faɪl/ *fail* 'file'	/ɸuaɪ/ *fuai* 'goanna type'
æɪ	/æɪ/ *ái* 'yikes!'	/ræɪs/ *ráis* 'rice'	/markæɪ/ *markái* 'white man'
əɪ	—	/mʷəɪɲ/ *mwèiny* 'sorcery type'	/wəɪ/ *wèi* 'now'
aʊ	/aʊsə/ *ausè* 'old woman'	/rabaʊl/ *Rabaul* 'place name'	/jaʊ/ *yau* 'no, not'
æʉ	/æʉjə/ *áuyè* 'cassowary'	—	/kʷæʉ/ *kwáu* 'tree type'
əʊ	/əʊmjoɣ/ *èumyogh* 'rely on others'	/wəʊmyoɣ/ *wèumyogh* 'allow'	/dəʊ/ *dèu* 'support stick'
ɔɪ	/ɔɪl/ *oil* 'cooking oil'	—	/dʒərɔɪ/ *jèroi* 'decorative frame for singsing'

Some minimal/subminimal pairs for Nama diphthongs are given in Table 2.17.

Table 2.17: Minimal/subminimal pairs for diphthongs in Nama.

Diphthongs	Minimal/subminimal pairs	
aɪ – æɪ	/jaɪ/ *yai* 'type of singsing'	/jæɪ/ *yái* 'type of plant'
aɪ – əɪ	/waɪ/ *wai* 'yes!'	/wəɪ/ *wèi* 'now'
aʊ – æʉ	/waʊ/ *wau* 'ripe'	/kʷæʉ/ *kwáu* 'tree type'
aʊ – əʊ	/kaʊkaʊ/ *kaukau* 'fish type'	/kəʊkəʊ/ *kèukèu* 'famine'
aʊ – əʊ	/jaʊ/ *yau* 'no'	/jəʊ/ *yèu* 'bottom'

2.3 Orthography

An orthography for Nama was first developed at a workshop at Morehead in the late 1990s, run by SIL International (Price 2000). This orthography was for teachers to use in elementary schools (first 4 years of education) in which the vernacular was then the medium of instruction. This orthography is basically what is currently used in this sketch grammar except for a few changes made in response to various issues. These changes were made in 2011–12 after discussions with consultants, teachers and other members of the community. The issues and resulting changes were as follows:

1. In the original orthography, <ñg> was used for /ᵑg/ and <ng> for /ŋ/. The first problem was confusion between the two symbols, which were used in opposite ways by many teachers. Another issue was that a diacritic (the tilde over the <n>) was necessary in one of the most common phonemes of the language /ᵑg/, and unnecessary in one of the rarest /ŋ/. The solution was to use <ng> for /ᵑg/ and only <ñ> for /ŋ/ – for example, *ngiri* /ᵑgiri/ 'small knife' and *ñoti* /ŋoti/ 'spotted cuscus'.

2. In the original orthography <à> was used for /ə/ and <è> for /ɪ/. The first problem was again confusion between symbols, and also confusion between <à> and <á>, the symbol for /æ/. Another problem was the variable pronunciation of /ə/, which sometimes extends to /ɪ/, especially in nominal and verb morphology where the two sounds are never distinctive. This led to great inconsistency in spelling – e.g. *yàfandam/yèfandam* 'he watched it'. In addition, there are only a very few lexical items in which the two phonemes are distinctive. The solution was to eliminate <à> and use <è> consistently for /ə/ and all its variant pronunciations in nominal and verb morphology. A new symbol <ì> was adopted for /ɪ/ to be used only where it contrasts with /ə/ – for example, *yìm* /jɪm/ 'tree type' and *yèm* /jəm/ 'he/she/it is'.

3. New symbols were created for the velarised bilabials, some of the labialised velars and the palatal nasal that were not originally recognised as separate phonemes: <mbw> /ᵐbʷ/, <fw> /ɸʷ/, <mw> /mʷ/, <gw> /gʷ/, <ngw> /ᵑgʷ/ and <ny> /ɲ/.

Consideration was given to change the digraph <gh> /ɣ/ to a single symbol, such as <q>. However, speakers are used to <gh>, and recognise its relationship to the cognate /ð/ in neighbouring languages (Namat and Komnzo), which is also represented by a digraph <th>.

On the word level, a decision was made regarding reduplication. If the reduplicated form is two or more syllables, then a hyphen is used – for example, *rokár* 'thing', *rokár-rokár* 'things' vs *tru* 'container', *trutru* 'small container'.

2.4 Phonotactics

2.4.1 Syllable structure

The maximal syllable in Nama is $[CCVC]_\sigma$, where 'V' stands for either a monophthong or a diphthong, and the minimal syllable is $[V]_\sigma$. $[V]_\sigma$ and $[VC]_\sigma$ have restricted distribution – occurring word-initially or in a monosyllabic word or following a syllable ending in a vowel. Otherwise, syllables require an onset, either complex or simple, but may or may not have a coda, which is always simple. The preferred or optimal syllable has both an onset and a coda. Examples are given in Table 2.18.

Table 2.18: Syllable types in Nama.

Syllable type	Monosyllabic word	Second syllable of disyllabic word
$[V]_\sigma$	/a/ *a* 'and'	/bu.a/ *bu.a* 'bird type'
$[VC]_\sigma$	/es/ *es* 'mist'	/dʒu.ar/ *ju.ar* 'tree type'
$[CV]_\sigma$	/ɸæ/ *fá* 'he/she'	/kʷa.sə/ *kwa.sè* 'hip'
$[CVC]_\sigma$	/søɸ/ *sóf* 'wave'	/ja.bun/ *ya.bun* 'big'
$[CCV]_\sigma$	/tru/ *tru* 'container'	/dʒarbry/ *jar.brú* 'healthy tree'
$[CCVC]_\sigma$	/kryt/ *krút* 'grey hair'	—

2.4.2 Consonant clusters within the syllable

As just shown, clusters of two consonants C_1C_2 can occur as the onset of a syllable. Thirteen intrasyllabic clusters occur in lexical items in the Nama corpus. A further 11 occur as the result of cliticisation of verbs (see below).

In 11 of the 13 clusters, C_2 is a liquid /r/ or /l/. In 9 of them, C_1 is a stop and in 4 a fricative /ɸ/ or /s/. These clusters and examples are shown in Table 2.19. (The first three occur frequently; the others are relatively rare. The /bl/ and /ɸl/occur only in loan words, and /sn/ occurs only in the speech of some speakers.)

Additional initial consonant clusters occur with the two proximal verb phrase clitics: /ɸ/ 'just now' and /s/ 'right here/there' (see section 5.4.1). The C_1 is one of these clitics and the C_2 one of the possible initial consonants of finite verbs: /n/, /k/, /w/, /j/, /kʷ/ and /t/. All possible combinations except */ɸkʷ/ are attested. See Table 2.20 for examples.

46 —— 2 Phonology

Table 2.19: Consonant clusters as onsets in Nama lexical items.

Consonant clusters	Examples as onsets
/tr/	/tro.rəɣ/ *tro.règh* 'squeeze'
/kr/	/kru.ɸər/ *kru.fèr* 'cold'
/ɸr/	/ɸro.ⁿde/ *fro.nde* 'first'
/br/	/bru/ *bru* 'fish type'
/ᵐbr/	/ᵐbraɪ.ᵐbraɪ/ *mbrai.mbrai* 'tree type'
/ⁿdr/	/ⁿdro.ⁿdro/ *ndro.ndro* 'bird type'
/ᵑgr/	/ᵑgru.ke/ *ngru.ke* 'sound type'
/kl/	/kli.sə/ *kli.sè* 'fresh feeling'
/bl/	/bli.bli/ *bli.bli* 'bouncing (baby) on knee'
/ɸl/	/ɸlaʊ.ə/ *flau.è* 'flour'
/dr/	/dren/ *dren* 'drain'
/st/	/səⁿdʒ.stə/ *sènj.stè* 'thick'
/sn/	/sno.ⁿdʒo/ *sno.njo* 'today'

Table 2.20: Consonant clusters with proclitic markers on verbs.

Consonant cluster	Examples with proclitic markers on verbs
/ɸn/	/ɸneɸɔɪ/ *fnefoi* 'it just finished'
/ɸk/	/ɸkuɸrotaʊ/ *fkufrotau* 'he just kept doing it'
/ɸw/	/ɸwəm/ *fwèm* 'just am'
/ɸj/	/ɸjaɪtote/ *fyaitote* 'he just told it'
/ɸt/	/ɸtənete/ *ftènete* 'she just ate it'
/sn/	/snuwanɔɪ/ *snuwanoi* 'he set off right there'
/sk/	/skaᵑgotaʊ/ *skangotau* 'he was going back right there'
/sw/	/swəmⁿdetat/ *swèmndetat* 'they're telling me right here'
/sj/	/sjeretat/ *syeretat* 'they're holding her right there'
/skʷ/	/skʷəmormən/ *skwèmormèn* 'I was right there'

2.4.3 Stress

Stress is a word-level phenomenon in Nama in which a particular syllable has prominence or emphasis. It is not predictable, but the clear tendency is that the penultimate syllable is given primary stress. This means that in words of two syllables, the initial syllable is stressed. Lexical items are rarely longer than three syllables, with four the maximum, except for those with complete reduplication. However, forms with derivational and inflectional morphology have as many as five syllables. Some examples are given in Table 2.21.

2.4 Phonotactics ——— **47**

Table 2.21: Examples of stress patterns in Nama.

Syllables	Examples of stress patterns
2 syllables	/ˈwa.giɸ/ *wa.gif* 'fish'
	/ˈɣər.næt/ *ghèr.nát* 'dry'
	/ˈkæn.kə/ *kán.kè* 'tongue'
3 syllables	/li.ˈmæ.nəɣ/ *li.ˈmá.nègh* 'pull'
	/ɣə.ˈra.re/ *ghè.ˈra.re* 'moon'
	/məɪ.ˈɲot.jo/ *mèi.ˈnyot.yo* 'all'
4 syllables	/æ.ɸi.ˈtræ.rəɣ/ *á.fi.ˈtrá.règh* 'come off'
	/ˌdʒi.ma.ˈka.ri/ *ji.ma.ˈka.ri* 'dance type'
5 syllables	/ˌɸə.jo.ta.ˈoɸ.nar/ *fè.yo.ta.ˈrof.nar* 'unfair'
	/jən.kə.ma.ˈⁿgər.mən/ *yèn.kè.ma.ˈngèr.mèn* 'we slept'

Exceptions in two syllable words include /wə.ˈrær/ *wè.ˈrár* 'wallaby'and /də.ˈgin/ *dè.ˈgin* 'fish type'. These are most often words in which the first syllable is open and the nucleus is /ə/. Examples of exceptions in three syllable words are /ˈga.si.jə/ *ˈga.si.yè* 'crocodile' and /ˈsæ.læ.me/ *ˈsá.lá.me* 'clothes' with initial stress, and /ˌmə.ⁿgo.ˈtu/ *mè.ngo.ˈtu* 'village' and /ˌa.ra.ˈkaɸ/ *a.ra.ˈkaf* 'sea' with final stress.

Also, in reduplicated forms, each element retains the same stress as it would have individually – e.g. /ˈfə.rən.ˈfə.rən/ *ˈfè.rèn-ˈfè.rèn* 'pattern' and /gi.ˈda.min.gi.ˈda.min/ *gi.ˈdá.min-gi.ˈdá.min* 'plant type'.

2.4.4 Vowel syncope and epenthesis

As mentioned in section 2.2.2, the mid central schwa vowel [ə] in neighbouring languages of the Yam family has been analysed as non-phonemic or marginally phonemic, occurring primarily as the result of epenthesis. Therefore, the absence of schwa between consonants in some contexts is seen to be the result of the lack of epenthesis according to the phonemic rules of the language. In Nama, however, the mid central schwa vowel [ə] is phonemic (section 2.2.2); therefore, the absence of /ə/ in an environment where it otherwise occurs is analysed as the result of vowel loss, or syncope, according to the rules of the language, as shown below. And as also shown below, it is not only /ə/ that is subject to syncope.

Nevertheless, epenthesis of /ə/ does occur in Nama in conjunction with nominal and verb morphology – i.e. with affixes that are single consonants. Schwa is inserted when a single consonant suffix follows a nominal base or a verb stem or another suffix ending in a consonant. Some examples follow:

48 —— 2 Phonology

/man/ *man* 'close' + nominalising suffix -/ɣ/ -gh → /manəɣ/ *manègh* 'closing'. (See section 4.1.)

/osær/ *osár* 'compete' + 3rd person nonsingular S$_A$-indexing suffix -/t/ -*t* (and /n/- *n*-, one of the two prefixes indicating lack of a P argument with an A-aligned intransitive verb) → /nosærət/ *nosárèt* 'they (2) competed'. (See section 4.2.2.)

Schwa is also inserted when a single consonant prefix precedes a verb stem or another prefix beginning with a consonant (with some exceptions). (See section 4.2.2.)

/ræɸær/ *ráfár* 'break' + 3rd person singular P-indexing prefix /j/- y- (and 1st person singular A- indexing suffix -/ən/ -*èn*) → /jəræɸærən/ *yèráfárèn* 'I just broke it'.

/fanda/ *fanda* 'look at' + 1st person singular P-indexing prefix /w/- w- → /wəfanda/ *wèfanda* 'look at me'.

(An alternative analysis would be that each of these morphemes has two allomorphs: for the suffixes, -C after vowels and -əC after consonants, and for the prefixes, C- before vowels and Cə- before most consonants.)

Syncope, however, is more widespread. It can occur in two contexts. The first is in lexical items where /ə/ occurs as the nucleus of an open syllable which has an onset C_1 that can form a consonant cluster with the onset of the following syllable C_2. If syncope takes place, /ə/ is deleted and the cluster C_1C_2 is formed. (This cannot occur when the vowel of the second syllable is also /ə/.)

$$[C_1ə]_\sigma[C_2V(C_3)]_\sigma \to [C_1C_2V(C_3)]_\sigma$$

An example is /ⁿgərake/ *ngèrake* 'sound type' becoming /ⁿgrake/ *ngrake*:

$$[^ngə]_\sigma[ra]_\sigma[ke]_\sigma \to [^ngra]_\sigma[ke]_\sigma$$

In this context, there is often variation between lexical items without and with syncope – for example: /təraktə/~/traktə/ *tèraktè~traktè* 'tractor', /kərærə/~/krærə/ *kèrárè~krárè* 'parrot' and /sənoⁿdʒo/~/snoⁿdʒo/ *sènonjo~snonjo* 'today'. This is similar to the syncope that occurs variably in English words such as *director* /dərɛktə/~/drɛktə/ and *police* /pəlis/~/plis/.

The second context for syncope occurs in a lexical item or base (a nominal or a verb stem) when /ə/ again occurs in an open syllable with an onset C_1, but here it is preceded by another open syllable and followed by a syllable with an onset C_2. Resyllabification can occur where the onset C_1 of the syllable with /ə/ becomes the coda of the preceding syllable, and the original syllable with /ə/ is lost.

2.4 Phonotactics — **49**

$$[(C)V]_\sigma,[C_1\partial]_\sigma,[C_2V(C)]_\sigma \rightarrow [(C)VC_1]_\sigma[C_2V(C)]_\sigma$$

An example of a lexical item is /dʒakərə/ *jakèrè* 'plant type', which may be pronounced /dʒakrə/ *jak*rè.

$$[dʒa]_\sigma,[kə]_\sigma[rə)]_\sigma \rightarrow [dʒak]_\sigma[rə]_\sigma$$

An example with a verb stem is /ætəno/ *átèno* 'hide' which becomes /ætnoɣ/ *átnogh* in its nominalised form (with the nominaliser -/ɣ/ -*gh*). This is similar to the syncope that occurs in English words such as *memory* /mɛməri/~/mɛmri/

This context for syncope also arises when an inflectional or derivational morpheme is added to a base with a particular structure.

In the case of the addition of a suffix, syncope can occur when the base ends in a closed syllable with an onset C_1, /ə/ as its nucleus and a coda C_2, and the preceding syllable is open – i.e. $[(C)V]_\sigma[C_1\partial C_2]_\sigma$. When a suffix with the first syllable beginning in a vowel -$[V(C)]_\sigma$ is added, resyllabification can occur, where the coda of the original final syllable of the base C_2 becomes the onset of the new final syllable with the suffix, and the onset of the original final syllable C_1 becomes the coda of the preceding syllable. The original final syllable with /ə/ is then lost.

$$[(C)V]_\sigma[C_1\partial C_2]_\sigma + [V(C)]_\sigma \rightarrow [(C_1)VC_2]_\sigma[C_3V(C)]_\sigma$$

An example with a nominal is /ɣakər/ *ghakèr* 'boy' plus the ergative suffix -/am/ resulting in /ɣakram/ *ghakram*:

$$[ɣa]_\sigma[kər]_\sigma + [am]_\sigma \rightarrow [ɣak]_\sigma[ram]_\sigma$$

Another is /mərən/ *mèrèn* 'family' plus the comitative suffix /aɸə/ -*afè*, which results in /mərnaɸə/ *mèrnafè* 'with the family'.

An example with a verb stem is /ætəm/ *átèm* 'move (next to someone or something)' with the 2|3NSG S_A-indexing suffix -/t/ (along with /k/-, one of prefixes indicating lack of a P argument with an A-aligned intransitive verb) resulting in /kætmət/ *kátmèt* 'they (2) moved'.

With verb stems it is not only /ə/ but also the near open near front vowel /æ/ <*á*> that is subject to syncope when particular verb suffixes are added – especially the imperfective suffix -/ta/. This is used for 2nd person singular imperatives (and, for A-aligned intransitives, along with the /k/- prefix). In example without syncope, the imperative form of /ar^mbo/ *armbo* 'climb' is /kar^mbota/ *karmbota*. However, this is not always straightforward suffixation, as complex morphophonemic changes may also be involved (see section 4.6.2), and in these contexts, /æ/ deletion most often occurs. One such change is that when a verb stem ends in /r/, this fuses with the /t/ of the -/ta/ suffix to become a single consonant /n/. For example, the verb stem /atær/

atár 'jump' plus the /k/- ØP prefix and the -/ta/ suffix becomes /katna/ *katna* 'jump (imperative)'. Here, as above, resyllabification occurs and the original syllable with /æ/ is deleted.

$$/k/ + [a]_\sigma[tær]_\sigma \to [ka]_\sigma[tær]_\sigma + [ta]_\sigma \to [ka]_\sigma[tæ]_\sigma[na]_\sigma \to [kat]_\sigma[na]_\sigma$$

Another example is with /ymbær/ *úmbár* 'bathe'. With the other ØP prefix used with A-aligned intransitive verbs, /n/-, the imperfective suffix -/ta/ and the 2nd/3rd person nonsingular S_A-indexing suffix -/t/, we get /ny^{m}bnat/ *númbnat* 'they are bathing'.

In the case of the addition of a prefix, syncope can occur when the first syllable of the base is open with an onset C_1 and a /ə/ nucleus, and the following syllable has an onset C_2 – i.e. $[C_1ə]_\sigma[C_2V(C)]_\sigma$. When a prefix ending in a vowel $[(C)V]_\sigma$ is added, resyllabification can occur where the onset C_1 of the original first syllable with /ə/ becomes the coda of the new first syllable with the prefix. The original first syllable with /ə/ is then lost.

$$[(C)V]_\sigma + [C_1ə]_\sigma[C_2V(C)]_\sigma \to [(C)VC_1]_\sigma[C_2V(C)]_\sigma$$

An example with a nominal is /nə^{n}gən/ *nèngèn* 'younger brother' plus the first person singular possessive prefix /ta/- resulting in /tan^{n}gən/ *tanngèn* 'my younger brother':

$$[ta]_\sigma + [nə]_\sigma[^ngən]_\sigma \to [tan]_\sigma[^ngən]_\sigma$$

An example with a verb stem is /ɸəno/ *fèno* 'rupture' with the 3rd nonsingular P-indexing prefix /e/-, resulting in /eɸno/ *efno* 's/he just ruptured them'.

Note that syncope can occur differently with the same base depending on the type of affixation. For example, the nominal /mərəs/ *mèrès* 'girl, sister' with the ergative suffix -/am/ becomes /mèrsam/ *mèrsam*. But with the 3rd person possessive prefix /jæ/- *yá-* it becomes /jæmrəs/ *yámrès* 'his sister'.

Similarly, the verb stem /ɸətərær/ *fètèrár* 'tear' with the nominalising suffix -/ɣ/ -*gh* becomes /ɸətrærəɣ/ *fètrárègh*. With the 3rd person singular P-indexing prefix /j/- *y-* and no suffix, it becomes /jəɸtrær/ *yèftrár* s/he just tore it'. But with the addition of the -/ta/ imperfective suffix (section 4.6.1) plus 1st person A-indeing suffix -/ən/ and the resultant morphophonemic changes (section 4.6.2), it becomes /jəɸtərnan/ *yèftèrnan* 'I am tearing it'.

3 Nominal morphology

One of the two major word classes in Nama is nominals. Its members are defined by their ability to act as the head of a nominal phrase and by their distinct morphological behaviour. This chapter examines the morphology of nominals, and introduces the subclass of pronouns.

The most common morphological process affecting nominals is affixation. Suffixation is most frequent and is used for case-marking (sections 3.3 and 3.4), and to a limited extent for number marking (3.2). Prefixation is used less frequently for possessive marking with a limited subclass of nominals (3.7). A small subset of these nominals are bound forms; otherwise the nominal bases to which affixes are attached can all occur as free forms. Another morphological process, restricted to nominal word formation, is reduplication, which has a variety of functions (3.8).

3.1 Morphophonemic processes

A consistent morphophonemic process occurs when a suffix follows a nominal base ending in schwa <è>. If the suffix begins with *a*, then the *è* is deleted – for example: *ághè* 'dog' plus the ergative suffix *-am* becomes *ágham*. If the suffix begins with any other phoneme, the final *è* changes to *a* – for example: *ághè* 'dog' plus the ablative suffix *-ta* becomes *ághata*. In addtion, when a base ends in a consonant and the suffix is a single a consonant, epenthesis of *è* occurs – for example: *Moet* 'Morehead' plus the allative suffix *-t* becomes *Moetèt.*

3.2 Number marking suffixes

The marking of number on nominals is marginal in Nama, with number normally expressed by mandatory verb morphology. Nominal suffixes marking both nonsingular and singular are described in detail by consultants; however, they occur only rarely in the recordings. It appears that they are vestiges of a once more productive system.

3.2.1 Nonsingular marking *-af, -f, -of*

Consultants provide nonsingular forms for most nominals with human referents (especially relations – i.e. terms for relatives and friends), and also for some referring to culturally important animals. These forms most commonly have the

https://doi.org/10.1515/9783111077017-003

52 —— 3 Nominal morphology

suffix *-af*. This suffix is used for all nominals ending in *è*, with the final *è* deleted. Examples are given in Table 3.1.

Table 3.1: Nonsingular nominals with *-af*.

Singular		Nonsingular	
amaf	'woman'	*amafaf*	'women'
ghakèr	'boy'	*ghakraf*	'boys'
ambum	'child'	*ambumaf*	'children'
mèrès	'girl'	*mèrsaf*	'girls'
taghar	'in-law'	*tagharaf*	'in-laws'
afè	'father'	*afaf*	'fathers'
mwitè	'exchange cousin'	*mwitaf*	'exchange cousins'
anè	'older sibling'	*ánaf*	'older siblings'
yare	'close friend'	*yáraf*	'close friends'

The suffix *-f* is also used with some nominals (with epenthetic *è* inserted after consonants). Examples are given in Table 3.2

Table 3.2: Nonsingular nominals with *-f*.

Singular		Nonsingular	
áki	'grandfather'	*ákif*	'grandfathers'
tagase	'my namesake'	*tagasef*	'my namesakes'
nakum	'brother-in-law'	*nakumèf*	'brothers-in-law'
yákál	'cousin'	*yákálèf*	'cousins'

There is some variation – e.g. *nèngnaf ~nèngnèf* 'younger siblings', *frenaf ~ frenèf* 'friends'.

Consultants also describe another nonsingular suffix *-of*, which refers to paucal number (2 or 3) as opposed to plural (4+). There is general agreement about its existence for a small number of nominals; however, none appear in the naturalistic recorded data. See Table 3.3 for examples.

Table 3.3: Plural vs paucal.

Singular		Paucal	Plural
ghakèr	'boy'	*ghakrof*	*ghakraf*
ambum	'child'	*ambumof*	*ambumaf*
mèrès	'girl'	*mèrsof*	*mèrsaf*
nèngèn	'younger sibling'	*nèngnof*	*nèngnaf/nèngnèf*
fren	'friend'	*frenof*	*frenaf/frenèf*

Two examples with plural marking from recordings of narratives follow:[13]

(5) *tane*   *nèngènèf*   *kètè*   *támorangèrwèn.*   (MK13(4))
    tane   nèngèn-**f**   kètè   támorangèrwèn
    1SG:GEN   younger.sibling-**PL**   there   stay:3NSG:REM.DUR
    'my (younger) brothers were staying there.'

(6) *áki-áki*   *kèraf*   *ndernae*   *támorangèrwèng.*   (ND11)
    ákiáki   kèr-**af**   ndernae   támorangèrwèng
    ancestor   dead-**PL**   how   stay:3NSG:REM.DUR
    'how our dead grandfathers lived.'

The plural and paucal markers occur most often with major case-marking suffixes (section 3.3), in which case, they are followed first by the dative suffix -*e* (section 3.3.2). In only one instance (example 7), the dative marker occurs with the plural marker without other case markers and without a dative function:

(7) *mèngotuan*   *ánde*   *tane*   *bafafe*   *emor...*   (YD11(2))
    mèngotu-an   ánde   tane   bafè-**af-e**   emor
    village-LOC1   where   1SG:GEN   uncle-**PL-DAT**   stay:3NSG:CUR
    'in the village where my uncles live ...'

3.2.2 Singular marking -*o*

There is also an optional singular marker -*o* which occurs mainly with nominals with human referents but also for *Ngánján* 'God' and significant animals such as *kimb* 'pig' and *wèrár* 'wallaby'. However, this marker only occurs preceding a case marker, as described below.[14]

---

13 Since the details of verb morphology are not introduced until the following chapter, an integrated glossing style is used for verbs in this chapter, rather than morpheme by morpheme glosses. The meaning of the verb stem is given along with the person and number of the arguments. The tense/aspect is also given, although the individual affixes indicating tense/aspect are not specified. Other individual affixes (such as deictic and applicative) and clitics (proximal) are indicated.
14 An alternative analysis of the -*of* paucal marker is that it is composed of the singular suffix -*o* plus the nonsingular suffix -*f*.

54 —— 3 Nominal morphology

3.3 Major case-marking suffixes

Case-marking suffixes in Nama indicate grammarical or semantic relationships beween nominals. Nama has 15 case-marking suffixes that are widely used and occur frequently in the recorded data. Each of these is decribed here. (Marginal case-marking suffixes are described in section 3.4.)

3.3.1 Ergative (ERG) -*am*, -*m*

The ergative suffix has a syntactic function, marking the A argument of a transitive verb (see section 1.4). It occurs only with the noun subclass of nominals (see section 5.1.1), attached to the final constituent of the nominal phrase making up the A argument.[15] It has different forms determined by number and the semantic subclass of the nominal.

The general ergative suffix is -*am*. It is used for nominals with both animate and inanimate referents, including the stimulus in P-focussed constructions (section 6.1.1). Some examples:

(8) *kambanam* *yènd* *fèfè* *wèrnam.* (YD11(3))
kamban-**am** yènd fèfè wèrnam
snake-**ERG** 1.ABS TA bite:3SG>1SG:REM.DLT
'The snake nearly bit me.'[16]

(9) *nèkwam* *yifoi.* (MD11(2))
nèkw-**am** yifoi
anger-**ERG** finish:3SG>3SG:REM.PUNC
'He got angry.' (lit. 'anger finished him.')

(10) *ágham* *wèrár* *yifnamènd.* (MD93)
ághè-**am** wèrár yifnamènd
dog-**ERG** wallaby chase:3SG>3SG:REM.DLT
'The dog chased the wallaby.'

---

15 The fact that the ergative suffix appears to have scope over the whole NP is an argument for analysing it as a clitic. However, since it is restricted to only nouns, and it does not follow nouns with other case suffixes, it is analysed here as a suffix.

16 The abbreviation TA is used to gloss preverbal tense/aspect markers until they are described in chapter 5, section 5.4. The symbol '>' in glosses in this chapter is used for transitive verbs. The person and number of the A argument precedes '>' and the person and number of the P argument follows '>'.

The suffix *-m* is used with personal names (with schwa inserted after a consonant), and with plural-marked nouns, following the dative suffix (see section 3.3.2).

(11) *Ausè*   *Kafukèm*   *yinjoi.*   (MD99(1))
ausè   Kafuk-**m**   yinjoi
old.woman Kafuk-**ERG** see:3SG>3SG:REM.PUNC
'Old woman Kafuk saw him.'

(12) *Gumam*   *yètkwèn*   *tosae*   *yaram.*   (NB2:80)
Guma-**m**   yètkwèn   tosè-e   yaram
Guma-**ERG** name   baby-DAT give:3SG>3SG:APP:CUR.INC
'Guma gave a name to his baby.'

The allomorph *-yam* follows nominals, including personal names, ending in *o* (but not nominals with the singular marker *-o*)

(13) Kwaro**yam**   yènmaramai.   (SK11(2))
Kwaro-**yam** yènmaramai
Kwaro-**ERG** give:1SG>1NSG:APP:REM.PUNC
'Kwaro gave it to us.'

The *-m* ergative suffix is also most commonly used with *ár* 'man, person'.

(14) *Yènè* *árèm*   *yèsmete.*   (PT:350)
yènè   ár-**m**   yèsmete
DEM   man-**ERG** hit:3SG>3SG:CUR
'This man hit her.'

The *-m* ergative suffix follows the singular suffix *-o* that optionally occurs with some nominals – e.g.:

(15) *Yènè* *árom*   *yènd* *wawèrsote.*   (NB4:41)
yènè   ár-o-**m**   yènd   wawèrsote
DEM   man-SG-**ERG** 1.ABS trouble:3SG>1SG:APP:CUR
'This man is troubling me.'

(16) *...tane* *ambumom* *nèmèf* *yaitote.*   (KS11)
tane   ambum-o-**m** nèmèf   yaitote
1SG:GEN child-SG-**ERG** just:what tell:3SG>3SG:CUR
'...what my son just told.'

The distinction between singular and nonsingular is illustrated in these examples:

(17) *Mèrsom   yame   yèrnyang.*   (NB2:2)
mèrès-**o-m**   yame   yèrnyang
girl-SG-**ERG**   mat   start:3SG>3SG:CUR.INC
'The girl started (making) the mat.'

(18) *Mèrsam   yame   yèrnyangi.*   (NB2:2)
Mèrès-**am**   yame   yèrnyangi
girl-**ERG**   mat   start:3PL>3SG:CUR.INC
'The girls started (making) the mat.'

(19) *ndernae   so   ghakrom   tèkafangè?*   (B&B:195)
ndernae   so   ghakèr-**o-m**   tèkafangè
how   FUT boy-**SG-ERG**   pick:3SG>3SG:REC.INC
'How will the boy choose her?'

(20) *ghakèramro   enetat.*   (PT:107)
ghakèr-**am**-ro   enetat
boy-**ERG**-RSTR   drink:3NSG>3NSG:CUR
'Only boys drink them.'

Absolutive vs ergative case

The P argument of tranisitve verbs and the S argument of intransitive verbs (both A- and P-aligned) have absolutive case. As opposed to ergative case, absolutive case is zero marked on nouns in Nama, as can be seen in example (10), where the P argument *wèrár* has no suffix. With regard to pronouns, however, there are distinct absolutive forms.

Ergative and absolutive pronouns

Pronouns make up a subclass of nominals (see section 5.1.2). Nama has a set of six independent ergative pronouns, three with singular (SG) reference and three with nonsingular (NSG) reference. It also has a distinct set of three independent absolutive pronouns that do not distinguish singular and nonsingular. (See Table 3.4.) The use of the 1st person absolutive pronoun can be seen in examples (8) and (15).

The combination of the plural marker -*f*, the dative marker -*e* (section 3.3.2) and the ergative suffix -*m* can be seen in the nonsingular ergative forms of the pronouns: *yèndfem* 'we', *fèmofem* 'you (NSG)' and *yèmofem* "they'. (This is considered

3.3 Major case-marking suffixes — **57**

Table 3.4: Ergative and absolutive pronouns in Nama.

	Ergative	Absolutive
1SG	*yèndo(n)*	*yènd*
1NSG	*yèndfem*	
2SG	*fèmo(n)*	*fèm*
2NSG	*fèmofem/fèmafem*	
3SG	*yèmo(n)*	*fá*
3NSG	*yèmofem*	

to be historical rather than prodcutive.) The distinction between paucal and plural seems to be maintained in the two 2nd person nonsingular ergative pronouns:

(21) **Fèmofem**   nèmè   ewafrot?   (DC01)
      2NSG.ERG   what   do:2DU>3NSG:CUR
      'What are you (two) doing?'

(22) ndenè   **fèmafem**   nèmè   wagh-e   welkèm   yènd
      like.this   **2NSG.PL**   what   dance-INST   welcome   1.ABS
      yènmafrotangi.   (B&B:143)
      do:3PL>1NSG:APP:INC.IPFV
      'like this you all welcomed us with whatever dance.'

3.3.2 Dative (DAT) -e

The dative suffix -e marks the recipient or beneficiary in ditransitive constructions which have the applicative prefix (APP) on the verb (see section 4.5.2). It is also used to mark the beneficiary or goal in some other contructions.

(23) *Yèmo*   *náifè*   *yaram*   *Mawaie.*   (NB3:12)
      yèmo   náifè   yaram   Mawai-**e**
      3SG.ERG   knife   give:3SG>3SG:APP:CUR.INC   Mawai-**DAT**
      'He just gave Mawai the knife.'

(24) *Yèndo*   *Marie*   *si*   *yawátemban.*   (NB3:38)
      Yèndo   Mari-**e**   si   yawátemban
      1SG.ERG   Murry-**DAT**   talk   send:3SG>3SG:APP:CUR.INC
      'I've just sent a message to Murry.'

58 —— 3 Nominal morphology

(25) *Bene*   *yaoroi.*   (NB9:7)
Ben-**e**   yaoroi
Ben-**DAT**   leave:3SG>3SG:APP:REM.PUNC
'He left it for Ben.'

The dative suffix can follow the optional singular suffix *-o* – e.g.:

(26) *Yèndo*   *yènè*   *ároe*   *yaulon.*   (NB6:95)
Yèndo   yènè   ár-**o-e**   yaulon
1SG.ERG   DEM   man-SG-**DAT**   lie.in.wait:3SG>3SG:CUR.INC
'I've just started lying in wait for that man.'

Most animate nominals and some inanimate ones as well occur with the singular
and dative suffixes – e.g.: *gharkèr-o-e* 'to/for the boy', *kimb-o-e* 'to/for the pig'. *skul-o-e* 'to/for the school' and *wem-o-e* 'to/for the yam'.

     Word final *è* changes to *a* preceding the dative suffix – e.g. *áuyè* 'cassowary' + *-e*
→ *áuyae* 'to/for the cassowary'; *bafè* 'uncle' + *-e* → *bafae* 'to/for the uncle'.

     Nama also has a distinct set of independent dative pronouns, as shown in
Table 3.5, followed by examples.

Table 3.5: Dative pronouns
in Nama.

1SG	*ta*
1NSG	*tèfe*
2SG	*fe*
2NSG	*fèfe*
3SG	*yá*
3NSG	*yèfe*

(27) *Yèmo*   **yá**   *sifayè*   *yawalo.*   (NB5:79)
yèmo   **yá**   sifayè   yawalo
3SG.ERG   **3SG.DAT**   place   give.way:3SG>3SG:CUR.INC
'He just gave way for a place for him.'

3.3 Major case-marking suffixes — 59

(28) *Jefene   flauamèn   ta   Marim   klind*
    Jefene   flauè-mèn   **ta**   Mari-m   klind
    Jeff:GEN donut-ORIG **1SG.DAT** Murry-ERG piece
    waram.   (NB10:15)
    waram
    give:3SG>1SG:APP:CUR.INC
    'Murry just gave me a piece from Jeff's donut.'

The dative suffix -*e* also acts as a ligature preceding other case suffixes. (Note that it is still glossed as DAT when it has this function.) As already mentioned, if a nominal is marked by the nonsingular suffix -*af*/-*f*, the dative suffix must precede the ergative suffix -*m*:

(29) *tanjo   ákifem ...*   *ndèrnae*
    tanjo   áki-**f-e**-m   ndèrnae
    1SG.own grandfather-**NSG-DAT**-ERG how
    kafrotawèt.   (TE13)
    kafrotawèt
    tell.stories:3NSG:REM.DUR
    'how my own grandfathers (i.e. ancestors) . . . were telling stories.'

(30) *Afafem   sái   yènè   ásáfogh*
    afè-**af-e**-m   sái   yènè   ásáfo-gh
    father-**PL-DAT**-ERG intend DEM   work-NOM
    ewafrotat.   (FD13)
    ewafrotat
    do:3NSG>3NSG:CUR
    'The fathers intend to do this work.'

The endings of the nonsingular dative pronouns (Table 3.5) also consist of the nonsingular marker -*f* plus the dative suffix -*e*, as in this example:

(31) *mer   si   ewaramwè*   *yèfe*   (PT:157)
    mer   si   ewaramwè   **yèfe**
    good talk give:3SG>3DU:APP:CUR **3NSG.DAT**
    'he's giving the two of them a good talk.'

The dative suffix precedes many of the other case markers, especially for nominals with animate referents. These are described below and summarised in Table 3.14. The dative pronouns also form part of other independent pronouns (see below).

60 —— 3 Nominal morphology

3.3.3 Instrumental (INST) *-e*

The instrumental suffix is *-e*. Here it is analysed as a separate suffix that is a homophone of the dative suffix. However an alternative analysis would be that it is actually the dative suffix which has an instrumental function with nominals with non-animate referents.

(32)   *Yèndo*   *mbilae*   *wèn*   *táfotawèn*   *áráke*
       Yèndo   mbilè-**e**   wèn   táfotawèn   áráke
       1SG.ERG   axe-**INST**   wood   cut:1SG>3NSG:REM.DUR   fence
       ramghèt.   (TE11(1))
       ramè-gh-t
       make-NOM-ALL
       'I cut the wood with an axe to make a fence.'

(33)   *Ámb kwèfon*   *sèkwe*   *nóngèfnamoyèm*   *witma*
       Ámb kwèfon   sèkw-**e**   nóngèfnamoyèm   witma
       next.day   canoe-**INST**   cross:1NSG:REM.PUNC   side
       karfot.   (YD11(3))
       karèf-ot
       side-PURP
       'The next day we crossed by canoe to get to the other side.'

The instrumental suffix is also used on nominals for an adverbial function:

(34)   *Mètare*   *ndenè*   *endmè*   *tèngmorwèn.*   (PT:699)
       mètar-**e**   ndenè   end-mè   tèngmorwèn
       quiet-**INST**   like.this   road-PERL   go:3SG:REM.DUR
       'Quietly, he walked down the road.'

(35)   *Fá*   *kwankwane*   *nuwanoi.*   (DR8/18)
       fá   kwankwan-**e**   nuwanoi
       3.ABS   secret-**INST**   set.off:3SG:REM.PUNC
       'She left secretly.'

3.3.4 Genitive (GEN) *-ne*

The genitive (or possessive) suffix is *-ne*, most commonly following the dative suffix *-e*:

3.3 Major case-marking suffixes —— **61**

(36) *Mariene*   *yáf*   (NB5:82)
Mari-e-**ne**   yáf
Murry-DAT-**GEN**   basket
'Murry's basket'

(37) *Mendwarene*   *tokèf*   *yèndfem*   *yènfandamèm.*   (MD99(1))
Mendwar-e-**ne**   tokèf   yèndfem   yènfandamèm
Mendwar-DAT-**GEN**   grave   1NSG.ERG   look:1NSG>3SG:VEN:REM.DLT
'Mendwar's grave, we came and looked at it.'

This includes plural marked nominals:

(38) *ákiafene*   *yam*   (DC15)
áki-af-e-**ne**   yam
grandfather-PL-DAT-**GEN**   custom
'grandfathers' (i.e. ancestors') custom'

(39) *Fá*   *yau*   *kèr*   *árfene*   *Ngánján*   *yèm.*   (Mark 12:27)
fá   yau   kèr   ár-f-e-**ne**   Ngánján   yèm
3.ABS   NEG   dead   person-PL-DAT-**GEN**   God   COP:3SG:CUR
'He is not the God of dead people.'

It also includes singular marked nominals:

(40) *mèrsoene*   *kimb*   (ND11)
mèrès-o-e-**ne**   kimb
girl-SG-DAT-**GEN**   pig
'the girl's pig'

(41) *Yènè*   *ároene*   *amaf*   *tándái*   *tè*
yènè   ár-o-e-**ne**   amaf   tándái   tè
DEM   man-SG-DAT-**GEN**   woman   long.ago   TA
nawangnai.   (NB10:73)
nawangnai
die.in.childbirth:3SG:REM.PUNC
'This man's wife died in childbirth long ago.'

62 —— 3 Nominal morphology

(42) *Pitoene*   *yènè fyèm.*   (DC14)
      Pitè-o-e-**ne**[17]   yènè fyèm
      Peter-SG-DAT-**GEN** DEM   COP:PROX1:3SG:CUR
      'It's just Peter's.'

Consultants report that for some nominals, there are four possible genitive forms, one general and three marked for specific number: singular, paucal and plural. Some examples are given in Table 3.6. [17]

Table 3.6: Number-marked genitives.

Nominals	General	Singular	Paucal	Plural
boy's/boys'	*ghakrene*	*ghakroene*	*ghakrofene*	*ghakrafene*
child's/children's	*ambumene*	*ambumoene*	*ambumofene*	*ambumafene*
girl's/girls'	*mèrsene*	*mèrsoene*	*mèrsofene*	*mèrsafene*
wallaby's/wallabies'	*wèrárene*	*wèrároene*	*wèrárofene*	*wèrárafene*
bandicoot's/bandicoots'	*ñárárene*	*ñárároene*	*ñárárofene*	*ñárárafene*
snake's'/snakes'	*kambanene*	*kambanoene*	*kambanofene*	*kambanafene*

The dative suffix is not used with the genitive suffix for bases ending in *è*, which changes to *a* when followed by *-ne*:

(43) *Ausane*   *mèrès yènè fyèm.*   (PT:6)
      ausè-**ne**   mèrès yènè fyèm
      old.woman-**GEN** girl   DEM   COP:PROX1:3SG:CUR
      'This is just the old lady's girl.'

(44) *Ghèrayane   fifi*   *mènè   yèm*   *fái.*   (KS11)
      Ghèrayè-**ne**   fifi   mènè   yèm   fái
      Daraia-**GEN**   meaning thing COP:3SG:CUR EMPH
      'It is Daraia's real meaning.'

The independent possessive (genitive) pronouns (Table 3.7) consist of the dative pronoun with the *-ne* genitive suffix – e.g. *tane* 'my' (*ta* 1SG.DAT + *-ne* GEN).

---

17 Note that the final *è* in *Pitè* is dropped when followed by the singular suffix.

Table 3.7: Possessive pronouns in Nama.

1SG	*tane*
1NSG	*tèfene*
2SG	*fene*
2NSG	*fèfene*
3SG	*yáne*
3NSG	*yèfene*

(45) *Fronde* **tane** *mèngo yènfandwèm.* (MD13(1))
First **1SG.GEN** house look:1DU>3SG:VEN:CUR
'First we (2) look at my house.'

(46) *Yèndo wèi* **yèfene** *ambum engfandan.* (FD13)
1SG.ERG now **3NSG.GEN** child look:1SG>3NSG:AND:CUR
'I watch their children now.'

3.3.5 Comitative (COM) *-afè, -fè*

The most common form of the comitative suffix 'with' is *-afè*. Again, final *è* is deleted when this suffix is added.

(47) *Ámb* *misimisiafè* *nuwanoyèm* *Kiungat.* (YD11(3))
ámb *misimisi-***afè** *nuwanoyèm* *Kiunga-t*
some pastors-**COM** set.off:1NSG:REM.PUNC Kiunga-ALL
'(We) set off with some pastors to Kiunga.'

(48) *Yènd* *a* *afafè* *álet* *nuwaneayèm.* (MD93)
Yènd *a* *afè-***afè** *ále-t* *nuwaneayèm*
1.ABS and father-**COM** hunting-ALL set.off:1DU:REM.PUNC
'I set off for hunting with father.'

(49) *ághafè* *tèmorwèn.* (MD13(1))
*ághè-***afè** *tèmorwèn*
dog-**COM** stay:3SG:REM.DUR
'he stayed with the dog.'

64 —— 3 Nominal morphology

An allomorph *-fè* is variably used with bases ending in *e* – for example: *sáláme-fè* ~ *sáláme-afè* 'with the clothes'. But it is always used with personal names and following the *-o* singular marker:

(50) *Seforfè*   *kokawend.*   (MD99(1))
      Sefor-**fè**   kokawend
      Sefor-**COM**   fight:RECIP:3DU:REM.DUR
      'he fought with Sefor.' (lit. 'with Sefor they (2) fought.')

(51) *Ámb káye*   *yènd*   *tanjo*   *amafofè*   *nuwaneayèm*
      ámb káye   yènd   tanjo   amaf-o-**fè**   nuwaneayèm
      one.day   1.ABS   1SG.own   woman-SG-**COM**   set.off:1NSG:REM.PUNC
      ndauot.   (TE11(1))
      ndau-ot
      garden-ALL
      'One day I set off to the garden with my wife.'

Some other examples of the singular/nonsingular distinction with the comitative suffix are shown in Table 3.8.

Table 3.8: Singular/nonsingular distinctions in comitative case.

Singular		Nonsingular	
mèrsofè	'with the girl'	*mèrsafè*	'with the girls'
ghakrofè	'with the boy'	*ghakrafè*	'with the boys'
ároefè	'with the man'	*árfè/árafè*	'with the men'
amafofè	'with the woman'	*amafafè*	'with the women'
ambumofè	'with the child'	*ambumafè*	'with the children'
kimbofè	'with the pig'	*kimbafè*	'with the pigs'
wèrárofè	'with the wallaby'	*wèrárafè*	'with the wallabies'

Nominals with the comitative suffix can function as adjectives or adverbs:

(52) *Dafi*   *nèkwafè*   *kamndangè.*   (YD11(3))
      Dafi   nèkw-**afè**   kamndangè
      Duffy   anger-**COM**   become:3SG:REC.INC
      'Duffy became angry.'

3.3 Major case-marking suffixes — 65

(53) *wèriafè*   *kár*   *tèm.*   (PT:241)
     wèri-**afè**   kár   tèm
     drunkenness-**COM**   might   COP:3SG:REC
     'he might have been drunk.'

(54) *áuwèghafè*   *tawambnangèn.*   (TE11(1))
     áuwè-gh-**afè**   tawambnangèn
     be.happy-NOM-**COM**   hug:1SG>3NSG:REC.INC
     'I hugged them with happiness.'

The independent comitative pronouns, (Table 3.9) differ from those of most other cases. First, there are no separate singular and nonsingular forms for each person. Second, the case suffix *-fe/-afe* is not appended to the dative pronoun, but rather to the absolutive pronoun for 1st and 2nd person and to the ergative singular pronoun for 3rd person:

Table 3.9: Comitative pronouns in Nama.

1SG	*yèndfè*
1NSG	
2SG	*fèmafè*
2NSG	
3SG	*yèmafè/yèmofè*
3NSG	

(55) *yènd*   **yèmafè**   *yèntèmangre.*   (KS11)
     1.ABS   **3SG.COM**   be.united:1DU:CUR.INC
     'I am united with him.'

3.3.6 Privative (PRIV) *-ofnar, -afnar*

The main privative suffix is *-ofnar*:

(56) *Matsiofnar*   *tè*   *yènfrangoyènd.*   (MD99(1))
     matsi-**ofnar**   tè   yènfrangoyènd
     report-**PRIV**   TA   leave:3NSG>1NSG:REM.PUNC
     'They left us without a report.'

66 —— 3 Nominal morphology

(57) *Árofnar*   *sifayan*   *mè*   *yámbya*   *ár*
ár-**ofnar**   sifayè-n   mè   yámbya   ár
person-**PRIV**   place-LOC1   still   3SG.alone   man
tèkmangèrwèn.   (MD13(1))
tèkmangèrwèn
sleep:3SG:REM.DUR
'The man was still sleeping by himself at a place without people.'

When the nominal ends in *o*, the initial *o* is deleted – e.g. *mèngo* 'house' + *-ofnar* becomes *mèngofnar* 'without a house'.

The allomoph *-afnar* occurs with bases ending in *è*, which is deleted:

(58) *yènd*   *waft**afnar***   *wèm*   *ramèghèt.*   (Mark 10:40)
yènd   waftè-**afnar**   wèm   ramè-gh-t
1.ABS   power-**PRIV**   COP:1SG:CUR   give-NOM-ALL
'I am without the power to grant it.'

With personal names, the dative suffix *-e* is required before the privative suffix – e.g. *Toni-e-ofnar* 'without Tony'. The singular and nonsingular markers are not used with the privative suffix.

The Namat (Mibini) form of the privative suffix *-ofnae* is also heard in Nama – e.g.:

(59) *si**ofnae***   *wèm.*   (B&B:159)
si-**ofnae**   wèm
talk-**PRIV**   COP:1SG:CUR
'I am without words.'

Nominals with the privative suffix can function as adjectives:

(60) *Yènd*   *yènamamèn*   *áuwègh**ofnar***   *wèm.*   (TE13)
yènd   yènamamèn   áuwè-gh-**ofnar**   wèm
1.ABS   because.of.this   be.happy-NOM-**PRIV**   COP:1SG:CUR
'I was unhappy because of this.'

(61) *Sáláme*   *njúnjún**ofnar***   *em.*   (NB4:56)
sáláme   njúnjún-**ofnar**   em
clothes   dirt-**PRIV**   COP:3PL:CUR
'The clothes are clean (lit. without dirt).'

Privative pronouns are formed by the addition of *-ofnar* to the dative pronoun – for example: *feofnar* 'without you (SG)'; *tèfeofnar* 'without us'.

3.3.7 Locative

Nama has two locative suffixes, one inessive (*-n*) and one both adessive and superessive (*-an*). Although similar in form and function, they do contrast in some contexts, as illustrated at the end of this subsection. A nominal with a locative suffix most often has an adverbial function, like a locative propositional phrase with *in* or *on* in English.

Locative 1 (LOC1) (inessive) *-n*

The inessive marker is *-n*, again with *è* inserted when it follows a consonant. Its meaning is mainly 'in' or 'inside', but its usage does not always correspond to the meanings of these prepositions in English. Some examples:

(62) *Yèmo* *yèlawete*   *mèngon.* (NB2:97)
Yèmo yèlawete   mèngo-**n**
1SG.ERG put.inside:3SG>3SG:CUR house-**LOC1**
'He's putting it in the house.'

(63) *Yènd* *kuflangèm*   *tèfenjo* *sèkwèn.* (YD11(3))
yènd kuflangèm   tèfenjo sèkw-**n**
1.ABS go.inside:1NSG:REC.INC 1NSG.own canoe-**LOC1**
'We got into our own canoe.'

(64) *yèkwèn* *tènyawangè.* (MD99(2))
yèkw-**n** tènyawangè
well-**LOC1** put.in:3SG>3SG:REC.INC
'he put him in the well.'

(65) *. . . yalitotam*   *Jonèm* *Jodan* *wanjen.* (Mark 1:9)
yalitotam   Jon-m Jodan wanje-**n**
submerge:3SG>3SG:REM.DLT John-ERG Jordan river-**LOC1**
'. . . he was baptised by John in the river Jordan.'

68 — 3 Nominal morphology

(66) *Fangarkemèn   ár   ewinjeayèm   endèn.*   (TE11(1))
Fangarke-mèn   ár   ewinjeayèm   end-**n**
Fangarke-ORIG   man   see:1DU>3NSG:REM.PUNC   road-**LOC1**
'We (2) saw people from Fangarke on the road.'

The *-n* locative suffix is used for most place names – for example *Sarghárèn* 'in Sarghar' and *Moetèn* 'in Morehead'.

Like other case markers, the final *è* in a base changes to *a* when followed by *-n*, as with *sifayè* becoming *sifayan* 'in a place' (example 57) and the following:

(67) *Bulu   Mato   Ghèráyan   fronde   fifi   nufáryèng.*   (SK11(1))
Bulu   Mato   Ghèráyè-**n**   fronde   fifi   nufáryèng
Bulu   Mato   Daraia-**LOC1**   first   real   arrive:3SG:REM.PUNC
'Bulu Mato at the very beginning arrived at Daraia.'

When a base ends in *i, u* or *au*, the allomorph *-an* is used – for example *kafusi-an* 'in the cup'. (Note that this is homophonous with the adessive/superessive marker (see below).) Other examples:

(68) *Mbi   nèitam   kwartruan.*   (NP11)
mbi   nèitam   kwartru-**an**
sago   settle:3SG:REM.DLT   palm.sheath.container-**LOC1**
'The sago settled in the palm sheath container.'

(69) *nuan   kèghárngè.*   (MD13(1))
nu-**an**   kèghárngè
water-**LOC1**   walk.in.water:3SG:REC.INC
'he went in the water.'

(70) *Fèyo   ndauan   ásáfogh   nufngeayèm.*   (TE11(1))
fèyo   ndau-**an**   ásáfo-gh   nufngeayèm
then   garden-**LOC1**   work-NOM   start:1DU:REM.PUNC
'Then we started work in the garden.'

Locative 2 (LOC2) (adessive/superessive) *-an*

The adessive/superessive marker *-an* means mainly 'at' or 'on', but again its usage does not always correspond to the meanings of these prepositions in English.

3.3 Major case-marking suffixes — **69**

(71) *Amam   njamke ewaflitam   merekinan.*   (MD93)
amè-m   njamke ewaflitam   merekin-**an**
mother-ERG food   put.on:3SG>3NSG:REM.DLT   plate-**LOC2**
'Mother put the food on a plate.'

(72) *mbwito tènèm   eee   tane   kafkafan*
mbwito tènèm   eee   tane   kafkaf-**an**
rat   come:3SG:REC.INC PROL 1SG.GEN foot-**LOC2**
nafai.   (MD11(1))
nafai
go.on.top:3SG:CUR.INC
'a rat came eeee and went on my foot.'

(73) *skulan   skwèmorwèn.*   (FD13)
skul-**an**   skwèmorwèn
school-**LOC2**   stay:PROX2:3SG:REM.DUR
'I stayed at the school.'

(74) *yènd so yènmor   yènè mbandan.*   (ND11)
yènd so yènmor   yènè mband-**an**
1.ABS fut stay:1NSG:CUR DEM   ground-**LOC2**
'we will stay on this land.'

(75) *yáf   tawasurngè   kèghatan.*   (YD11(3))
yáf   tawasurngè   kèghat-**an**
basket pour:3SG>3NSG:REC.INC bush-**LOC2**
'she poured out the contents of the basket in the bush.'

An allomorph *-yan* follows word final *e* and word final *è*, which changes to *a* – e.g. *bèlè* 'grill' + *-an* is *bèlayan* 'on the grill'. Other examples:

(76) *kwèmb yáf kafweyan   yámayèn.*   (NP11)
kwèmb yáf kafwe-**yan**   yámayèn
basket.type branch-**LOC2** hang:1SG>3SG:REM.PUNC
'I hung the basket on the branch.'

(77) *diburayan yèlau   kwèfkwèfan.*   (PT:185)
diburè-**yan** yèlau   kwèfkwèf-**an**
prison-**LOC2** be.inside:3SG:CUR.INC darkness-**LOC2**
'he was inside the prison, in darkness.'

(78) *Mwighayan   kemnangè   emo   wèláite?*   (MD13(1))
mwighè-**yan**   kemnangè   emo   wèláite
mind-**LOC2**   think:3SG:REC.INC   who   shoot.at:3SG>1SG:CUR
'In his mind, he thought who's shooting at me?'

(79) *Kándè-kándè   efayotam   kitárayan.*   (MK13(4))
kándè-kándè   efayotam   kitárè-**yan**
yam.pudding   put.on.top:1NSG>3NSG:CUR   shelf-**LOC2**
'We put the yam pudding on the shelf.'

The *-an/-yan* locative morpheme is used with time and date expressions with an adverbial function – e.g. *yúnjèf-an* 'in the year'and *Sande-an* 'on Sunday'. Other examples:

(80) *yènè   sèmbáran ... fá   engèrmèn   Mátè.*   (MD99(1))
yènè   sèmbár-an   fá   engèrmèn   Mátè
DEM   night-**LOC2**   3.ABS   go:3DU:REM.DLT   Mata
'On that night ... they (2) went to Mata.'

(81) *yènè   akwan   kimb   yèsmáyènd.*   (MK13(4))
yènè   akw-**an**   kimb   yèsmáyènd
DEM   morning-**LOC2**   pig   kill:3NSG>3SG:REM.PUNC
'that morning they killed the pig.'

Locative 1 vs Locative 2
Because of the homophony of the inessive suffix occurring as *-an* following some vowels and the adessive/superessive suffix *-an*, the meaning may be ambiguous. For example, does *mèngotu* 'village' + *-an* mean 'in the village' or 'at the village'? However, that the two suffixes are distinct is clear, as shown by the following examples:

(82) *fèsèn   tè   náwúároi.*   (ND13)
fès-**n**   tè   náwúároi
fire-**LOC1**   TA   catch.on.fire:3SG:REM.PUNC
'it already burnt in the fire.'

(83) *Yèmo   brubru   yèfale   fèsan.*   (NB2:65)
yèmo   brubru   yèfale   fès-**an**
3SG.ERG   kundu.drum   warm.up:3SG>3SG:CUR   fire-**LOC2**
'He's warming up the kundu drum on the fire.'

3.3 Major case-marking suffixes — **71**

Note that neither locative suffix occurs with nominals with inanimate referents, and therefore neither cooccurs with number markers or the dative suffix. To express locative functions with nominals with animate referents, the associative case suffix -faf is used (section 3.3.12). Also, no locative forms of pronouns have been recorded; instead the associative forms of pronouns are used with locative functions.

3.3.8 Ablative (ABL) -ta

The ablative suffix indicating 'from' is -ta:

(84) *Alam*   *yame* *mèngota*
      Alè-am   yame mèngo-**ta**
      grandmother-ERG   mat   house-**ABL**
      ewewone.   (NB4:74)
      ewewone
      carry.down:3SG>3NSG:CUR
      'Grandmother is carrying mats down from the house.'

(85) *Farasi*   *ár*   *a*   *nák* *watamegh* *ár*   *efe*   *enmormèn*
      Farasi   ár   a   nák   watame-gh   ár   efe   enmormèn
      Pharisee man and law   teach-NOM   man who   come:3NSG:REM.DLT
      Jerusalemta   *Yesufè*   *náumendayènd.*   (Mark 7:1)
      Jerusalem-**ta** Yesu-fè   náumendayènd
      Jerusalem-**ABL** Yesu-COM   gather:3NSG:REM.PUNC
      'The Pharisees and some of the teachers of the law who had come from Jerusalem gathered around Jesus.'

(86) *áuwèghafè*   *wèi* *kènangotawèt*   *yènè*
      áuwè-gh-afè   wèi   kènangotawèt   yènè
      be.happy-NOM-COM now   come.back:3NSG:REM.DUR DEM
      waghta.   (ND11)
      wagh-**ta**
      singsing-**ABL**
      'happily now they were coming back from the singsing.'

As when other case markers are added, final *è* in a base becomes *a* – e.g. *ághè* 'dog' + -*ta* becomes *ághata* 'from the dog'.

72 —— 3 Nominal morphology

(87) *Mátata*   *nèngusaryèn*   *kètan...*   (FD13)
      Mátè-**ta**   nèngusaryèn   kètan
      Mata-**ABL** move:1SG:AND:REM.PUNC there
      'From Mata, I moved there...'

With personal names, the dative suffix -*e* precedes the ablative suffix – e.g. *Toni-e-ta* 'from Tony'. The dative suffix preceding the ablative suffix can also sometimes occur with other nominals with human referents, leading to variable forms – e.g. *áki-ta* ~ *áki-e-ta* 'from grandfather'; *fren-ta* ~ *fren-e-ta* 'from the friend'.

The dative suffix is required following a number-marking suffix, and thus also preceding the ablative suffix. Some examples are given in Table 3.10.

Table 3.10: Ablative suffix with number-marked nominal.

Nominals	General	Singular	Paucal	Plural
from the boy/boys	*ghakèrta*	*ghakroeta*	*ghakrofeta*	*ghakrafeta*
from the girl/girls	*mèrèsta*	*mèrsoeta*	*mèrsofeta*	*mèrsafeta*
from the pig/pigs	*kimbta*	*kimboeta*	*kimbofeta*	*kimbafeta*
from the friend/friends	*fútárta*	*fútároeta*	*fútárofeta*	*fútárafeta*

The ablative suffix -*ta* also indicates position or location with a subset of nominals, as shown in the following example and Table 3.11.

(88) *Fèm*   *tane*   *sèmta*   *kokái.*   (NB10:29)
      Fèm   tane   sèm-**ta**   kokái
      2.ABS   1SG.GEN   back-**ABL** stand:2SG:REC.INC
      'Stand behind me (lit. from my back)'

Table 3.11: Locational/positional nominals marked by -*ta*.

Nominals		With -*ta*	
tambèn	'side, right (direction)'	*tambènta*	'right side'
fátrè	'left (direction)'	*fátrata*	'left side'
sèm	'back'	*sèmta*	'back (of a person or house)'
tikèf	'heart'	*tikèfta*	'front (of a person)'
sèsáfne	'door'	*sèsáfneta*	'side of the house where the door is'
tèndo	'side'	*tèndota*	'side by side position'
kwèmb	'bottom'	*kwèmbta*	'back (inside a vehicle)'
sènko	'head'	*sènkota*	'front (inside a vehicle)'

There are no ablative pronouns. Rather, the orientational noun *tambèn* 'side' (which also means 'right [direction]' as shown in Table 3.11) follows the dative pronoun to indicate ablative – for example:

(89)   *Fá*   *nákárote*   *tèfe*   **tambèn.**   (NB6:14)
       fá   nákárote   tèfe   **tambèn**
       3.ABS cut.oneself.off:3SG.CUR 1NSG.DAT **side**
       'He's cutting himself off from us.'

(For other uses of *tambèn* as an orientational noun, see chapter 5, examples 424–426.)

3.3.9 Perlative (PERL) -*mè*

The perlative suffix -*mè* indicates 'along' or 'through' – for example:

(90)   *Yènd*   *nuwaneayèm*   *Suki*   *endmè.*   (YD11(3))
       Yènd   nuwaneayèm   Suki   end-**mè**
       1.ABS set.off:1DU:REM.PUNC Suki   road-PERL
       'We two set off along the Suki road.'

(91)   *Fèyo*   *tètkafwemè*   *susi*   *wigh*
       fèyo   tètkafwe-**mè**   susi   wi-gh
       then   creek-**PERL**   fishing.line throw-NOM
       engwafngeayèm.   (TE11(2))
       engwafngeayèm
       start:1DU>3NSG:AND:REM.PUNC
       'Then along the creek (we two) started throwing the fishing lines.'

The allomorph -*amè* is used after bilabial consonants, the *au* diphthong, final *n* and final *è* (which is deleted):

(92)   *Wagif* *wanje* *kènjkamamè*   *narende.*   (NB6:80)
       Wagif wanje kènjkam-**amè**   narende.
       fish   river deep.part-**PERL** move.around:3SG:CUR
       'The fish is swimming along the deep part of the river.'

74 —— 3 Nominal morphology

(93) *tafarorngai*   *nèndkèfamè.*   (MD99(2))
tafarorngai   nèndkèf-**amè**
split:3SG>3NSG:REM.INC middle-**PERL**
'he divided them along the middle.'

(94) *tènfangèm*   *yabun*   *sauamè.*   (YD11(3))
tènfangèm   yabun   sau-**amè**
cut:1NSG>3SG:REC.INC big   swamp-**PERL**
'we cut through a big swamp.'

The perlative marker does not appear to occur with number-marked nominals. And there are no perlative pronouns.

3.3.10 Allative (ALL) -*t*

The allative suffix, meaning 'to', is -*t*, with epenthetic *è* inserted following a consonant:

(95) *Yènd* *yènngèm*   *Moetèt.*   (TE11(1))
yènd yènngèm   Moet-**t**
1.ABS go:1NSG:CUR.INC Morehead-**ALL**
'We're going to Morehead.'

(96) *kangotawèn*   *mèngot.*   (NP11)
kangotawèn   mèngo-**t**
return:1SG:REM.DUR house-**ALL**
'I returned home.'

(97) *tanangè* . . .   *yánjo* *kèmègh* *sifayat.* (ND13)
tanangè . . .   yánjo kèmègh sifayè-**t**
take:3SG>3NSG:REC.INC 3SG.own sleeping place-**ALL**
'she took them . . . to her own sleeping place.'

An allomorph -*ot* occurs when the base ends in *u*, *au* or *kw*:

(98) *kufarèn*   *mèngotuot.*   (MK13(2))
kufarèn   mèngotu-**ot**
arrive:1SG:REC.INC village-**ALL**
'I arrived at the village.'

3.3 Major case-marking suffixes — **75**

(99) *nuot   nègháryèng.*   (MD13(1))
     nu-**ot**   nègháryèng
     water-**ALL**   go.in.water:3SG:REM.PUNC
     'he went to/in the water.'

(100) *Amè   ndauot   tèyotang.*   (MB13)
      amè   ndau-**ot**   tèyotang
      mother   garden-**ALL**   travel:3SG:REC.INC
      'Mother travelled to the garden.'

With nominals marked for plural, the allative suffix preceded by the dative suffix can have a benefactive function – e.g.:

(101) *tane   futárfet   wèi   táwánfetawèt.*
      tane   futár-f-e-**t**   wèi   táwánfetawèt
      1SG.GEN   friend-PL-DAT-**ALL**   now   cut:3NSG>3NSG:APP:REM.DUR
      'they were cutting them for my friends now.'

(102) *ághafet   ne   ewatárnan.*   (MK13(2))
      ághè-f-e-**t**   ne   ewatárnan
      dog-PL-DAT-**ALL**   guts   throw:1SG>3NSG:APP:CUR
      'I threw the guts to/for the dogs.'

Consultants report that a paucal/plural distinction exists for at least two nominals, as shown in Table 3.12.

Table 3.12: Paucal/plural distinction with allative *-t.*

	paucal	plural
for the boys	*ghakrofet*	*ghakrafet*
for the girls	*mèrsofet*	*mèrsafet*

With nominalised verbs (section 4.1), the allative suffix indicates purpose or intention, like the English infinitive marker *to* – e.g.:

76 —— 3 Nominal morphology

(103) *fèyo yènè seyotamèng fái sau*
fèyo yènè seyotamèng fái sau
then DEM travel:PROX2:3NSG:REM.DLT right swamp
unghèt. (ND13)
unè-gh-**t**
net.fish-NOM-**ALL**
'Then they travelled right to the swamp to net fish.'

(104) *Ndambrorarèt yènngre ásáfoghèt* (TE11(1))
Ndambrorar-**t** yènngre ásáfo-gh-**t**
Ndambrorar-**ALL** go:1DU:CUR work-NOM-**ALL**
'We (2) are going to Ndambrorar to work.'

(105) *Mèrsom yánjo mèngot tènamnde*
mèrès-o-m yánjo mèngo-t tènamnde
girl-SG-ERG 3SG.own house-ALL tell:3SG>1NSG:CUR
kèmghèt. (YD11(3))
kèmè-gh-**t**
lie.down-NOM-**ALL**
'The girl told us [to go] to her own house to sleep.'

(106) *Lea yèngmorman fès rèsaghèt.* (MD99(1))
Lea yèngmorman fès rèsa-gh-**t**
Lea go:3SG:REM.DLT firewood carry-NOM-**ALL**
'Lea went to carry firewood.'

The allative suffix is also used to indicate purpose or intention with some action nominals, although most of them take the purposive suffix *-ot* (see section 3.11).

(107) *Fá náfrende álet.* (NB2:88)
fá náfrende ále-**t**
3.ABS prepare.self:3SG:CUR hunting-**ALL**
'He prepared himself for hunting.

(108) *Fá náunotat waghèt.* (NB10:37)
fá náunotat wagh-**t**
3.ABS get.ready:3NSG:CUR singsing-**ALL**
'They're getting ready for the singsing.'

3.3 Major case-marking suffixes — **77**

(109) *Yènd farat   nuwanotan.*   (ND2:43)
    yènd farè-**t**   nuwanotan
    1.ABS hunt.by.self-**ALL** set.off:1SG:CUR
    'I set off for hunting by myself.'

No allative forms of pronouns have been recorded. Rather, allative functions are expressed with associative forms of pronouns (section 3.3.12).

3.3.11 Purposive (PURP) -*ot*

The purposive suffix -*ot* is found most commonly with action nominals (see section 5.1):

(110) *Yènd a   tanjo   amaf susiot*
    yènd a   tanjo   amaf susi-**ot**
    1.ABS and 1SG.own wife fishing-**PURP**
    nuwaneayèm.   (TE11(2))
    nuwaneayèm
    set.off:1DU:REM.PUNC
    'My wife and I set off for fishing.'

(111) *Anna Forak nuwanoi   wanjet anuot.*   (YD11(2))
    Anna Forak nuwanoi   wanje-t anu-**ot**
    Anna Forak set.off:3SG:REM.PUNC river-ALL bathing-**PURP**
    'Anna Forak set off to the river to bathe . . . .'

(112) *Yèmo   nánji   fèsot   yafngo.*   (NB7:96)
    yèmo   nánji   fès-**ot**   yafngo
    3SG.ERG banana cooking-**PURP** start:3SG>3SG:REC.INC
    'She started to cook the banana.'

(113) *Árèm   wèrár   ghèman   ngiot*
    ár-m   wèrár   ghèm-an   ngi-**ot**
    man-ERG wallaby river.bend-LOC2 killing-**PURP**
    yarendangi.   (NB9:30)
    yarendangi
    surround:3PL>3SG:CUR.INC
    'The men surrounded the wallaby on the river bend to kill it.'

78 —— 3 Nominal morphology

But it is also used with other nominals:

(114) *Fá wèn fènean fámgharot nomanyote.* (NB6:60)
fá wèn fène-an fámghar-ot nomanyote
3.ABS tree hole-LOC2 honey-**PURP** put.hand.inside:3SG:CUR
'She put her hand in the hole in the tree to get honey.'

(115) *Yèmo sau wagifot yèrote.* (NB5:16)
yèmo sau wagif-ot yèrote
3SG.ERG swamp fish-**PURP** poison:3SG>3SG:CUR
'He's poisoning the swamp for fish.'

The purposive suffix is also used to mark a nominal that another has changed into or replaced:

(116) *Ár kètè ngafurnawèt kariot.* (SK11(2))
Ár kètè ngafurnawèt kari-ot
person there change:3NSG:AND:REM.DUR pandanus-**PURP**
'People there changed into pandanus.'

(117) *Fá ámb amafot awerghèt*
fá ámb amaf-ot awerè-gh-t
1.ABS another woman-**PURP** marry-NOM-ALL
nawalote. (NB5:67)
nawalote
change:3SG:CUR
'He changed to another woman to get married.'

It is sometimes difficult to distinguish the purposive suffix -*ot* from the homophonic allomorph of the allative suffix -*t*, as it occasionally appears to be used with allative functions with nominals not ending in *u* or *au*:

(118) *yènè áuyè ... so yèrsan Moet*
yènè áuyè ... so yèrsan Moet
DEM cassowary FUT carry:1SG>3SG:CUR Morehead
tesenot. (MD99(1))
tesen-ot
station-**ALL**
'I'll take this cassowary ... to Morehead station.'

3.3 Major case-marking suffixes — **79**

However, the distinction between the allative and the purposive suffixes can be seen with the word *fès* 'fire, cooking': *fèsèt* 'to the fire', *fèsot* 'for cooking' (see example 112).

An allomorph *-yot* occurs with nominals ending in *o*:

(119) *ndènd*   *so*   *tatarwèm*   *kwakoyot.*   (TE11)2))
       ndènd   so   tatarwèm   kwako-**yot**
       worm   FUT   dig:1DU>3NSG:CUR   bait-**PURP**
       'we'll dig worms for bait.'

This allomorph is also used with personal names, following the dative suffix *-e*, with a benefactive function:

(120) *Yèmo*   *njamke*   *Beneyot*   *yèrse.*   (LN99:11)
       yèmo   njamke   Ben-e-**yot**   yerse
       3SG.ERG   food   Ben-DAT-**PURP**   carry:3SG>3SG:CUR
       'He's getting food for Ben.'

This allomorph is also appended to dative pronouns to have a benefactive function.

(121) *si*   *so*   *yaitotan*   *fèfeyot* ...   (FD13)
       si   so   yaitotan   fèfe**yot**
       talk   FUT   tell:1SG>3SG:CUR   2NSG.**PURP**
       'I will tell the story for you all ...'

(122) ... *mènamèn*   *totèr*   *yam*   *tèmorwèn*   *yèfeyot.*   (MD13(1))
       mènamèn   totèr   yam   tèmorwèn   yèfe**yot**
       because   new   custom   COP:3SG:REM.DUR   3NSG.**PURP**
       '... because it was a new way for them.'

Consultants also report special dual forms that appear to use the allative suffix *-t*: *tèfet* 'for us 2', *fèfet* 'for you 2', *yèfèt* 'for those 2'. However no examples have been recorded.

3.3.12 Associative (ASS) *-faf*

The associative suffix *-faf* indicates 'at a place of' or 'associated with' – for example: *amjogh-faf* 'at a sitting place', *susi-faf* 'at a fishing place', *mbèr-faf* 'at a muddy place'. Other examples:

80 — 3 Nominal morphology

(123) *fèyo kètan ndaufaf nufareayèm.* (TE11(1))
fèyo kètan ndau-**faf** nufareayèm
then there garden-**ASS** arrive:1DU:REM.PUNC
'then we arrived there at the garden place.'

(124) *Sèkw namghetam yabun wènfaf.* (YD11(3))
sèkw namghetam yabun wèn-**faf**
canoe go.alongside:3SG:REM.DLT big tree-**ASS**
'The canoe went to the side where there were big trees.'

(125) *Sènonjo yènè yam syènèm totèr árfaf.* (ND11)
sènonjo yènè yam syènèm totèr ár-**faf**
today DEM custom come:PROX2:3SG:CUR young person-**ASS**
'Today this custom continues with the young people.'

For nominals with animate, mainly human, referents, and personal names, the associative suffix *-faf* generally follows the dative suffix *-e*. (An exception is *ár* 'man, person', as in the preceding example.) Some examples are *taghar-e-faf* 'where my in-law is' or 'associated with my in-law' and *Fransis-e-faf* 'where Francis is' or 'associated with Francis'. Other examples:

(126) *Yaufi ghèngan enjnefè ár yènmormèn*
yaufi ghèngan enjne-fè ár yènmormèn
bad skin sickness-COM man come:3SG:REM.DLT
Yesuefaf... (Mark 1:40)
Yesu-e-**faf**...
Jesus-DAT-**ASS**
'A man with leprosy came to where Jesus was...'

(127) *Fekifefaf kèngèm fès*
feki-f-e-**faf** kèngèm fès
your.grandfather-PL-DAT-**ASS** go:2SG:REC.INC firewood
tènmanang. (MD99(1))
tènmanang
take:3SG>3SG:VEN:REC.INC
'Go to where your grandfathers are and get firewood.'

When a nominal, inanimate or animate, ends in *è*, the dative suffix is not used and the usual change to *a* occurs, as in *ághè* + *-faf* becoming *ághafaf* 'where the dog is' and *Gaitè* + *-faf* becoming *Gaitafaf* 'where Gaita is' or 'associated with Gaita.'

For nominals with human referents, the associative suffix is generally used instead of the allative suffix to indicate 'to' – for example:

(128) *nuwanoyènd* Moet *ADCefaf*
nuwanoyènd Moet ADC-e-**faf**
set.off:3NSG:REM.PUNC Morehead ADC-DAT-**ASS**
nafroyènd. (MD99(1))
nafroyènd
report:3NSG:REM.PUNC
'they set off to Morehead to the ADC [Assistant District Commisioner] and reported.'

(129) *Si yátembayèn* *Mátè ár fútárfefaf.* (MK13(4))
si yátembayèn Mátè ár fútár-f-e-**faf**
talk send:1SG>3SG:REM.PUNC Mata man friend-PL-DAT-**ASS**
'I sent word to Mata people, to friends.'

The associative suffix may also indicate other meanings:

(130) *Fèyo Erodias nèkwafè tèmorwèn Jonefaf...* (Mark 6:19)
fèyo Erodias nèkw-afè tèmorwèn Jon-e-**faf**
so Herodias anger-COM COP:3SG:REM.DUR John-DAT-**ASS**
'So Herodias nursed a grudge against John...'

(131) *Danielèm mèrès yúmnjo*
Daniel-m mèrès yúmnjo
Daniel-ERG girl give.for.marriage:3SG>3SG:CUR.INC
Kwambarafaf. (NB5:82)
Kwambarè-**faf**
Kwambarè-**ASS**
'Daniel gave his daughter to Kwambarè for marriage.'

The associative suffix is also occasionally used with the singular suffix *-o* followed by the dative suffix *-e*:

(132) *fronde ghakroefaf tèram.* (B&B:284)
fronde ghakèr-o-e-**faf** tèram
first boy-SG-DAT-**ASS** give:3SG>3SG:CUR.INC
'first she gave it to the boy.'

82 —— 3 Nominal morphology

(133) *Nèmè yèndon so yafrotan fèfeyot*
nèmè yèndon so yafrotan fèfeyot
what 1SG.ERG FUT do:1SG>3SG:APP:CUR 2NSG.PURP

yènè ároefaf... (Mark 15:12)
yènè ár-o-e-**faf**
DEM man-SG-DAT-**ASS**
'What shall I do for you to this man ...'

Examples (127) and (129) show the use of *-faf* with the plural suffix *-f*. Consultants also report a paucal/plural distinction with at least two nominals (as shown in Table 3.13), but again, these do not occur in recordings:

Table 3.13: Number-marked associatives.

	General	Singular	Paucal	Plural
to/associated with the boy/boys	*ghakèrfaf*	*ghakroefaf*	*ghakrofefaf*	*ghakrafefaf*
to/associated with the girl/girls	*mèrsèsfaf*	*mèrsoefaf*	*mèrèsofefaf*	*mèrèsafefaf*

Independent associative pronouns are composed of the dative pronoun plus the associative suffix – e.g. *tafaf* 'to me/associaed with me', *tèfefaf* 'to us/associated with us' and *yèfefaf* 'to them/associated with them.

3.3.13 Originative (ORIG) *-mèn*

The originative suffix *-mèn* indicates 'from' or 'because of' – i.e. that the nominal is the origin or cause of something:

(134) *Fangarkemèn ár ewinjeayèm endèn.* (TE11(1))
(=66) Fangarke-**mèn** ár ewinjeayèm end-n
Fangarke-**ORIG** man see:1DU>3NSG:REM.PUNC road-LOC1
'We saw two men from Fangarke on the road.'

(135) *Fá sègerú wèrimèn nawakáyote.* (NB7:27)
fá sègerú wèri-**mèn** nawakáyote
1.ABS alcohol drunkenness-**ORIG** stagger:3SG:CUR
'He's staggering from drunkenness.'

3.3 Major case-marking suffixes — **83**

(136) *Ben áuyamèn káuwetau.* (KK11)
Ben áuyè-**mèn** káuwetau
Ben cassowary-**ORIG** be.happy:3SG:REM.DUR
'Ben was happy because of the cassowary.'

(137) *Fá nawarse wem yabunmèn.* (NB10:52)
fá nawarse wem yabun-**mèn**
3.ABS extol.self:3SG:CUR yam largeness-**ORIG**
'He is extolling himself because of the large size of his yams.'

(138) *Mèrès ndèrmbu ghènj rárèghmèn nawambetam.* (ND13)
mèrès ndèrmbu ghènj rár-gh-**mèn** nawambetam
girl sugar.ant bite-NOM-**ORIG** cry.out:3SG:REM.DLT
'The girl cried out because of the ants biting.'

The originative suffix can also mean 'about':

(139) *Yèndon mbi táfèghmèn si so yaitotan.* (NP11)
yèndon mbi táf-gh-**mèn** si so yaitotan
1SG.ERG sago chop-NOM-**ORIG** talk FUT tell:1SG>3SG:CUR
'I'll tell a story about making sago.'

(140) *Ár Ghèrayè mbandmèn namndèt.* (NB9:51)
ár Ghèrayè mband-**mèn** namndèt
man Daraia ground-**ORIG** argue:3DU:CUR
"The two Daraia men are arguing about land.'

(141) *Darfi tárfár maimèn nawatete.* (NB9:16)
Darfi tárfár mai-**mèn** nawatete
Darfi many long.yam-**ORIG** boast:3SG:CUR
'Darfi is boasting about the many yams.'

(142) *Ghèrayamèn tè yèngwafwerèn.* (KS11)
Ghèrayè-**mèn** tè yèngwaferèn
Daraia-**ORIG** TA explain:1SG>3SG:AND:REM.PUNC
'I've just explained about Daraia.'

The suffix also indicates other relationships relating to origin, as in the following examples:

(143)   *Yèmo*   *ambum*   *olitoghmèn*   *wanjen*
       yèmo   ambum   olito-gh-**mèn**   wanje-n
       3SG.ERG   child   drown-NOM-**ORIG** river-LOC1
       yanjane.   (NB5:77 rev 8/18)
       yanjane
       save:3SG>3SG:CUR
       'He's saving the child from drowning in the river.'

(144)   *Wár mèngo*   *yúsmèn*   *tèmorwèn.*   (DC37)
       Wár mèngo   yús-**mèn**   tèmorwèn
       initiation.house grass-**ORIG** COP:3SG:REM.DUR
       'The initiation house was made of grass.'

When the originative suffix is used with nominals with animate, mainly human, referents, it normally follows the genitive suffix -*ne* which follows the dative suffix -*e* – e.g. *Toni-e-ne-mèn* 'because of/about Tony', *taghar-e-ne-mèn* 'because of/about my in-law'. An example from the recordings:

(145)   *Yènè* *si*   *sáf...*   *fèrèt-fèrètenemèn*   *so*
       yènè   si   sáf   fèrètfèrèt-e-ne-**mèn**   so
       DEM   talk kind naive.bachelor-DAT-GEN-**ORIG** FUT
       yèm.   (MD13(1))
       yèm
       COP:3SG:CUR
       'This kind of story... is about a naive bachelor. '

According to consultants, this combination of suffixes can also follow the singular marker *o* for some animates – e.g. *mèrès-o-e-ne-mèn* 'because of/about the girl'; *kimb-o-e-ne-mèn* 'because of/about the pig'.

    Originative pronouns are composed of the dative pronoun -*e*, the genitive suffix -*ne* and the originative suffix -*mèn*: for example: *yánemèn* 'because of/about him/her' and:

(146)   *tukènro*   *yakái*   *tèfenemèn.*   (DC13)
       tuk-n-ro   yakái   tèfene**mèn**
       top-LOC1-RSTR stand:3SG:CUR.INC 1NSG.ORIG
       'it's standing up on top because of us.'

3.3.14 Semblative (SEMB) -*nit*

The semblative suffix -*nit* indicates 'like' or 'similar to' – for example" *onjonit* 'like a dream'; *bru wagifnit* 'like a catfish', *ághanit* 'like a dog'. Other examples:

(147) *árnit*   *kènáyátotau*   *tane* *mènè* *mwighè*
      ár-**nit**   kènáyátotau   tane mènè mwighè
      person-**SEMB** become:3SG:REM.DUR 1SG.GEN thing mind
      kènjún.   (FD13)
      kènjún
      ainside
      'this became like a person in my mind'

(148) *PNG* *Austrelianit* *yau* *yèm.*   (NB7:14)
      PNG Austreliè-**nit** yau yèm
      PNG Australia-**SEMB** NEG COP:3SG:CUR
      'PNG isn't like Australia.'

3.3.15 Summary

Table 3.14 lists the core case-marking suffixes and their allomorphs, and indicates whether or not they co-occur with the plural suffix –*(a)f* and the dative suffix -*e*.

3.4 Peripheral case-marking suffixes

Four other morphemes reported by consultants appear to function as case-marking suffixes but are either very rare or absent in the recorded data and Bible translation. Two of these are attributive and two are temporal.

3.4.1 Attributive

Each attributive suffix indicates that the nominal base is an attribute or characteristic of the nominal with which it cooccurs, thus taking on an adjectival function.

86 —— 3 Nominal morphology

Table 3.14: Summary of major case-marking suffixes in Nama.

case	suffix	allomorph(s)	following plural + dative suffix?	preceded by dative suffix without singular or plural suffix?
ergative	*-am*	*-m/-yam*	*-(a)fem*	no
dative	*-e*		*-(a)fe*	
instrumental	*-e*		*-(a)fe*	no
genitive	*-ne*		*-(a)fene*	always (*-ene*), except for bases ending in *è*
comitative	*-afè*	*-fè*	*-(a)fefè*	no
privative	*-ofnar*	*-afnar*	*-(a)feofnar*	no
locative 1	*-n*	*-an*	——	no
locative 2	*-an*	*-yan*	——	no
ablative	*-ta*		*-(a)feta*	with personal names (*-eta*); optional for other animates
perlative	*-mè*	*-amè*	——	no
allative	*-t*	*-ot*	*-(a)fet*	no
purposive	*-ot*	*-yot*	——	with personal names (*-eyot*)
associative	*-(a)faf*		*-(a)fefaf*	with personal names and nominals with human referents (*-efaf*); optional for some other animates
originative	*-mèn*			no
semblative	*-nit*			no

Attributive 1 (ATTR1) *-are*

The first attributive suffix *-are* is most often appended to nominalised verbs (section 4.1) but may occur on other nominals as well. Some examples are:

(149)   *Saf*   *umbghare*   *yèm.*   (NB4:57)
       saf   umbè-gh-**are**   yèm
       face   wrinkle-NOM-**ATTR1**   COP:3SG:CUR
       'The face is wrinkled.'

(150)   *Sódè*   *áftrárèghare*   *yèm.*   (NB2:53)
       sódè   áfètèrár-gh-**are**   yèm
       shirt   tear-NOM-**ATTR1**   COP:3SG:CUR
       'The shirt is torn.'

(151) *Ambum   simare   yèm.*   (NB6:47)
   ambum   sim-**are**   yèm
   child   snot-**ATTR**   COP:3SG:CUR
   'The child has a runny nose.'

Attributive 2 (ATTR2) -*kaf*

The suffix -*kaf* also indicates having a particular attribute or characteristic, but one that is not intrinsic; rather it is the result of a recent action or process, or of an initiation of a state. It most often follows a nominalised verb (section 4.1) or an action or stative nominal (5.1.1), as in these examples:

(152) *Wem   ánèghkaf   yèm.*   (NB7:29)
   Wem   ánè-gh-**kaf**   yèm
   yam   be.cooked-NOM-**ATTR2**   COP:3SG:CUR
   'The yam is cooked.'

(153) *Tembè   wutárèghkaf   em.*   (NB2:9)
   Tembè   wutár-gh-**kaf**   em
   timber   join-NOM-**ATTR2**   COP:3PL:CUR
   'The timbers are joined.'

(154) *Yènè   matkaf   yènd   so   yènmor   yènè*
   Yènè   mat-**kaf**   yènd   so   yènmor   yènè
   DEM   knowing-**ATTR2**   1.ABS   FUT   stay:1NSG:CUR   DEM
   mbandan.   (ND11)
   mband-an
   earth-LOC2
   'Having come to know this, we will stay on this earth.'

It is also found in some lexicalised phrases, such as *rendaghkaf wem* 'fried yam', *nánji árau sútkaf* 'wrapped in banana leaves' and *fèskaf* 'cooked in the fire'. This construction occurs only twice in the naturalistic recordings.

3.4.2 Temporal

According to consultants, there are two temporal case markers that can be appended to activities or temporal nominals. However, in addition to not occurring

88 —— 3 Nominal morphology

in the recorded data or the Bible translation, they behave differently from other case suffixes in some word formations.

Temporal 1 (TEMP1) -tau

The temporal suffix -tau indicates during an activity or for a period of time in the present or future – e.g. tèmènd-tau 'during the feast', kembne-tau 'during the game'. It can also mean 'next' as in this example:

(155) *Wem ghèraretau   taoro.*   (NB6:1)
  wem ghèrare-**tau**   taoro
  yam month-**TEMP1** leave:2SG>3NSG:REC.INC
  'Leave some yams for next month.'

There are two unique forms with this suffix: *káye* 'tomorrow' + *-tau* → *káytau* 'during/for tomorrow'; *nambèt* 'a day or two after tomorrow' + *-tau* → *nambtau* 'during/for the next day or two'.

Temporal 2 (TEMP2) -táwár

The temporal suffix -táwár indicates during or for a period of time that began in the past – e.g. *ghèrare-táwár* 'for a month'. There is also a unique form *káytáwár* 'since yesterday'. Some examples:

(156) *Yúnjèftáwár   Darun   yèmor.*   (NB6:42)
  yúnjèf-**táwár** Daru-n   yèmor
  year-**TEMP2**   Daru-LOC1   stay:3SG:CUR
  'He's been staying in Daru for a year.'

(157) *Fá   yèsio   káytáwár.*   (NB6:89)
  fá   yèsio   káye-**táwár**
  3.ABS be.away:3SG:CUR yesterday-**TEMP2**
  'He's been away since yesterday.'

3.5 Exclusive and restrictive suffixes

Nama has two suffixes that mark nominals with many of the meanings of the multifunctional and hard to define English words *just* and *only*. They occur at the end of a nominal, and can follow case suffixes. However, they cannot co-occur with each other. These suffixes behave morphophonemically like case suffixes – i.e.

3.5 Exclusive and restrictive suffixes —— **89**

when following a base ending in *è*, the *è* changes to *a*, as in examples (173) and
(176) below.

3.5.1 Exclusive (EXCL) *-yo*

The sufix *-yo* indicates that the referent or meaning of the nominal is exclusive to
that referred to before or to that and none other. In other words, it excludes a possible different identity or meaning. In this way, it is like *just* in English in sentences
such as:

> *Who's there? Just me.*
> *It was just what I wanted.*
> *Isn't that just the car you were driving last year?*
> *I finished the job just yesterday.*

In many cases *-yo* has the meaning 'the same':[18]

(158) *Yènè* *áryo* *fètè* *yèm.* (PT:323)
　　　 yènè ár-**yo** fètè yèm
　　　 DEM man-**EXCL** really COP:3SG:CUR
　　　 'It's really the same man.'

(159) *Yènè* *mèrèsyo* *yèna* *fiya* *tèm.* (PT:631)
　　　 yènè mèrès-**yo** yèna fiya tèm
　　　 DEM girl-**EXCL** here MOD COP:3SG:REC
　　　 'This had to be the same girl here.'

(160) *Fáyo* *fètè* *yèm.* (PT:526)
　　　 fá-**yo** fètè yèm
　　　 3.ABS-**EXCL** really COP:3SG:CUR
　　　 'He's actually the same one.'

(161) *Tárfáryo* *fès* *tánman.* (NB7:85)
　　　 tárfár-**yo** fès tánman
　　　 plenty-**EXCL** firewood bring:2SG>3NSG:REC.INC
　　　 'Bring more firewood.' (lit. 'bring plenty of the same')

---

18 The first three of these examples are from the "Picture Task", described in section 1.1.3, in which
two consultants had to put in order pictures forming parts of a story. To do so, they had to recognise
the characters that appeared in other pictures.

90 —— 3 Nominal morphology

This suffix is frequently used to mark the S$_A$ argument in reflexive sentences (section 4.5.1) and sometimes in anticausative sentences (4.12.2):

(162) *Fáyo   númbne.*   (NB6:19)
    fá-**yo**   númbne
    3.ABS-**EXCL**   wash:3SG:CUR
    'He's washing himself.'

(163) *yèndyo   so   nawasindan.*   (FD13)
    yènd-**yo** so   nawasindan
    1.ABS-**EXCL** FUT   look.after:1SG:CUR
    'I'll look after myself.'

(164) *Sèsafne   fáyo   namne.*   (NB6:81)
    sèsafne   fá-**yo**   namne
    door   3SG.ABS-**EXCL**   close:3SG:CUR
    'The door closed by itself.'

The *-yo* suffix is frequently used on temporal nominals to exclude other possible times:

(165) *nambètyo   tèrametan.*   (DC28)
    nambèt-**yo**   tèrametan
    day.before.yesterday-**EXCL**   make:1SG>3SG:REC
    'I made it just the day before yesterday.'

(166) *Yabun   tè   namndè   tekyo*   (DC20)
    yabun   tè   namndè   tek-**yo**
    big   TA   become:3SG:CUR.INC   a.while-**EXCL**
    'It got big in just a while.'

(167) *kènwanongèm   aligh   táfyo.*   (B&B:184)
    kènwanongèm   aligh   táf-**yo**
    set.off:1NSG:VEN:REC.INC   walking   then-**EXCL**
    'we set off walking just then.'

As already mentioned, the exclusive suffix -*yo* can follow case-marking suffixes:

(168)　*Yèmo*　*ndene*　***yameyo***　　　　*mèngo*　*so*
　　　yèmo　ndene　yam-e-**yo**　　　mèngo　so
　　　3SG.ERG　the.same　custom-INST-**EXCL**　house　FUT
　　　yèramete.　　　　　　　　　　　　　　　　　　(NB9:8)
　　　yèramete
　　　make:3SG>3SG:CUR
　　　'He will build the house in the same way.'

(169)　*Fèm*　*yènè*　***mèngonyo***　　*nèmor*　　*ánde*　*fronde.*
　　　Fèm　yènè　mèngo-n-**yo**　　nèmor　　ánde　fronde
　　　2.ABS　DEM　house-LOC1-**EXCL**　stay:2SG:CUR.INC　where　first
　　　kènákmangèrwèn?　　　　　　　　　　　　　　(NB7:25)
　　　kènákmangèrwèn
　　　sleep:2SG:REM.DUR
　　　'Do you live in the same house where you slept before?'

(170)　*Yèndfem*　*Sefor*　*mbilae*　*tè*　*yènèsmangèm.*
　　　yèndfem　Sefor　mbilè-e　tè　yènèsmangèm
　　　1NSG.ERG　Sefor　axe-INST　TA　hit/kill:1NSG>3SG:VEN:CUR.INC
　　　krotyo.　　　　　　　　　　　　　　　　　(MD99(1))
　　　kèr-ot-**yo**
　　　dead-PURP-**EXCL**
　　　'We killed Sefor with an axe.'

In addition, -*yo* is attached to bases that normally do not take suffixes, such the quantifier *tárfár* 'plenty' (section 5.1.4), as shown in example (161), to free forms of pronouns, as in examples (162) and (163), and to temporal and orientational nouns (section 5.1), such as *kènjún* 'inside':

(171)　*yènè*　*wèri*　　　*kènjúnyo*　*kès*　　*wèi*　*námnjongi.*
　　　yènè　wèri　　　kènjún-**yo**　kès　　wèi　námnjongi
　　　DEM　drunkenness　inside-**EXCL**　right.there　now　sit:3PL:CUR.INC
　　　'They're sitting down now inside the same drinking place.'

(172)　*tanjoyo*　　*wem*　*tón*　*ewafrotan.*　　　　(DC12)
　　　tanjo-**yo**　　wem　tón　ewafrotan
　　　1SG.own-**EXCL**　yam　mound　make:1SG>3NSG:CUR
　　　'I'm making my own yam mounds.'

The addition of the -*yo* suffix sometimes leads to a change of meaning of the nominal. This may be the result of lexicalised rather than "live" affixation. It may be the case in example (170) with *krotyo* and the following (in which the analysis shows the origin, not necessarily the online derivation, of the word):

(173) *soramayo*   *áram*   *tumbange*   *yènamndamènd.*   (YD11(3))
      soramè-**yo**   ár-am   tumbang-e   yènamndamènd
      later-**EXCL**   man-ERG   spear-INST   shoot:3NSG>3SG:REM.DLT
      'Afterwards the men speared it.'

(174) *frondeyo*   *wasurègh*   *tane*   *fifian*   *mer*
      fronde-**yo**   wasur-gh   tane   fifi-an   mer
      first-**EXCL**   pour-NOM   1SG.GEN   body-LOC2   good
      wimafè   *wèkwère.*   (Mark 14:8)
      wim-afè   wèkwère
      smell-COM   liquid
      'Beforehand she poured perfume on my body . . .'

(175) *fenjo*   *tèfnanyo*   *yènè*   *syèmon*   *tè*
      fenjo   tèfèn-an-**yo**   yènè   syèmon   tè
      2SG.own   basis-LOC2-**EXCL**   DEM   COP:PROX2:3SG:CUR   TA
      nèsmetat.   (ND13)
      nèsmetat
      hit:3NSG>2SG:CUR
      'It's your own fault that they're attacking you.'

(176) *Fá*   *fènatayo*   *yèm*   *ta*   *tambèn.*   (NB1:22)
      fá   fènatè-**yo**   yèm   ta   tambèn
      3.ABS   sort-**EXCL**   COP:3SG:CUR   1SG.DAT   side
      'He is different from me.'

Similarly, the occurrence of -*yo* with the comitative case marker – *(a)fè* has the meaning of 'with' or 'together' on a more permanent basis, as with ownership, permanent attributes and family relationships. So for example, *Yènd ághafè wèm.* 'I am with a dog.' versus *Yènd ághafayo wèm.* 'I have a dog.'

    Other examples are:

(177) *yánjo*   *ambumafayo*   *amafafayo*   *wèi*   *kètè*
      yánjo   ambum-afè-**yo**   amaf-afè-**yo**   wèi   kètè
      3SG.own   child-COM-**EXCL**   wife-COM-**EXCL**   now   there

tek-ofnar-e	*nafrendamènd*	*nuwanoyènd*
tek-ofnar-e	nafrendamènd	nuwanoyènd
long.time-PRIV-INST	get.ready:3NSG:REM.DLT	set.off:3NSG:REM.PUNC

Moet. (MD99(1))

Moet

Morehead

'His own children and wife quickly got ready and set off to Morehead.'

(178)
tafandawèm		*bout*	*tanmorwèn*	*yèfenjo*
tafandawèm		bout	tanmorwèn	yèfenjo
watch:1NSG>3NSG:REM.DUR		boat	come:3NSG:REM.DUR	3NSG.own

diyáriafayo. (YD11(3))

diyári-afè-**yo**

light-COM-**EXCL**

'we watched the boats coming with their own lights.'

(179)
femeliafayo	*yènd*	*yènre.*	(MK13(2))
femeli-afè-**yo**	yènd	yènre	
family-COM-**EXCL**	1.ABS	COP:1DU	

'we (2) are family.'

3.5.2 Restrictive (RSTR) *-ro*

The restrictive suffix *-ro* generally has another meaning of 'just' – i.e. 'only'. The following example contrasts with the use of *-yo* in example (162) above:

(180)
Fáro	*númbne.*	(NB6:19)
fá-**ro**	númbne	
3.ABS-**RSTR**	wash:3SG:CUR	

'Only he's washing.'

Other examples are:

(181)
Nambyo	*árro*	*álet*	*tangèm.*	(NB7:70)
nambyo	ár-ro	ále-t	tángèm	
three	man-**RSTR**	hunting-ALL	go:3NSG:REC	

'Only three men went hunting.'

94 —— 3 Nominal morphology

(182) *Yèndo*   *tónro*   *ewafrotan.*   (DC12)
yèndo   tón-**ro**   ewafrotan
1SG.ERG   yam.mound-**RSTR**   make:1SG>3NSG:CUR
'I'm making only yam mounds.'

(183) *wemro*   *fètè*   *nèmè*   *ámb*   *yau?*   (DC26)
wem-**ro**   fètè   nèmè   ámb   yau
yam-**RSTR**   right   what   else   NEG
'only yams, nothing else?'

(184) *Mbilaro*   *tanangayèng*   *yawar*   *fogh*
mbilè-**ro**   tanangayèng   yawar   fo-gh
axe-**RSTR**   take:3SG>3SG:VEN:REM.INC   black palm   chop-NOM
tafngongè.   (MD99(2))
tafngongè
start:3SG>3NSG:REC.INC
'He took only an axe and started chopping the black palm.'

Like *-yo*, the restrictive suffix differs from case suffixes in attaching to nominals that do not normally take case markers, such as free forms of pronouns (example 180 above), and to nominals already marked with a variety of case suffixes:

(185) *nèmè*   *wèt*   *ghakèramro*   *enetat*   (PT:107)
nèmè   wèt   ghakèr-am-**ro**   enetat
what   so   young.man-ERG-**RSTR**   drink:3NSG>3NSG:CUR
'so what are just the young men drinking?'

(186) *wamaneamè*   *merero*   *mè*   *karmbotawèt.*   (MK13(3))
wamane-amè   mer-e-**ro**   mè   karmbotawèt
yam.stick-PERL   good-INST-**RSTR**   still   climb:3NSG:REM.DUR
'they were climbing along the yam stick nicely.

(187) *yákè túfèr*   *njam*   *akwotro*   *kufnawèng . . .*   (DC33)
yákè túfèr   njam   akw-ot-**ro**   kufnawèng
coucal   when   morning-ALL-**RSTR**   call:3SG:REM.DUR
'When the coucal was calling in the early morning . . .'

An argument against *-ro* being a clitic is that it cannot always have scope over a larger phrase, such as in the following, where a pronoun must be used:

3.5 Exclusive and restrictive suffixes — **95**

(188) *Kambèl a yánjo amaf fáro*
Kambèl a yánjo amaf fá-**ro**
Kambel and 3SG.own wife 3.ABS-**RSTR**
kafrangend. (MD99(2))
kafrangend.
remain:3DU:REC.INC
'Only Kambel and his wife remained.'

Note that -*ro* can only modify constituents of a NP. When it is used to modify an action, a nominalised verb or action nominal must be used:

(189) *Fá átánèghro kátne wende yau* (NB7:71)
fá átán-gh-**ro** kátne wende yau
3.ABS jump-NOM-**RSTR** jump:3SG:REC but NEG
nawatemb.
nawatemb
race:3SG:CUR.INC
'He only jumped; he didn't race.'

(190) *Yèmo waghro taufane, wende yau* (NB7:71)
yèmo wagh-**ro** taufane wende yau
3SG.ERG singsing-**RSTR** sing.out:3SG>3SG:REC but NEG
keremnde.
keremnde
dance:3SG:REC
'He sang only but he didn't dance.'

The suffix -*ro* is also used with nominals that have attributive meanings to indicate an intrinsic rather than temporary characteristic:

(191) *Ambum sènkèrro yèm.* (NB3:39)
ambum sènkèr-**ro** yèm
child cheeky-**RSTR** COP:3SG:CUR
'The child is cheeky (in nature).'

(192) *Gaita njamkean kwènjtaro yèm.* (PC/MD 1.5.19)
Gaita njamke-an kwènjtè-**ro** yèm
Gaita food-LOC2 greedy-**RSTR** COP:3SG:CUR
'Gaita is greedy with food.'

96 —— 3 Nominal morphology

(193)   *Fá*   *múyaro*   *yèm.*   (NB5:39)
       fá   múyè-**ro**   yèm
       3.ABS   strong-**RSTR**   COP:3SG:CUR
       'He's strong.'

Also like *-yo*, the addition of *-ro* may be lexicalised. creating a word with a different meaning – for example, *sènjtè* 'sharp' + *-ro* → *sènjtaro* 'vicious', as in *sènjtaro ághè* 'vicious dog'. This is illustrated in the following examples (in which the analysis shows the origin, not necessarily the online derivation, of the word):

(194)   *Yènd* *mbandro*   *wèm*   *ndau*   *ásáfoghmèn.*   (NB7:89)
       yènd mband-**ro**   wèm   ndau   ásáfo-gh-mèn
       1.ABS ground-**RSTR** COP:1SG garden work-NOM-ORIG
       'I'm dirty from working in the garden.'

(195)   *Fá*   *Darut*   *ámb*   *káyenro*   *yèngèm.* (NB7:69)
       fá   Daru-t   ámb   káye-n-**ro**   yèngèm
       3.ABS Daru-ALL some yesterday/tomorrow-LOC1-**RSTR** go:3SG:CUR
       'He sometimes goes to Daru.'

(196)   *Sènonjonro*   *Darut*   *tèngèm.*   (NB7:93)
       sènonjo-n-**ro**   Daru-t   tèngèm
       today-LOC1-**RSTR** Daru-ALL go:3SG:REC
       'He went to Daru for the first time.'

(197)   *fá*   *mat*   *ere*   *ámbiroero.*   (B&B:283)
       fá   mat   ere   ámbiro-e-**ro**
       3.ABS knowing COP:3DU:CUR one-INST-**RSTR**
       'they (2) know only one at a time.'

In addition, *-ro* appended to names of fauna indicates a place where many can be found – for example: *wèrárro* 'place where there are many wallabies' and *wagifro* 'place where there are many fish'.

3.6 Affixes in word formation

3.6.1 Case suffixes used in word formation

It has already been mentioned that the addition of the exclusive and restrictive suffixes can be lexicalised, forming new lexical items. The same phenomenon may have occurred historically with case markers, for example in the creation of nominals with adverbial functions, as shown in Table 3.15. An alternative analysis would be that inflection for case in these forms occurs synchronially. But here the use of hyphens does not indicate productive synchronic processes.

Table 3.15: Adverbs formed by the use of case markers.

Morphological structure	Adverb	
yènè-n DEM-LOC1	*yènan*	'here, at this place'
yènè-n-mè DEM-LOC1-PERL	*yènanmè*	'(from) along here; this way'
yènè-n-ot-yo DEM-LOC1-PURP-EXCL	*yènanotyo*	'right up to; all the way to'
kètè-n there-LOC1	*kètan*	'there, at that place'
kètè-n-mè there-LOC1-PERL	*kètanmè*	'(from) along there; through there; that way; previous'

Examples can be seen with *kètan* in (87) and (123) and with others following:

(198) **yènan**   *tane*   *ambum*   *syakái.*   (B&B:277)
      yènè-n   tane   ambum   syakái
      DEM-LOC1   1SG.GEN   child   be.standing:PROX2:3SG:CUR
      'here my child is standing.'

(199) . . . *kènangotau*   **yènanotyo**   *Sarghár.*   (SK11(2))
      kènangotau   **yènè-n-ot-yo**   Sarghár
      return:3SG:REM.DUR   **DEM-LOC1-PURP-EXCL**   Sarghar
      '. . . he was returning all the way to Sarghar.'

98 — 3 Nominal morphology

(200) *Fèyo ambum **kètanmè**   wúnjègh*
      fèyo   ambum   **kètè-n-mè**   wúnj-gh
      then   child   **there-LOC1-PERL** raise-NOM
      ewafngeayènd.   (MD13(1))
      ewafngeayènd
      start:3DU>3NSG:REM.PUNC
      'Then that way they (2) started raising children.'

Many conjunctive adverbs and adverbial subordinating conjunctions are also derived by the addition of case markers. Several are based on *yènè* 'here' and *kètè* 'there', as well as the nominal *mènè* 'thing' (see Table 3.16). Again, whether these forms are the result of historical word formation process or productive affixation is a topic for further research.

Table 3.16: Conjunctive adverbs formed by the use of case markers.

Morphological structure	Conjunctive adverb	
yènè-mè-mèn DEM-PERL-ORIG	*yènamamèn*	'for this reason; this is why'
kètè-n-mè-yo there-LOC1-PERL-EXCL	*kètanmayo*	'since'
kètè-n-ot-yo there-LOC1-PURP-EXCL	*kètanotyo*	'until'
mènè-e thing-INST	*mènae*	'with this purpose'
mènè-t thing-ALL	*mènat*	'so that'
mènè-yan thing-LOC2	*mènayan*	'that's why'
mènè-mèn thing-ORIG	*mènamèn*	'because'
njam-an-yo when-LOC2-EXCL	*njamanyo*	'just when, as soon as'

Examples are given in sections 6.8 and 7.2.

3.6.2 Derivational nominal affixes

Several other affixes are used in word formation. However, only two of these appear to be currently productive.

3.6 Affixes in word formation — **99**

Productive derivational suffix: Characterised (CHAR) -*aro*
Related to the function of -*ro* to mark an intrinsic attribute or a place characterised by an animal is the suffix -*aro*. This also indicates 'place of', or 'place characterised by' (and glossed CHAR) but it is used for flora rather than fauna – for example, *yús-aro* 'grassy place'. Other examples are:

(201) *Yèmofem   náifae   wènaro   yarotat.*   (NB5:80)
      yèmofem   náifè-e   wèn-**aro**   yarotat
      3NSG.ERG knife-INST tree-**CHAR** cut.through:3NSG>3SG:CUR
      'They're cutting through the bush with a bushknife.'

(202) *yènd   fwèm   fái   kendaro   yèmafè.*   (KS11)
      yènd   fwèm   fái   kend-**aro**   yèmafè
      1.ABS COP:PROX1:1SG:CUR EMPH   melaleuca-**CHAR** 3SG.COM
      'I am really with him at the melaleuca tree place.'

An allomorph -*waro* occurs after vowels:

(203) *yènèm   ee   nánjiwaro*   (ND13)
      yènèm   ee   nánji-**waro**
      come:3SG:CUR   PROL   banana
      'she came (a long way) to the banana place.'

Other derivational suffixes
Four other suffixes that appear to have been involved historically in word formation. One of these is -*tè*, which occurs on the end of a large number of nominals that can have attributive functions. These include *kèrtè* 'heaviness, heavy', *sèntè* 'sharpness, sharp', *senjtè* 'thickness, thick', *waftè* 'heat, hot', *ghaftè* 'bitterness, bitter' and *kwefitè* 'darkness, dark, black'.

The other suffixes apply to small closed sets of lexical items. The first, -*notmè* (which could possibly be derived from three case markers – -*n* LOC1, -*ot* PURP and -*mè* PERL), is found only with three numerals, with their final syllable or final *o* deleted. It means 'time(s)': *ámbiro* 'one', *ámbinotmè* 'one time'; *sembyo* 'two', *sembinotmè* 'two times'; *nambyo* 'three/several', *nambinotmè* 'three/several times'.

The remaining two non-productive suffixes have been appended to the dative pronouns. The first -*njo* gives an exclusive or emphasised possessive meaning: *tanjo* 'my own', *fenjo* 'your own', *yánjo* 'his/her own', *tèfenjo* 'our own' , etc. (This suffix is clearly related to exclusive -*yo* and restrictive -*ro* [section 3.5].) It can be seen in examples (175), (177) and (178), and the following:

100 —— 3 Nominal morphology

(204) *Yènd kuflangèm* **tèfenjo** *sèkwèn.* (YD11(3))
yènd kuflangèm **tèfenjo** sèkw-n
1.ABS go.inside:1NSG:REC.INC **1NSG.own** canoe-LOC1
'We got inside our own canoe.'

The second suffix on dative pronouns, -*mbya*, has a different exclusive meaning: 'alone' or 'by oneself': *tambya* 'I alone, by myself'; *fembya* 'you alone, by yourself', *yámbya* 'she/he alone, by herself/himself'; *tèfembya* 'we alone, by ourselves', etc. – for example:

(205) *Ár tèkmangèrwèn* **yámbya.** (MD13(1))
ár tèkmangèrwèn **yámbya**
man stay:3SG:REM.DUR **3SG.alone**
'The man was staying by himself.'

Derivational prefix

The prefix *mámá-* 'exceedingly' has been involved in the formation of many lexical items – for example: *mámátèkrèt* 'exceedingly painful', *mámáminde* 'very fast or strong'and *mámáfi* 'very bad'. This prefix is still productive, as evidenced by its use with a Tok Pisin loanword, *ambag* 'bighead' (originally from English *humbug*): *mámáambag* 'real bighead'.

3.7 Possessive prefixing

For most nominals that reference relationships (relatives and close friends), possession is shown not by using independent possessive pronouns (Table 3.7) but by prefixing the dative form of the pronoun (Table 3.5). For a small number of relationship nominals the dative pronoun is prefixed to the full free form of the nominal – for example: *ta-* + *ghakèr* → *taghakèr* 'my brother', *yá-*+ *mèrès yámrès* 'his/her sister', *tèfe* + *fútár* → *tèfefútár* 'our friend', *fe-* + *mwitè* → *femwitè* 'your exchange cousin'.

More commonly, however, there is a special bound form of the nominal, to which the dative form of the pronoun is prefixed to show possession. Some of these are shown in Table 3.17, followed by examples.

Note that some of these bound forms are not generally used with the first person singular pronoun *ta*; rather the usual independent possessive pronoun is used – e.g. *tane afè* 'my father', *tane ambum* 'my child', *tane áki* 'my grandfather', *tane amaf* 'my wife'.

Table 3.17: Free and possessed bound forms of relationship nominal.

Free form		Bound form	Example	
afè	'father'	-rafè	yárafè	'his/her father'
amè	'mother'	-tmè	yátmè	'his/her mother'
áki	'grandfather'	-ki	yáki	'his/her grandfather'
alè	'grandmother'	-kealè	yákealè	'his/her grandmother'
ambum	'child'	-mbum	yámbum	'his/her child'
anè	'older sibling'	-nè	yánè	'his/her older sibling'
bafè	'uncle'	-njafè	yánjafè	'his/her uncle'
bafalè	'aunt'	-njafalè	yánjafalè	'his/her aunt'
ár	'man/husband'	-njár	yánjár	'her husband'
amaf	'woman/wife'	-tèn	yátèn	'his wife'

(206)   *kètè*   **yátèn**   *wèmènj*   *fèrèk*   *kèfnongè.*   (MD99(2))
        kètè   **yá-tèn**   wèmènj   fèrèk   kèfnongè
        there   **3SG-wife**   menstruation   blood   flow:3SG:REC.INC
        'There his wife's menstruation blood flowed out.'

(207)   ***Fèferafè***   *mbanè*   *yèm.*   (DC20)
        fèfe-rafè   mbanè   yèm
        1NSG-father   unfortunate   COP:3SG:CUR
        'Our father is unfortunate.'

(208)   ***yánjár***   *kès*   *yèkèmang.*   (ND13)
        yá-njár   kès   yèkèmang
        3SG-husband   just.there   be.lying.down:3SG:CUR.INC
        'her husband slept just there.'

Some relationship nominals have only a bound possessed form and no free form (Table 3.18).

Table 3.18: Relationship nominals that have no free form.

Bound form		Example	
-gase	'namesake'	tagase	'my namesake'
-mbrè	'exchange man'	tambrè	'my exchange man'
-blalè	'exchange girl'	tablalè	'my exchange girl'
-nat	'father-in-law'	tanat	'my father-in-law'
-natalè	'mother-in-law'	tanatalè	'my mother-in-law'
-wèn	'type of friend'	tawèn	'my friend'
-ngar	'type of friend'	tangar	'my friend'

102 —— 3 Nominal morphology

Some examples are:

(209) ***Tagase***　　　　*fèmafayof*　*tángre?*　　　　　　　(DC29)
　　　　ta-gase　　　　fèmafayof　tángre
　　　　1SG-namesake　2NSG.COM　go:3DU:REC
　　　　'Did my namesake go with you?'

(210) ***Fembrè***　　　　　*kès*　　　*yèmong.*　　　*wèi.*　（B&B:260)
　　　　fe-mbrè　　　　　kès　　　yèmong　　　wèi
　　　　2SG-exchange.man　just.there　COP:3SG:CUR.INC　now
　　　　'Your exchange man is just there now.'

(211) *PNG*　*kènjún*　*yènè*　*yam*　　*yèmon*　　*awerègh*
　　　　PNG　kènjún　yènè　yam　　yèmon　　awer-gh
　　　　PNG　inside　DEM　custom　COP:3SG:CUR　marry-NOM
　　　　yámblalè.　　　　　　　　　　　　　　　　（ND11)
　　　　yá-mblalè
　　　　3SG-exchange.girl
　　　　'In PNG the custom is to marry one's exchange girl.'

If the possessor of a nominal with no free form is a personal name rather than a pronoun, the bound nominal is prefixed with the 3rd person singular dative pronoun *yá* and the personal name may be marked with the dative suffix – for example: *Jefe yágase* 'Jeff's namesake'. If the possessor is another common nominal, it is marked by the genitive suffix – e.g. *afane yágase* 'father's namesake'.

3.8 Reduplication

The morphological process of reduplication is not productive in Nama but is evident in the word formation of a limited number of nominals. It is almost exclusively complete reduplication and has several functions: indicating plural, diminutive, intensification, associative and naming of plants and animals.

For a small number of nominals, reduplication indicates plural – for example:

rokár 'thing'　　　　*rokár-rokár* 'things'
mèrès 'girl'　　　　　*mèrès-mèrès* 'girls'
ghakèr 'boy'　　　　*gharkèr-ghakèr* 'boys'
yondèr 'sandfly'　　*yondèr-yondèr* 'sandflies'

3.8 Reduplication — **103**

karèf 'large piece' *karèf-karèf* 'many largepieces'
ndimbul 'small piece' *ndimbul-ndimbul* 'small pieces'
ausè 'old woman' *ausè-ausè* 'old women'
áki 'ancestor' *áki-áki* 'ancestors'
fèrèn 'pattern' *fèrèn-fèrèn* 'patterns'

For another small group, reduplication has a diminutive function:

kafwe 'branch (of tree)' *kafwe-kafwe* 'small branch (of tree)'
kiánj 'grass type' *kiánj-kiánj* 'smaller variety of *kiánj*'
sangu 'planting yam' *sangu-sangu* 'small planting yam'
sèmbár 'night' *sèmbár-sèmbár* 'dusk'
tru 'container' *trutru* 'small container'
yáf 'basket, bag' *yáfyáf* 'small basket, bag'
sèki 'floor' *sèki-sèki* 'small platform'
námb 'bow' *námbnámb* 'small bow'

For an even smaller group, reduplication indicates intensification:

sún 'startled' *súnsún* 'very startled'
ofè 'light' *ofè-ofè* 'very light'
kèrtè 'heavy' *kèrtè-kèrtè* 'very heavy'
krúfèr 'cold' *krúfèr-krúfèr* 'extreme cold'

Reduplication can sometimes be used to create another lexical item for something that is a bit similar to the unreduplicated item:

ághè 'dog' *ághè-ághè* 'bronze quoll'
gásiyè 'crocodile' *gásiyè-gásiyè* 'stick insect'
mai 'long yam' *maimai* 'wild bush yam type'
ndúmáng 'yam type' *ndúmáng-ndúmáng* 'wild bush yam type'
kend 'melaleuca tree' *kendkend* 'tree similar to *kend*'

Reduplication can also be used to create another lexical item for something associated with the unreduplicated item:

karèf 'termite mound' *karèf-karèf* 'goanna that lives around termite mounds and eats termites'
filè 'pillow' *filè-filè* 'tree with cotton-like seed pods used for pillow stuffing'

mai 'long yam'	*maimai* 'plant used to make *mai* yams grow larger'
yam 'louse'	*yamyam* 'plant with seeds that when bitten make the same sound as when a louse is bitten'

Most often, nominals (especially names of animals) are reduplicated to provide names for plants and sometimes fish. There may have originally been some associations with the unreduplicated items, as above, but they are not currently apparent.

kimb 'pig'	*kimbkimb* 'tree type'
áuyè 'cassowary'	*áuyè-áuyè* 'tree type'
kárimam 'pademelon'	*kárimam-kárimam* 'tree type'
wèrár 'wallaby'	*wèrár-wèrár* 'tree type'
wagif 'fish'	*wagif-wagif* 'tree type'
tèngan 'ear'	*tèngan-tèngan* 'tree type'
ñárár 'bandicoot'	*ñárár-ñárár* 'grass type'
yátmè 'her/his mother'	*yátmè-yátmè* 'grass type'
kankè 'tongue'	*kankè-kankè* 'fish type'
ninyè 'witch'	*ninyè-ninyè* 'banana type'
efogh 'sun, day'	*efogh-efogh* 'tree type' and 'fish type'

The only example found of partial reduplication is *sikwál* 'rainbow' → *sikwálkwál* 'lighter rainbow in a double rainbow'. (A completely reduplicated form also exists: *sikwál-sikwál* 'halo around the moon'.)

4 Verbal morphology

Verbs, like nominals, are defined as a class by their morphology. This chapter examines verbal morphology – exclusively affixation, as shown in the grammatical overview (section 1.4). Prefixes and suffixes are added to a bound stem, which never occurs on its own (i.e. without at least one prefix or suffix). (In the description below, verb stems are marked with slanted lines – \. . ./.) The verb stem may undergo morphophonemic changes with the addition of particular affixes.

The grammatical overview also showed that there are three subclasses of verbs: transitive, A-aligned intransitive and P-aligned intransitive. In the data, the subclasses have the following number of verb stems: transitive: 392; A-aligned intransitive: 370; and P-aligned intransitive: 35.

The stems of all transitive and P-aligned intransitive verbs begin with a consonant and the two classes are distinguished by their morphological behaviour. The stems of all A-aligned intransitive verbs begin with a vowel – most commonly *a* or *á*. The majority of transitive verb stems in Nama (approximately 220 out of 392) have a corresponding A-aligned intransitive verb that is identical except for the addition of the initial vowel. Whether or not this vowel should be considered a productive detransitivising prefix is discussed below in section 4.12.6.

4.1 Nominalising suffix (nom) *-gh*

The stems of transitive and A-aligned intransitive verbs can take the nominalising suffx *-gh* – for example, \olito/ 'dive' + *-gh* → *olitogh* 'to dive' or 'diving' and \faror/ 'split' + *-gh* → *farorègh* 'to split' or 'splitting". This nominalised form, called the "big word" by consultants, is considered to be an infinitive form and is used as the citation form in the Nama dictionary. Nominalised verbs act as other nominals and can take case markers – for example:

(212) *yènd* *yènè* *ásmèghan*   *kwèm*   (PT:759)
        yènd   yènè ásèmè-**gh**-an kw-m
        1.ABS DEM fight-**NOM**-LOC2 1SG$_\text{P}$-COP
        'I was (involved) in that fighting.'

(213) *Lea* *yèngmormèn*   *fès*   *rèsaghèt.*   (MD99(1))
        Lea   y-ngèmo-èrmèn   fès   rèsa-**gh**-t
        Lea   3SG$_\text{P}$-go-REM.DLT.PA firewood carry-**NOM**-ALL
        'Lea went to get firewood.'

https://doi.org/10.1515/9783111077017-004

106 —— 4 Verbal morphology

(214) | *Tane* | *mbi* | *táfèghmèn* | *si* | *nefo.* | (NP11)
| tane | mbi | táf-**gh**-mèn | si | n-efo |
| 1SG.GEN | sago | chop-**NOM**-ORIG | story | ØP-finish |

'My story about making sago is finished.'

(See also examples 138, 139, 149, 150, 152 and 153 in chapter 3.)
Nominalised verbs can also modify other nominals:

(215) | *ufrogh* | *si* | *so* | *yaitotan.* | (MD11(2)) |
| ufro-**gh** | si | so | y-waito-ta-èn |
| originate-**NOM** | story | FUT | 3SGP-tell-IPFV-1SGA |

'I will tell an origin story.'

(216) | *watamegh* | *mèngo* | *so* | *yèfrendan.* | (FD13) |
| watame-**gh** | mèngo | so | y-freng-ta-èn |
| teach-**NOM** | house | FUT | 3SGP-fix.up-IPFV-1SGA |

'I will fix up the school house.'

(Note that further distinctions exist in perfectivity, some Ø marked. These are discussed in section 4.6.1, but are not glossed in examples until that section. Also, many morphophonemic changes occur with the addition of perfective suffixes, as in the preceding example. These are described in section 4.6.2.)

The nominalised form of the verb occurs in an inchoative construction with the verbs meaning 'to start' or 'be about to' *wafngo*/ for transitives and *ufngo*/ for A-aligned intransitives. (This inchoative construction was used to elicit the citation form from consultants.) Some example are:

(217) | *árèm* | *yèfenjo* | *yu* | *ronjagh* | |
ár-m	yèfenjo	yu	ronja-**gh**	
person-ERG	3NSG.own	place	look.for-**NOM**	
ewafngoyènd.				(MD11(2))
e-wafngo-ay-nd				
3NSGP-start-REM.PUNC-2|3NSGA				

'The people started looking for their own places.'

(218) | *awasogh* | *nufngoyèn.* | (NP11) |
| awaso-**gh** | n-ufngo-ay-èn | |
| pack.up-**NOM** | ØP-start-REM.PUNC-1SG$_A$ | |

'I started packing up.'

Some nominals appear to be the result of nominalisation, but are not, at least not synchronially. For example, *kèmègh* is the word for 'sleep' or 'sleeping', but the nominalised form *kèmè* + *-gh* means 'laying something/someone down'. In addition, *aligh* 'walking' and *siogh* 'hunger' do not have corresponding verb stems.

4.2 Core argument indexing affixes

4.2.1 A- and S$_A$-indexing suffixes

As pointed out in the grammatical overview in chapter 1, the same suffixes index the person and number (singular vs nonsingular) of both the A argument of transitive verbs and the S argument of A-aligned intransitive verbs (S$_A$). There are different sets of A/S$_A$-indexing suffixes for different tenses. The set for imperfective CURRENT and RECENT imperfective tenses is shown in Table 4.1.

Table 4.1: A/S$_A$-indexing suffixes (imperfective current and recent tenses).

1SG	*-èn*
2/SG	*-e*
1 NSG	*-m*
2|3 NSG	*-t-*

For other tenses and aspects, the 1st person suffixes remain *-èn* 1SG and *-m* 1NSG, but 2nd/3rd person suffixes have different forms: Ø, *-è* and *-ng* for 2\|3 SG and *-nd* and *-i* for 2\|3NSG. (See sections 4.7.2, 4.8 and 4.9 below.)

4.2.2 P- and S$_P$-indexing prefixes

The same prefixes index the person and number (singular vs nonsingular) of both the P argument of transitive verbs and the S argument of P-aligned intransitive verbs (S$_P$). With transitive verbs, the P argument is generally the semantic patient. However, with ditransitive verbs it is the semantic recipient or beneficiary.

There is also a prefix that indicates the absence of a P argument in A-aligned intransitive verbs. This is glossed as ØP. This prefix is also used when two arguments of a verb are simultaneously both a semantic agent and semantic patient – i.e. in reflexive and reciprocal constructions. (See sections 4.5.1 and 4.12.3.)

108 —— 4 Verbal morphology

The P/S$_P$-indexing and ØP-indexing prefixes also play a vital role in indicating tense and aspect. In Nama, there are two sets of prefixes. In some cases, it is the choice of prefix from one set or the other in combination with particular suffixes that determines tense and aspect (see Siegel 2015). This is because some tense/aspect suffixes can have two possible values, and the particular prefix set that is used serves as a switch to determine one value or the other. Each prefix set is used for a variety of tenses and aspects, and therefore it is difficult to assign them precise functions. Therefore, the two sets are distinguished by the Greek letters α and β, following the practice used by Evans (2012a, 2015a) for the closely related language of Nen. The two prefix sets in Nama are shown in Table 4.2.

Table 4.2: P/S$_P$-indexing and ØP prefixes.

	α	β
1SG	w-	kw-
2SG	n-/nèn-	kèn-
3SG	y-	t-
1NSG	yèn-	tèn-
2|3NSG	e-	ta-/tá-[19]
ØP	n-	k-

Examples (219a) and (219b) illustrate how the different prefixes distinguish imperfective current tense from imperfective recent tense. (The scope of these tenses is discussed in section 4.7.1.) Both are indicated by the imperfective suffix *-ta* followed by one of the A/S$_A$-indexing suffixes shown above – here *-t*, indicating 2nd or 3rd person nonsingular. A prefix from the α set indicates current tense while a prefix from the β set indicates recent tense:

(219) a. *mèrsam   ghakèr efrangotat.*   (based on YD11(3))
         mèrès-am *ghakèr* **e**-frango-ta-t
         girl-ERG boy   α.2\|3NSGP-leave-IPFV-2\|3NSGA
         'The girls **are** leaving the boys.'

      b. *mèrsam   ghakèr tafrangotat.*   (based on YD11(3))
         mèrès-am *ghakèr* **ta**-frango-ta-t
         girl-ERG boy   β.2\|3NSGP-leave-IPFV-2\|3NSGA
         'The girls **were** leaving the boys.'

---

19 There are two subclasses of verbs that determine which alternative *ta-* or *tá-* is used. These are discussed below in section 4.3.

4.2 Core argument indexing affixes — **109**

This is an example of DISTRIBUTED EXPONENCE – a characteristic of other Yam languages, as described, for example, by Carroll (2016, 2020) and Döhler (2018). There is no single inflection that is the exponent of the grammatical meaning imperfective current tense as opposed to recent tense. Rather, this meaning is distributed over more than one inflection and each of these has to be taken into account for it to be derived. The inflections here are the imperfective suffix *-ta*, the following A-indexing suffix *-t*, and the P-indexing prefix. The combination of suffixes indicates an imperfective tense but does not distinguish between current versus recent. The particular set of the P-indexing prefix (α versus β) in combination with these suffixes gives the specific meaning of either current or recent. For example, the combination of *e-* (series α) with *-ta* and *-t* expresses imperfective current tense in example (219a).

Thus, prefixes in both sets have two functions. The first is to index the person and number of the P argument of a transitive verb or S argument of a P-aligned intransitive verb. This function can be specified. The second is to signal particular tenses or aspects according to which prefix set the prefix is in. The set can be specified, but not the specific function because this depends what other inflections are involved. This can be seen in examples (220a) and (220b). The same prefixes that are used to distinguish current and recent tenses are used here to distinguish aspect within the remote tense – the α set for remote delimited (DLT) and the β set for remote durative (REM.DUR). (See section 4.7.2.)

(220) a. *Yèndo*   *rokár-rokár*   *eflitamèn.*   (based on NP11)
           yèndo   rokár~rokár   e-fèli-ta-**m**-èn
           1SG.ERG   thing~PL   α.2|3NSGP-put.inside-IPFV-**REM.DLT**-1SGA
           'I put the things inside (long ago).'

       b. *Yèndo*   *rokár-rokár*   *taflitawèn.*   (based on NP11)
           yèndo   rokár~rokár   ta-fèli-ta-**w**-èn
           1SG.ERG   thing~PL   β.2|3NSGP-put.inside-IPFV-**REM.DUR**-1SGA
           'I was putting the things inside (long ago).'

Several morphophonemic changes are involved when some of the prefixes are joined to the many transitive and P-aligned verb stems beginning with *w*. First, the 3rd person singular and 1st person singular prefixes of both the α and β sets – i.e. *y-*, *t-*, *w-*, and *kw-* – replace the initial *w* of the stem. (Of course, for 1st person singular *w-* there is no apparent change.) Some examples:

110 — 4 Verbal morphology

(221) *Yèndo yènè ár   yeghetan.*   (NB5:59)
      yèndo yènè ár   y-weghè-ta-èn
      1SG.ERG DEM man α.3SGP-avoid-IPFV-1SGA
      'I avoid that man.'

(222) *Mbormboram yènè wèn endèn*   (NB7:40)
      mbormbor-am yènè wèn end-n
      wind-ERG   DEM   tree road-LOC1
      táfmote.
      t-wáfmo-ta-e
      β.3SGP-flatten-IPFV-2|3SGA
      'The wind has flattened this tree on the road.

(223) *Yènè árèm   wafalngote.*   (NB10:79)
      yènè ár-m   w-wafalngo-ta-e
      DEM   man-ERG α.1SGP-curse-IPFV-2|3SGA
      'That man is cursing me.'

(224) *Kwáumyo   trak   wawerghèt.*   (NB7:86)
      kw-wáumyo   trak   wawerè-gh-t
      β.1SGP-allow vehicle drive-NOM-ALL
      'Let me drive the vehicle.'

There is also a general rule in prefixing that when *n* precedes *w*, the *w* changes to *m*. (This rule is also relevant to the deictic and applicative prefixes (sections 4.4 and 4.5.2).) The change occurs with the 1st person nonsingular prefixes *yèn-* and *-tèn*:

(225) *fèmofem welkèm   yènmafrotati.*   (B&B:69)
      fèmofem welkèm   yèn-wafro-ta-t-i
      2NSG.ERG welcome α.1NSGP-do-IPFV-2|3NSGA-2PL
      'you all are welcoming us.'

(226) *siogham   tènmifongè.*   (YD11(3))
      siogh-am   tèn-wifo-ang-è
      hunger-ERG β.1NSGP-finish-INC-2|3SGA
      'we were hungry.' (lit. 'hunger finished us.')

It is more complicated with 2nd person singular prefixes. The β set prefix generally occurs as *kèn-*. An example is:

(227) *Yèndo   fèm   kènfandan.*   (NB6:34)
yèndo   fèm   kèn-fanda-ta-èn
1SG.ERG 2.ABS β.2SGP-look.at-IPFV-1SGA
'I was looking at you.'

When it occurs before a *w*-initial stem, the *w* changes to *m* as expected. However, the final *n* of the suffix does not appear – e.g.:

(228) *yèmo   kèmerete.*   (NB6:34)
yèmo   kèn-werè-ta-e
3SG.ERG β.2SGP-hold-IPFV-2|3SGA
'he held you.'

(229) *fèyo   yènandmè   so   kèmauyaufon.*   (PT:691)
fèyo   yènandmè   so   kèn-wauyaufo-èn
then   this.way   FUT   β.2SGP-show-1SGA
'then I'll show you this way.'

The α set 2nd person singular prefix generally occurs as *n-*. An example is:

(230) *Yèndo   fèm   nèfrendan.*   (NB4:2)
yèndo   fèm   n-freng-ta-èn
1SG.ERG 2.ABS α.2SGP-greet-IPFV-1SGA
'I'm greeting you.'

However, when it precedes a *w*-initial stem, the *w* changes to *m* as if the prefix were *nèn*, but like with *kèn-*, the final final *n* does not appear:

(231) *yèndo   fèm   nèminjon.*   (NB6:34)
yèndo   fèm   n-winjo-èn
1SG.ERG 2.ABS α.2SGP-see-1SGA
'I see you.'

(232) *Yèmo   nèmaitote.*   (SK11-trans-nb:17)
yèmo   n-waito-ta-e
3SG.ERG α.2SGP-follow-IPFV-2|3SGA
'He's following you.'

4.3 Morphological subclasses

Nama verbs are divided into two morphological subclasses that are distinguished by the vowel *a* /a/ versus *á* /æ/ in the β set 3rd person nonsingular P- or S$_P$-indexing prefix – i.e. *ta-* vs *tá-*. Examples are *a*-subclass *faro*/ 'write' vs *á*-subclass *fenè*/ 'feed' – e.g. *tafarota* 'write them' vs *táfeneta* 'feed them'. The *a-/á-* distinction is also found with the reflexive/reciprocal prefix (section 4.5.1) and the applicative prefix (*wa-/wá-*, section 4.5.2).

The existence of the *á* vowel in these prefixes can be attributed partially to phonological conditioning. If the first vowel of the verb stem is *á* or the front rounded vowels *ú* or *ó*, then the verb belongs to the *á*-subclass – for example: *káyo*/ 'attach' *tákáyota* 'attach them', and *wúyo*/ 'lift' *táwúyota* 'lift them'. However, the majority of *á*-subclass verbs do not have *á, ú* or *ó* in the stem (e.g. the example with *fenè*/ above), and thus lexical conditioning is more significant. Further evidence of lexical conditioning can be seen in pairs such as *a*-subclass *ro*/ 'poison' vs *á*-subclass *so*/ 'grate', and *a*-subclass *fèli*/ 'put inside' vs *á*-subclass *mèli*/ 'disturb (water)'.

Out of the total of 392 recorded transitive verbs, 245 are in the *a*-subclass and 147 are in the *á*-subclass. Out of 35 P-aligned intransitive verbs, 15 are in the *a*-subclass and 20 in the *á*-subclass.

With regard to A-aligned intransitive verbs, of the 370 stems recorded, 144 begin with *a* and 111 with *á*. (The rest [115] begin with other vowels.) As mentioned above, a total of 220 A-aligned intransitive verbs have corresponding transitive forms (i.e. without the initial vowel). For those A-aligned intransitives beginning with *a* or *á* that have corresponding transitive forms, the initial vowel depends on the subclass of the transitive form. For example, *a*-subclass *falo*/ 'divide up', *afalo*/ 'become divided up', and *á*-subclass *káyo*/ 'attach', *ákáyo*/ 'become attached'. There are 96 A-aligned intransitives with initial *a* that have corresponding *a*-subclass transitive forms, and 77 with initial *á* that have corresponding *á*-subclass transitive forms. Of the remaining 84 that begin with *a* or *á*, the vowel may be conditioned by the following vowel – e.g. *átán*/ 'jump' and *áfála*/ 'squat' – but the majority are lexically conditioned.

4.4 Deictic prefixes

Most Nama verbs, other than P-aligned ones, can optionally have one of two prefixes indicating direction of movement either towards or away from the speaker or deictic centre. The conventional labels are used here: VENITIVE (VEN) 'towards' and ANDATIVE (AND) 'away'; however the function of the prefixes is wider, as they can also indicate location ('close' or 'far') in space or time. The prefixes are *n-* (VEN) and *ng-* (AND). They occur with transitive and A-aligned intransitive verbs after the P/ØP-marking prefix.[20] The following examples show the use of the verb stem \ango/ 'return' without a deictic prefix, and then with the prefixes.

(233) a. *wèikor   nangote.*   (MD13(1))
      wèikor   n-ango-ta-e
      again   α.ØP-return-IPFV-2|3SGS$_A$
      'he returned again.'

   b. *sèite   so...   nènangotati.*   (DC27)
      sèite   so...   n-**n**-ango-ta-t-i
      afternoon FUT   α.ØP-**VEN**-return-IPFV-2|3NSGS$_A$-2PL
      'you all will return (here) in the afternoon.'

   c. káye   *akwan*   so   *nèngangom*
      káye   akw-an   so   n-**ng**-ango-m
      tomorrow   morning-LOC2 FUT   α.ØP-**AND**-return-1NSGS$_A$
      mèngotuot.   (MK13(1))
      mèngotu-ot
      village-ALL
      'tomorrow morning we (2) will return to the village.'

With A-aligned intransitive verbs, the ØP prefix *n-* or *k-* may be deleted preceding the andative prefix *ng-* – for example:

(234) *Dafifè   fá   **ng**angongè.*   (YD11(3))
      Dafi-fè   fá   **ng**-ango-ang-è
      Duffy-COM   3.ABS   **AND**-return-INC-2|3SGS$_A$
      'He returned with Duffy.'

(235) *kokái,   yame   tèkaf   **ng**uwano   mèngot.*   (Mark 2:11)
      kokái,   yame   t-kafè   **ng**-uwano   mèngo-t
      stand.up mat   β.3SGP-pick.up AND-set.off   house-ALL
      'get up, pick up (your) mat and go home.'

(236) *ámb   **ng**ákmetawèt.*   (ND13)
      ámb   **ng**-ákèmè-ta-w-t
      some   **AND**-lie.down-IPFV-REM.DUR-2|3NSGS$_A$
      'some were going to lie down.'

---

20 The P-aligned intranstive verbs for 'come' and 'go' (section 4.11.2) appear to be exceptions, but they are more likely the result of historical word formation rather than productive affixation.

114 —— 4 Verbal morphology

Deictic prefixes also occur with transitive verbs – e.g. with \wi/ 'throw':

(237) *susi...*   *yèngwitan.*   (MK13(1))
      susi...   y-**ng**-wi-ta-èn.
      fishing.line   α.3SGP-**AND**-throw-IPFV-1SGA
      'I threw the fishing line out.'

Note that with venitive *n-* the morphophonemic rule still applies where a following stem initial *w* becomes *m*:

(238) *mèrèsom*   *táf*   *ladè*   *yènmitam.*   (MD13(1))
      mèrès-o-m   táf   ladè   y-**n**-wi-ta-m
      girl-SG-ERG   then   ladder   α.3SGP-**VEN**-throw-IPFV-REM.DLT
      'the girl then threw the ladder in.'

(239) *nu*   *wèikor*   *yènmukayèn.*   (NP11)
      nu   wèikor   y-**n**-wuk-ay-èn
      water   again   α.3SGP-**VEN**-fetch-REM.PUNC-1SGA
      'I fetched water again.'

As mentioned above, when used with a verb that does not involve motion in a particular direction the *n-* venitive prefix can indicate closeness – for example:

(240) *yèmo*   *wèikor*   *enènfetam.*   (FD13)
      yèmo   wèikor   e-**n**-nèfè-ta-m-Ø
      3SG.ERG   again   α.2|3NSGP-**VEN**-cut-IPFV-REM.DLT-2|3SGA
      'he cut them again (nearby).'

And the *ng-* andative prefix can indicate farness or distance:

(241) *Mer*   *kakayam*   *ke*   *tángyárwèm*
      mer   kakayam   ke   tá-**ng**-yárè-Ø-w-m
      good   bird.of.paradise   sound   β.2|3NSGP-**AND**-hear-IPFV.DU-REM.DUR-1NSGA
      men   *ámb*   *ngutnawèt*   *wènaro*   *men.*   (TE11(1))
      men   ámb   **ng**-utár-ta-w-t   wènaro   men
      bird   other   **AND**-call.out-IPFV-REM.DUR-2|3NSGA   forest   bird
      'We heard nice bird-of-paradise cries and other forest birds calling (in the distance).'

4.5 Other verbal prefixes —— **115**

The *ng-* andative prefix can also indicate distance in time. For example, in the following, the speaker is clearing land and answers a question about what it will eventually be used for:

(242) *Mekswel* *yárafam*   *so* *yè**ng**èite.*   (DC24)
       Mekswel   yá-rafè-am   so   y-**ng**-y-ta-e
       Maxwell   3SG-father-ERG FUT α.3SGP-**AND**-plant-IPFV-2|3SGA
       'Maxwell's father will plant it.'

Note that in the inchoative construction with a nominalised verb, the deictic marker occurs on the finite verb:

(243) *mèrès* *eworègh*   *kènufngongai.*   (MD13(1))
       mèrès   ewor-gh   k-**n**-ufngo-ang-ay-Ø
       girl   come.down   β.ØP-**VEN**-start-INC-REM.PUNC-2|3SGS_A
       'the girl started coming down.'

(244) *waorogh*   *yèngwafngongi.*   (ND11)
       waoro-gh   y-**ng**-wafngo-ang-i
       leave-NOM α.3SGP-**AND**-start-INC-2|3PLA
       'they are starting to leave it.'

4.5 Other verbal prefixes

4.5.1 Reflexive/reciprocal prefix

In reflexive and reciprocal propositions, the semantic agent and the semantic patient are the same entity. For these constructions in Nama, instead of having an A argument and P argument with the same referent, there is only an S argument with this one referent. The reflexive/reciprocal prefix (REFL), *a-* or its allomorph *á-*, occurs before the verb and the construction is similar to that of an A-aligned intransitive. The ØP prefix is used insead of the P-indexing prefix and there is a single S_A (in the absolutive case) instead of the two arguments normally found with transitive verb.[21] Examples follow. (Note that the REFL prefix can have either a reflexive or a reciprocal interpretation according to context.)

---

21 An alternative analysis is that *a-/á-* is a general valency reducing prefix that changes transitive verbs to intransitive, and that this intransitive form is simply used for reflexive/reciprocal constructions (see section 4.12).

116 —— 4 Verbal morphology

(245) *yèndyo*   *so*   *náwásindan.*   (FD13))
yènd-yo   so   n-**á**-wásing-ta-èn
1.ABS-EXCL   FUT   α.ØP-**REFL**-take.care.of-IPFV-1SGS$_A$
'I will take care of myself.'

(246) *kawayamngotau*   *fáyo.*   (MD13(1)))
k-**a**-wayamngo-ta-w-Ø   fá-yo
β.ØP-**REFL**-look.closely-IPFV-REM.DUR-2|3SGS$_A$ 3.ABS-EXCL
'she was looking closely at herslf.'

(247) so   fètè   kèngawinjem   soramè.   (FD13)
so   fètè   k-ng-**a**-winjo-e-m   soramè
FUT   VAL   β.ØP-AND-**REFL**-see-PFV.DU-1NSGS$_A$ later
'we (2) should see each other later.'

4.5.2 Applicative prefix

Nama has an applicative prefix *wa-* (APP) (with the allomorph *wá-*) which converts transitive verbs to ditransitive verbs with an additional core argument. This argument, which would correspond to the indirect object in English, is semantically the recipient or beneficiary of the action of the verb. When the applicative prefix is employed, this becomes the P argument. Therefore P-indexing prefix that precedes the applicative prefix indexes the person and number of what corresponds to the the indirect object in English, not of the direct object that is normally indexed in a transitive verb. The recipient or beneficiary P argument does not have to be overt, but if it is, it is marked with the dative nominal suffix *-e* or (for nonsingular marked nominals) both this and the allative suffix *-t*. The same morphophonemic changes described above that occur with the P-indexing prefixes before stems beginning with *w* apply before the *wa-* prefix, resulting in the forms given in Table 4.3.

Table 4.3: P-indexing prefixes plus the applicative prefix.

	α	β	
1SG	*wa- / wá-*	*kwa- / kwá-*	'for/to me'
2SG	*nèma- / nèmá-*	*kènma / kènmá-*	'for/to you'
3SG	*ya- / yá-*	*ta- / tá-*	'for/to him/her/it'
1NSG	*yènma- / yènmá-*	*tènma- / tènmá-*	'for/to us'
2|3NSG	*ewa- / ewá-*	*tawa- / táwá-*	'for/to you/them'

4.5 Other verbal prefixes — **117**

The following examples illustrate the use of the applicative prefix:

(248) a. *Yèndo   yame   yarametan*   *yáyot.*   (NB7:36)
           yèndo   yame   y-**wa**-ramè-ta-èn   yáyot
           1SG.ERG mat   α.3SGP-**APP**-make-IPFV-1SGA 3SG.PURP
           'I'm making him a mat.'/'I'm making a mat for him.'
       b. *Yèndo   yame   ewarametan*   *yèfeyot.*   (NB7:36)
           yèndo   yame   e-**wa**-ramè-ta-èn   yèfeyot
           1SG.ERG mat   α.2|3NSGP-**APP**-make-IPFV-1SGA 3NSG.PURP
           'I'm making them a mat.'/'I'm making a mat for them.'

The *y*- and *e*- prefixes before the *wa*- applicative prefix index the person and number of recipient or beneficiary indirect object (*him* and *them*), not the direct object *yame* 'mat'. So the meanings of the preceding examples could be 'I'm making a mat for him/them.' or 'I'm making mats for him/them.' This contrasts to normal intransitive sentences in which the direct object (the P argument) is indexed:

(249) a. *Yèndo   yame   yèrametan.*   (DC13)
           yèndo   yame   y-ramè-ta-èn
           1SG.ERG mat   α.3SGP-make-IPFV-1SGA
           'I'm making a mat.'
       b. *Yèndo   yame   erametan.*   (based on DC13)
           yèndo   yame   e-ramè-ta-èn
           1SG.ERG mat   α.2|3NSGP-make-IPFV-1SGA
           'I'm making mats.'

For the vast majority of sentences with verbs that have the applicative prefix, the indexed P argument is the beneficiary of the action, as in the preceding examples and the following:

(250) *nu fès   wèi   kènmawanan.*   (MM13)
        nu fès   wèi   kèn-**wa**-wanè-èn
        boiled.water now   β.2SGP-**APP**-get-1SGA
        'I got boiled water for you.'

(251) *Yèmo   botru   yènmasauèr.*   (NB2:83)
        yèmo   botru   yèn-**wa**-sauèr
        3SG.ERG bottle   α.1NSGP-**APP**-open
        'He just opened the bottle for us.'

118 —— 4 Verbal morphology

(252) *tane*   *futárfet*   *wèi*
(=101) tane   futár-f-e-t   wèi
        1SG.GEN   friend-NSG-DAT-**ALL**   now
        táwánfetawèt.   (MK13(4))
        tá-wá-nèfè-ta-w-t
        β.2|3NSGP-**APP**-cut-IPFV-REM.DUR-2|3NSGA
        'they were cutting them for my friends now.'

However, the use of the applicative prefix can express meanings different from the
core meaning of the verb. For example, the applicative prefix used with verb *wi*/
'throw' (literally 'throw for someone') can have the meaning 'forgive':

(253) *Yèmo*   *yaufi* *yam*   *yawite.*   (NB6:95)
        yèmo   yaufi   yam   y-**wa**-wi-ta-e
        3SG.ERG   bad   action   α.3SGP-**APP**-throw-IPFV-2|3SGA
        'He's forgiving his bad deed.'

And in other examples, the indexed P argument is not so clearly the beneficiary:

(254) *Daianèm*   *so*   *buk*   *yawatembete*
        Daianè-m   so   buk   y-**wa**-watembè-ta-e
        Diana-ERG   FUT   book   α.3SGP-**APP**-send-IPFV-2|3SGA
        tiksae.   (NB2:82)
        tiksa-e
        teacher-DAT
        'Diana will send books [for the students] to the teacher.'

(255) *tèngwawitawèn*   *yáne*   *yèmo*
        t-ng-**wa**-wi-ta-w-èn   yáne   yèmo
        β.3SGP-AND-**APP**-throw-IPFV-REM.DUR-1SGA   3SG.GEN   3SG.ERG
        nèmè   *yam*   *kwawafrotau*   (FD13)
        nèmè   yam   kw-**wa**-wafro-ta-w-Ø
        what   action   β.1SGP-**APP**-do-IPFV-REM.DUR-2|3SGA
        'I was forgiving whatever action he did to me.'

(256) *Yèmo*   *merekin*   *kwáráfne.*   (PC/BS:31.7.21)
        yèmo   merekin   kw-**wá**-ráfár-ta-e.
        3SG.ERG   plate   β.1SGP-**APP**-break-IPFV-2|3SGA
        'He broke my plate.' (lit. 'He broke the plate, affecting me.')

4.5 Other verbal prefixes — **119**

Like the deictic prefixes, the applicative prefix occurs on the finite verb in inchoative constructions:

(257) *Yèmo*   *yaufi*   *yam*   *wigh*   *yawafngo.*   (NB8:1)
       yèmo   yaufi   yam   wi-gh   y-**wa**-wafngo
       3SG.ERG   bad   action   throw-NOM   α.3SGP-**APP**-start
       'He's about to forgive his bad deed.'

(258) *Yèmo*   *yame*   *ta*   *ramghèt*   *wawafngo.*   (NB8:21)
       yèmo   yame   ta   ramè-gh-t   w-**wa**-wafngo
       3SG.ERG   mat   1SG.DAT make-NOM-ALL   α.1SGP-**APP**-start
       'She's about to make a mat for me.'

Note that the andative deictic prefix *ng-* (AND) can be used along with the applicative prefix, preceding it, as in example (255) and the following:

(259) *yáf*   *yèngwárse.*   (NB7:41)
       yáf   y-**ng-wá**-rèsa-ta-e
       basket   α.1SGP-**AND-APP**-carry-IPFV-2|3SGA
       'He's carrying the basket for her (as they go).'

The venitive deictic prefix *n-* (VEN) can also be used, as shown in example (783).

The applicative prefix *wa-* can also be used with A-aligned intransitive verbs. But in such cases, it must be followed by the prefix *w-* (TR). This changes intransitive to transitive verbs with the recipient or beneficiary as the P argument, indexed by the P-indexing prefix that normally indexes P arguments in transitive verbs.[22] Again, the experiencer or beneficiary argument does not have to be overt, but if it is, it is marked with the dative nominal suffix *-e* or (for nonsingular marked nominals) both this and the allative suffix *-t*. Some examples:

(260) *Yèmo*   *ás*   *ta*   *kwawarmbote.*   (NB5:72)
       yèmo   ás   ta   kwa-**wa-w**-armbo-ta-e
       3SG.ERG   coconut   1SG.DAT   β.1SGP-**APP-TR**-climb-IPFV-2|3SGA
       'He was climbing the coconut tree for me.'

---

22 The *w-* (TR) prefix appears to be productive only in this context. However, it may have been involved in a historical word formation process, deriving transitive from intransitive verbs. This would explain why approximately 206 out of 392 Nama transitive verb stems begin with *w*.

(261) káye   so   ewáwásáfote.   (NB5:74)
     káye   so   e-**wá-w**-ásáfo-ta-e
     yesterday FUT   α.2|3NSGP-**APP-TR**-work-IPFV-2|3SGA
     'you will work for them tomorrow.'

4.5.3 Autobenefactive prefix

In autobenefactive constructions, the action of the verb benefits the agent. They are similar to constructions with an applicative prefix in that verb is ditransitive and the P-argument is the recipient or beneficiary rather the than the direct object. However, they are also semantically similar to reflexive constructions in that the agent and the beneficiary are the same entity. Thus, in autobenefactive constructions, like reflexive constructions, a single S_A argument refers to this entity, and the ØP prefix is used, as with A-aligned intransitives. But here, the autobenefactive prefix o- (AUTO), rather than the reflexive prefix a/-á, precedes the verb. Also, unlike the reflexive construction, an overt S_A argument in an autobenefactive construction has ergative case, like the A argument with a transitive verb. Some examples:

(262) Ausam   yame noramete.   (NB7:41)
     ausè-am   yame n-**o**-ramè-ta-e
     old.woman-ERG mat   α.ØP-**AUTO**-make-IPFV-2|3SGS$_A$
     'The old woman is making the mat for herself.'

(263) Yèndo mènde so kowifon.   (NB7:48)
     yèndo mènde so k-**o**-wifo-èn
     1SG.ERG wish   FUT β.ØP-**AUTO**-finish-1SGS$_A$
     'I'll get what I want.' (lit. 'I'll finish the wish for myself.')

The autobenefactive prefix can also be used with A-aligned intransitive verbs. Like the reflexive/reciprocal prefix with intransitives, it is then followed by the w- (TR) prefix. Again, an overt S_A argument in this construction has ergative case, like the A argument with a transitive verb.

(264) Fèmo ás   kowarmbota.   (NB7:41)
     fèmo ás   k-**o-w**-armbo-ta-Ø
     2SG.ERG coconut β.ØP-**AUTO-TR**-climb-IPFV-2|3SGA.IRR
     'Climb the coconut tree for yourself.'

(265) *Yèmofem   nowáwásáfotat.*   (NB7:41)
  yèmofem   n-**o-w**-ásáfo-ta-t
  3NSG.ERG   a.ØP-**AUTO-TR**-work-IPFV-2|3NSGA
  'They're working for themselves.'

4.5.4 Ordering of prefixes

The ordering of prefixes in different verbal constructions is shown in Table 4.4. (In the table, ØP = *n*- or *k*-; deictic = *n*- or *ng*-; applicative = *wa-/wá*; transitive = *w*-; reflexive = *a-/á*-; autobenefactive = *o*-.)

Table 4.4: Prefix ordering in Nama.

TRANSITIVE	P-indexing	(deictic)	(applicative)	verb stem
A-ALIGNED INTRANSTIVE	ØP	(deictic)	(applicative + transitive)	verb stem
REFLEXIVE/RECIPROCAL (TRANSITIVE)	ØP	(deictic)	reflexive	verb stem
REFLEXIVE/RECIPROCAL (A-ALIGNED INTRANSITIVE)	ØP	(deictic)	reflexive + transitive	verb stem
AUTOBENEFACTIVE (TRANSITIVE)	ØP		autobenefactive	verb stem
AUTOBENEFACTIVE (A-ALIGNED INTRANSITIVE)	ØP		autobenefactive + transitive	verb stem

4.6 Perfectivity marking

4.6.1 Imperfective vs perfective aspect

The distinction between imperfective and perfective aspect is indicated on every transitive and A-aligned intransitive verb in Nama.[23] As in other languages, the category of IMPERFECTIVE focuses on the internal structure of an event. In Nama it marks ongoing events, encompassing the progressive, habitual and iterative aspects. In contrast, PERFECTIVE focuses on an event or a state as a whole or on its boundaries. In Nama it marks punctual events and the commencement of non-punctual events, encompassing inceptive and inchoative aspects.

---

23 Most transitive and A-aligned intransitive verbs can vary in their perfectivity; P-aligned intransitives, however, are always perfective.

122 —— 4 Verbal morphology

This focus on the inception rather than completion in perfective marking is unusual cross-linguistically (see section 8.2.2). Even more unusual in Nama (and closely related languages such as Nen), however, is that marking for imperfective-perfective depends on the grammatical category of dual number with reference to the referents of the verb's nominal arguments. This is unusal since nominals themselves generally do not distinguish number, and pronouns distinguish only between singular and nonsingular (which includes dual and plural). As shown above, the verbal prefixes that index the person and number of the P argument also distinguish only between singular and nonsingular. And of the suffixes indexing the person and number of the A or S_A argument, only the suffixes for 2nd/3rd person have different forms for dual versus plural. But even more unusual, there are not three number categories with regard to perfectivity marking – singular, dual and plural – but rather two categories: dual versus nondual (which encompasses singular and plural – i.e. more than two). (See section 8.2.3.)

The imperfective and perfective markers are shown in Table 4.5.

Table 4.5: Perfectivity markers in Nama.

	Imperfective	Perfective
NONDUAL	-ta	Ø
DUAL	Ø	-e

One other important characteristic of perfectivity marking is that transitive verbs take the dual perfectivity marker when either one of the core arguments (A or P) has a dual referent (and also when both of them do). This is illustrated in section 4.7.1 below.

Perfectivity marking in Nama is also involved in the distributed exponence of number marking (section 4.2.2). Verbs in both imperfective current tense and imperfective recent tense end with the imperfective suffix and an A/S_A-indexing suffix. It is the particular set of the P-indexing prefixes (α versus β) that signals which tense (see example 219) . The grammatical meaning "imperfective current tense", for example, is distributed between the prefix and the final suffixes – e.g. e- (series α), -ta and -t in examples (219a) and (219b).

The perfective-imperfective marker is also a necessary component in signalling whether the referent for an argument of the verb is dual (2) or plural (3+), as illustrated in examples (266a) and (266b) below. In both, the A argument is indexed by the suffix -t, which indicates nonsingular 2nd or 3rd person. Thus, the A argument could be either dual or plural. The pronominal argument yèmofem does not provide any clue, as it also is nonsingular. It is the combination with the nondual imperfective suffix -*ta* (IPFV.ND) that gives the plural (3+) reading in (266a) whereas the Ø dual imperfective suffix (IPFV.DU) indicates dual in (266b).

(266) a. *Yèmofem*   *yètáftat.*   (based on NB2:65)
          yèmofem   y-táf-**ta**-t
          3NSG.ERG   α.3sGP-chop.up-**IPFV.ND**-2|3NSGA
          'They (3+) are chopping it up.'
     b. *Yèmofem*   *yètáfèt.*   (based on NB2:65)
          yèmofem   y-táf-**Ø**-t
          3NSG.ERG   α.3sGP-chop.up-**IPFV.DU**-2|3NSGA
          'They (2) are chopping it up.'

In the following examples, the S$_A$ argument is indexed by the nonsingular 1st person suffix -*m*. The 1st person absolutive pronoun *yènd* is absent in the example, but even if it was there, it would not provide information about dual versus plural as number is not distinguished at all in absolutive pronouns. Here the dual perfective suffix -*e* (PFV.DU) gives the dual reading in (267a) whereas its absence indicates plural in (267b).

(267) a. *mèngotu*   *kufarem.*   (YD11(3))
          mèngotu   k-ufar-**e**-m
          village   β.ØP-arrive-**PFV.DU**-1NSGS$_A$
          'we (2) arrived at the village.'
     b. *Kiunga*   *kufarngèm.*   (YD11(3))
          Kiunga   n-ufar-**Ø**-ang-m
          Kiunga   β.ØP-arrive-**PFV.ND**-INC-1NSGS$_A$
          'we (3+) arrived at Kiunga.'

The presence of the inceptive suffix -*ang* also indicates nondual (see section 4.8.1).

4.6.2 Morphophonemic changes

Many morphophonemic changes occur as a consequence of perfectivity marking. We have already seen in example (1) in the grammatical overview, and in many other examples earlier in this chapter, that the final *a* of the nondual imperfective marker -*ta* is deleted when it is followed by the 2|3sG A/S$_A$-indexing suffix -*e*. Another change is that for verb stems ending in *m*, the initial *t* in the nondual imperfective suffix becomes *nd* resulting in -*nda* – for example:

(268) *Yèmofem* *mèngotu* *yèyamndat.* (SK11(1))
yèmofem mèngotu y-yam-**ta**-t
3NSG.ERG village α.3SGP-scout.out-**IPFV.ND**-2|3NSGA
'They're scouting out the village.'

Other changes occur in the verb stem. (Whether they are actually morphophonemic is discussed below.) In fact, the majority of the 735 stems of verbs that can be marked for imperfective are affected by the addition of imperfective suffix -*ta*. The only ones not affected are those ending in *o* (approximately 280), *f* (6) and some ending in *r* (10).[24]

For verb stems ending in *è*, the *è* changes to *e* when followed by -*ta*:

(269) *áuyè* *yèfenetan.* (DC20)
áuyè y-fenè-**ta**-èn
cassowary α.3SGP-feed-**IPFV.ND**-1SGA
'I'm feeding the cassowary.'

(270) *Susi* *naumbete.* (NB5:38)
susi n-aumbè-**ta**-e.
fishing.line α.ØP-get.tangled-**IPFV.ND**-2|3SGS_A
'The fishing line is getting tangled up.'

When a stem ending in *è* is preceded by a bilabial – i.e. *mbè, fè* or *mè* – the bilabial can be velarised and rounded – i.e. pronounced as the corresponding velarised bilabial phoneme *mbw, fw* or *mw* – when the final *è* changes to *e* before the -*ta* imperfective nondual suffix. For example, in example (249a) above, *yèrametan* may be pronounced as *yèramwetan* and in (270) *naumbete* as *naumbwete*. Another example with the stem *nèfè* 'cut' would be *yènfetat/yènfwetat* 'they are cutting it'. Impressionistically, this variation apprears to occur more frequently with older speakers, but more rigorous research is needed.

Other changes are categorical and affect both the verb stem and the -*ta* suffix. Approximately 140 verb stems end in *r.* With the exception of 10 of these,[25] the final *r* changes to *n* for nondual imperfective and the initial *t* of the -*ta* suffix is deleted. This can be seen with the verb *wifár*/ 'chase', in which vowel elision also occurs:

---

24 Note that only a limited number of phonemes occur finally in verb stems: six consonants (*f, m, n, ng, ny, r*) and six vowels (*a, e, è, i, o, u*).
25 exceptions are *awar*/ 'plant oneself firmly on the ground', *er*/ 'leak out', *fìr*/ 'get dark' and *war*/ 'hollow out', as well as *fèr*/ 'repeat', *sèr*/ 'hump' and *tèr*/ 'cut hair' and their corresponding intransitive forms: *afèr*/, *ásèr*/ and *átèr*/.

(271) *Ágham* *wèrár* *yifnat.*   (VN:4)
    ághè-am wèrár y-wifá**r-ta**-t
    dog-ERG wallaby α.3SGP-chase-**IPFV.ND**-2|3NSGA
    'The dogs (3+) are chasing the wallaby.'

This change and others make nondual imperfective forms of verbs very different from others – as can be seen by comparing the verb in the following example with that in the preceding one.

(272) *Ágham* *wèrár* *yifárèt.*   (VN:4)
    ághè-am wèrár y-wifár-**Ø**-t
    dog-ERG wallaby α.3SGP-chase-**IPFV.DU**-2|3NSGA
    'The dogs (2) are chasing the wallaby.'

Another change is that verb stem final *ng* becomes *nd* for nondual imperfective and again the initial *t* of the *-ta* suffix is deleted – for example with the verb *úmeng*/ 'gather together (e.g. for a meeting)':

(273) *Ár* *tèmèndot* *númendat.*   (NB6:8)
    ár tèmènd-ot n-úme**ng-ta**-t
    people feast-PURP α.ØP-gather.together-**IPFV.ND**-2|3NSGS$_A$
    'People are gathering together for the feast.'

Verb stems ending in *ny* become *ind*, followed again by *-a* instead of *-ta* – for example with the verb *kuny*/ 'poke':

(274) *Yèmofem* *karèf*   *wène*
    yèmofem karèf   wèn-e
    3NSG.ERG termite.mound stick-INST
    *yèku**inda**t.*   (NB4:42)
    y-ku**ny-ta**-t
    α.1SGP-poke-**IPFV.ND**-2|3NSGA
    'They're poking the termite mound with a stick.'

In verb stems ending in *a*, the final *a* is preceded by a limited set of consonants: *gh, l, n, nd, nj, s* and *y*. When the verb is marked nondual imperfective, the final *a* is deleted and again the initial *t* of the *-ta* suffix is deleted – e.g.:

126 —— 4 Verbal morphology

(275) *Fá   náfálat.*   (NB2:32)
       fá   n-áfá**la-ta**-t
       3.ABS   α.ØP-squat-**IPFV.ND**-2|3NSGS$_A$
       'They're squatting.'

(276) *Yèndo   wèi   yèfene   ambum*
       yèndo   wèi   yèfene   ambum
       1SG.ERG now   3NSG.GEN child
       *engfa**nda**n.*   (FD13)
       e-ng-fa**nda-ta**-èn
       α.2|3NSGP-AND-look.at-**IPFV.ND**-1SGA
       'I look at (i.e. look after) their children now.'

(277) *Yèndo   sèrásèr   esnan.*   (NB5:55)
       yèndo   sèrásèr   e-sè**na-ta**-èn
       1SG.ERG post   α.2|3NSGP-line.up-**IPFV.ND**-1SGA
       'I'm lining up the posts.'

(278) *Fá   námnjat*   *waghfaf.*   (NB8:41)
       fá   n-ám**nja-ta**-t   wagh-faf
       3.ABS   α.ØP-go.for.special.purpose-**IPFV.ND**-2|3NSGS$_A$ singsing-ASS
       'They're going to the singsing place.'

For verbs stems ending in *a*, the final *a* is also deleted when the referent of an argument is dual and thus the imperfective marker is Ø (with epenthetic *è* preceding the consonant-initial suffixes *-m* and *-t*) – for example:

(279) *káye táf   yèronjèt.*   (NB5:10)
       káye táf   y-ronja-**Ø**-t
       3.ABS OBL2 α.3SGP-look.for-**IPFV.DU**-2|3NSGA
       'you should look for it tomorrow.'

(280) *nafalèm.*   (YD11(3))
       n-a-fala-**Ø**-m
       α.ØP-REFL-warm-**IPFV.DU**-1NSGS$_A$
       'we (2) warmed ourselves.'

(281) *kimb   társwè.*   (NB5:10)
       kimb   tá-rèsa-Ø-wè-Ø
       pig   β.2|3NSGP-carry-**IPFV.DU**-DA-2|3NSGA
       'Carry the pigs (2).'[26]

In multi-syllabic verb stems ordinarily ending in *man*, the stem to which the *-ta* suffix is added behaves as if it ends in *mè* – for example, *fenman*/ 'suck' (which has the nominalised citation form *fènmanègh*):

(282) a. *Yèmo   wiwi   wèkwèr   yèfèn**metat**.*   (NB2:54)
           yèmo   wiwi   wèkwèr   y-fènman-**ta**-t
           3SG.ERG mango   juice   α.3SGP-suck-**IPFV.ND**-2|3NSGA
           'They (3+) are sucking mango juice.'
       b. *Yèmofem wiwi   wèkwèr   yèfènmanèt.*   (NB2:54)
           yèmofem wiwi   wèkwèr   y-fènman-**Ø**-t
           3NSG.ERG mango   juice   α.3SGP-suck-**IPFV.DU**-2|3NSGA
           'They (2) are sucking mango juice.'

This raises the question of whether there are actually two separate verb stems, one only for nondual imperfective and one for dual imperfective and both nondual and dual perfective, similar to what is found in Komnzo (Döhler 2018). In other words, there would be, for example, two stems for 'chase' (examples 271and 272), *wifán*/ and *wifár*/, and two for 'gather together' (example 273), *úmend*/ and *úmeng*/. This could be a possible explanation, as the changes to the verb stem, described above (i.e. $r \rightarrow n$ and $ng \rightarrow nd$ when followed by *-t*) do not seem to be motivated by common morphophonemic principles. On the other hand, there are other unusal morphonemic changes in the language, such as *w* becoming *m* when preceded by *n* (section 4.2.2).

4.7 Imperfective tenses and aspects

The three tenses in Nama – current, recent and remote – have been mentioned in the grammatical overview. (As will become apparent below, terms commonly used to label tense – such as "present", "past" and "nonpast" – do not work for Nama.) These tenses occur for all transitive and A-aligned intransitive verbs, but are indicated differently depending on whether their aspect is imperfective or perfective.

---

26 See section 4.7.1 below for an explanation of the dual argument (DA) suffix *-wè.*

128 —— 4 Verbal morphology

Aspects other than imperfective-perfective are also marked. This section describes how these tenses and aspects are indicated on transitive and A-aligned intransitive verbs with imperfective aspect.

4.7.1 Imperfective current and recent tenses

As mentioned in the introduction (section 1.4.1), CURRENT TENSE covers progressive, habitual or iterative events that are occurring at the time of speaking or that have occurred earlier in the day. (This unusual conflation of present and hodiernal past tenses into one category is discussed in section 8.2.1.) In contrast, RECENT TENSE covers events that occurred the day before or sometimes two or three days before.

As we have seen, in both the current and recent tenses, the verb stem is followed by the imperfective suffix -ta for nondual arguments and Ø for dual, and then the A- or S_A-indexing suffix, as shown in Table 4.6 below.

The combined set of verb endings (perfectivity suffix plus A- or S_A-indexing suffix) for the current and recent tenses is also shown in Table 4.6. Note that, as already desscribed, the vowel a in the imperfective suffix -ta is dropped when followed by the $2^{nd}/3^{rd}$ person singular A- or S_A-indexing suffix -e, and because of the other morphonemic changes described above, the initial t in the -ta suffix may be realised as n or deleted. Note also that in the imperfective current and recent tenses, there is an additional A/S_A-indexing suffix -i for 2nd person plural (i.e. 3+). This distinguishes it from 2nd person dual and from 3rd person nonsingular.[27]

Table 4.6: Suffixes for the imperfective current and recent tenses.

Imperfective		A- or S_A-indexing		Combined	
1|2|3SG	-ta	1SG	-èn	1SG	-tan
		2|3SG	-e	2|3SG	-te
1DU	Ø	1NSG	-m	1DU	-m
1PL	-ta			1PL	-tam
2DU	Ø	2DU	-t	2DU	-t
2PL	-ta	2PL	-t-i	2PL	-tati
3DU	Ø	3NSG	-t	3DU	-t
3PL	-ta			3 PL	-tat

---

27 The -i (2PL) suffix can be seen in examples (225) and (233b) above.

4.7 Imperfective tenses and aspects — 129

These combined endings clearly indicate that the verb has imperfective aspect and the tense is either current or recent.

Also as mentioned in section 4.2, current and recent tenses are distinguished by which of the two prefix sets is used: current indicated by the α set, and recent by the β set. Other examples follow.

(283) a. *Yèmo   sèrásèr   yèrèrmete.*   (NB4:65)
         yèmo   sèrásèr   y-rèrmè-**ta**-e
         3SG.ERG   post   α.**3SGP**-straighten-**IPFV.ND**-2|3SGA
         'He's straightening the post.'

      b. *Yèmo   sèrásèr   tèrèrmete.*   (NB4:65)
         yèmo   sèrásèr   t-rèrmè-**ta**-e
         3SG.ERG   post   β.**3SGP**-straighten-**IPFV.ND**-2|3SGA
         'He was straightening the post (yesterday).'

The imperfective current tense is also used with the preverbal tense marker *so* (section 5.3.1) to indicate imperfective events in the future:

(284) *Yèmofem   fifi   so   yúrtotat.*   (MD99(1))
       yèmofem   fifi   so   y-wúrto-**ta**-t
       3NSG.ERG   body   **FUT**   α.3SGP-take.out-**IPFV.ND**-2|3NSGA
       'They (3+) will be taking out the body.'

(285)   *káye   akwan   so   nèngangom*
(=233c) káye   akw-an   so   n-ng-ango-**Ø**-m
        tomorrow   morning-**LOC2** **FUT**   α.ØP-**AND**-return-**IPFV.DU**-1NSGS$_A$
        mèngotuot.   (MK13(1))
        mèngotu-ot
        village-**ALL**
        'tomorrow morning we'll return to the village.'

As noted above, in transitive verbs the imperfective suffix is Ø when the number of either the A argument or P argument is dual. For verbs with a 2nd or 3rd person singular A argument or a 1st person singular or plural A argument, when the indexed P argument is dual, a dual argument suffix *-wè* (DA) follows the verb stem. Following this suffix, the 2nd and 3rd person singular A-indexing suffix is Ø:

(286) a. *Ágham   wèrár   ewifne.*   (VN:4)
          ághè-am   wèrár   e-wifár-**ta**-e
          dog-ERG   wallaby α.2|3NSGP-chase-**IPFV.ND**-2|3SGA
          'The dog is chasing the wallabies (3+).'
       b. *Ágham   wèrár   ewifárwè.*   (VN:4)
          ághè-am   wèrár   e-wifár-**Ø-wè-Ø**
          dog-ERG   wallaby α.2|3NSGP-chase-**IPFV.DU-DA**-2|3SGA
          'The dog is chasing the wallabies (2).'

However, the usual A-indexing suffix remains for 1st person – e.g.:

(287) *Yèndo   yame   tawaramwan.*   (NB7:36)
       yèndo   yame   ta-wa-ramè-**Ø-wè**-èn
       1SG.ERG mat   α.2|3NSGP-APP-make-**IPFV.DU-DA**-1SGA
       'I'm made a mat for the two of them.'

(288) *Yèndfem   ambum   tafandwèm.*   (based on VN:37)
       yèndfem   ambum   ta-fanda-**Ø-wè**-m
       1NSG.ERG child   β.2|3NSGP-look.at-**IPFV.DU-DA**-1NSGA
       'We (3+) looked at the children (2).'

Note that the dual argument suffix followed by the 1st person singular A-indexing suffix is -*wan* because when -*èn* follows a suffix or verb stem ending in *é*, the sequence *èèn* becomes *an*. (However, the dual argument suffix followed by the 1st person plural A-indexing suffix -*m* can similarly sometimes become -*wam* [see below].)

In contrast, the dual argument suffix does not usually occur when the A argument is nonsingular. If the P argument is also nonsingular, there is no indication, then, of which argument is dual. This can lead to ambiguity; for example in the following:

(289) *Yèmofem   ambum   tafandèt.*   (based on VN:37)
       yèmofem   ambum   ta-fanda-**Ø**-t
       3NSG.ERG child   β.2|3NSGP-look.at-**IPFV.DU**-2|3NSGA
       'They (3+) looked at the children (2).' or
       'They (2) looked at the children (3+).' or
       'They (2) looked at the children (2).'

It is only the context that gives the meaning indicated in the following:

(290) *wagif*   *tèfè*   *enèrnyèt.*   (TE11(2))
       wagif   tèfè   e-n-rèny-Ø-t
       fish   INCH2 α.2|3NSGP-VEN-catch-**IPFV.DU**-2|3NSGA
       'you (2) have caught fish.'

In a different context, the meaning could be 'you (3+) caught two fish'.

Ambiguity can also occur with 1st person plural A arguments. In the following, the meaning appears to be clear because of the absence of the dual argument suffix:

(291) *mèrèkin*   *ewúmbárèm.*   (DC01)
       mèrèkin   e-wúmbár-Ø-m
       plate   α.2|3NSGP-wash-**IPFV.DU**-1NSGA
       'we (2) are washing the dishes.'

However, the dual argument suffix is also sometimes used when the 1st person A argument, not the P argument, is dual:

(292) *mato*   *kor...*   *efandwèm.*   (PT:563)
       mato   kor...   e-fanda-Ø-**wè**-m
       INV   actually α.2|3NSGP-look.at-**IPFV.DU-DA**-1NSGA
       'let's (the 2 or us) try to look at them.'

For unknown reasons, the dual argument suffix and 1st person nonsingular suffix are most often realised as -*wam*:

(293) *fèyo*   *yáf-èn*   *efliwam.*   (TE11(2))
       fèyo   yáf-n   e-fèli-Ø-**wè**-m
       then   basket-LOC1 α.2|3NSGP-put.inside-**IPFV.DU-DA**-1NSGA
       'then we (2) put them in the basket.'

(294) *fronde*   *so*   *ekmèwam.*   (PT:197)
       fronde   so   e-kèmè-Ø-**wè**-m
       first   FUT α.2|3NSGP-lay.down-**IPFV.DU-DA**-1NSGA
       'first, we (2) will lay them down.'

132 — 4 Verbal morphology

(295) *fèyo   susi   ekènwam.*   (TE11(2))
fèyo   susi   e-kèna-**Ø**-**wè**-m
then   fishing.line   α.2|3NSGP-roll.up-**IPFV.DU**-**DA**-1NSGA
'then we (2) rolled up the fishing lines.'

Again, ambiguity is possible; for example, the meaning of (295) in a different context could be 'we (3+) rolled up the two fishing lines.'

The dual argument suffix also appears to occur sometimes with A-aligned intransitive verbs with a 1st person dual S argument. (However, as the only instances are with a verb stem ending in *o*, it may be that *-wèm* is an allomorph of 1NSGS$_A$ *-m* following *o*.)

(296) *so   nangowèm?*   (TE11(1))
so   n-ango-**Ø**-**wè**-m
fut   α.ØP-return-**IPFV.DU**-**DA**-1NSGS$_A$
'shall we (2) return?

(297) *Awe,   narmbowam.*   (MD13(1))
awe   n-armbo-**Ø**-**wè**-m
come   α.ØP-climb.up-**IPFV.DU**-**DA**-1NSGS$_A$
'Come, the two of us will climb up.'

A realis-irrealis distinction also occurs in the current tense but only with 2nd or 3rd person singular A or S$_A$ arguments. Irrealis (IRR) is indicated by the absence of the A- or S$_A$-indexing suffix *-e*. The realis (REAL) reading of the *-e* suffix is not generally indicated in the glosses, but it is in the following examples:

(298) a. *Fèm   tè   násáfote?*   (DC27)
fèm   tè   n-ásáfo-**ta-e**
2.ABS   PRF   α.ØP-work-**IPFV.ND-2**|3SGS$_A$.**REAL**
'Have you finished working?'

b. *Fèm   ndauan   fiya   násáfota.*   (NB5:57)
fèm   ndau-an   fiya   n-ásáfo-**ta-Ø**
2.ABS   garden-LOC   OBL1   α.ØP-work-**IPFV.ND-2**|3SGS$_A$.**IRR**
'You should work in the garden.'

(299) a. *Fá   ndauan   mè   násáfote.*   (NB5:57)
fá   ndau-an   mè   n-ásáfo-**ta-e**
3.ABS   garden-LOC   CONT   α.ØP-work-**IPFV.ND-2**|3SGS$_A$.**REAL**
'He's still working in the garden.'

4.7 Imperfective tenses and aspects — **133**

b. *Fá   ndauan   mètè   násáfota.*   (NB5:57)
   fá   ndau-an   mètè   n-ásáfo-**ta-Ø**
   3.ABS   garden-LOC1   POT1   α.ØP-work-**IPFV.ND-2|3SGS$_A$.IRR**
   'He might work in the garden.'

The Ø marking of the 2nd person singular A or S$_A$ argument to indicate irrealis is a feature of imperfective imperative sentences (see section 6.5.1).

4.7.2 Imperfective remote tense

For imperfective verbs, the remote tense is usually used to indicate progressive, habitual and completed events that occurred long ago – or at least more than a few days ago. Two different aspects are distinguished in this tense: delimited (DLT) and durative (DUR).[28]

The REMOTE DELIMITED is used for distant past events that have been completed or that lasted a limited amount of time. It is marked by the α set prefixes and a suffix -*m* (REM.DLT) that follows the imperfective marker -*ta* in the case of nondual arguments and directly follows the verb stem in the case of duals. This is followed by one of the A- or S$_A$-indexing suffixes. These are the same as those for the current and recent tenses for the 1st person singular and nonsingular (i.e. -*èn* and -*m*) but different for the 2nd and 3rd person singular (both either Ø or-*ng*) and for the 2nd and 3rd person nonsingular (-*nd*). The remote delimited verb endings are shown in Table 4.7, followed by examples.

Table 4.7: Verb endings with remote delimited suffixes.

Imperfective		REM.DLT	A- or SA-indexing		Combined	
1|2|3SG	-*ta*	-*m*	1SG 2|3SG	-*èn* Ø/-*ng*	1SG 2|3SG	-*tamèn* -*tam*/-*tamèng*
1DU 1PL	Ø -*ta*	-*m*	1NSG	-*m*	1DU 1PL	-*mèm* -*tamèm*
2|3DU 2|3PL	Ø -*ta*	-*m*	2|3NSG	-*nd*	2|3DU 2|3PL	-*mènd* -*tamènd*

---

28 Note that in Siegel (2015) these were referred to as "completive" and "continuous".

134 —— 4 Verbal morphology

(300) *Yèmo*   *wèn*   *yitrotam*   *endta.*   (NB4:68)
      yèmo   wèn   y-itro-ta-**m**-Ø   end-ta
      3SG.ERG tree   α.3SGP-move-IPFV.ND-**REM.DLT**-2|3SGA   road-ABL
      'He moved the tree from the road.'

(301) *nangotamèm*   *Wamamèngo.*   (MD99(2))
      n-n-ango-ta-**m**-m   Wamamèngo
      α.ØP-VEN-return-IPFV.ND-**REM.DLT**-1NSGS_A   Wamamèngo
      'we returned to Wamamèngo.'

(302) *yèmofem*   *mèngotu*   *ámbiro*
      yèmofem   mèngotu   ámbiro
      3NSG.ERG   village   one
      yafromènd.   (MD13(1))
      y-wafro-Ø-**m**-nd
      α.3SGP-make-IPFV.DU-**REM.DLT**-2|3NSGA
      'They (2) made one village.'

The REMOTE DURATIVE is used for distant past events that were continuing or lasted a long time, often focusing on their start. It is marked by the β set prefixes and a suffix -*w* (REM.DUR) that follows the imperfective marker -*ta* for nondual arguments and directly follows the verb stem for duals. This is followed by one of the A- or S_A-indexing suffixes. These are the same as those for the remote delimited except that the 2nd and 3rd person nonsingular suffix is -*t* rather than -*nd*. The remote durative verb endings are shown in Table 4.8, followed by examples. (When the -*w* remote durative marker follows *a* at the end of a word, the two sounds form the diphthong *au*.)

Table 4.8: Verb endings with remote durative suffixes.

Imperfective		REM.DUR	A- or SA-indexing		Combined	
1|2|3SG	-*ta*	-*w*	1SG	-*èn*	1SG	-*tawèn*
			2|3SG	Ø/-*ng*	2|3SG	-*tau*/-*tawèng*
1DU	Ø	-*w*	1NSG	-*m*	1DU	-*wèm*
1PL	-*ta*				1PL	-*tawèm*
2|3DU	Ø	-*w*	2|3NSG	-*t*	2|3DU	-*wèt*
2|3PL	-*ta*				2|3PL	-*tawèt*

(303) *Fèyo tánetawèm* yènè *wagif.* (TE11(2))
fèyo tá-ne-ta-**w**-m yènè wagif
then β.2|3NSGP-eat-IPFV.ND-**REM.DUR**-1NSGA DEM fish
'Then we started eating these fish.'

(304) *Bulu Mato sènko kufrotau.* (SK11(3))
Bulu Mato sènko k-ufro-ta-**w**-Ø
Bulu Mato head β.ØP-become-IPFV.ND-**REM.DUR**-2|3SGS$_A$
'Bulu Mato was becoming leader.'

(305) *Ágham wèrár tifárwèt.* (VN4)
ághè-am wèrár t-wifár-Ø-**w**-t
dog-ERG wallaby β.3SGP-chase-IPFV.DU-**REM.DUR**-2|3NSGA
'The (2) dogs were chasing the wallaby.'

4.8 Perfective tenses and aspects

The three Nama tenses also occur with verbs with perfective aspect. The perfective current and recent tenses are grouped together under the heading of INCEPTIVE to distinguish them from the imperfective tenses. Unlike the imperfective tenses, some of the perfective ones can occur with P-aligned intransitive verbs as well as with transitive and A-aligned intransitive verbs (see section 4.10 below).

4.8.1 Inceptive tenses

The inceptive tenses indicate a punctual event or focus on the start of a non-punctual event. CURRENT INCEPTIVE is similar to what is called the "immediate perfect", indicating "just now". It is indicated by the α set prefixes and either by the inceptive suffix *-ang* (inc) or by the lack of any other tense-aspect marking other than perfective. The inceptive suffix *-ang* is obligatory for plural (i.e. 3+) arguments, but it is absent when there is a dual argument, in which case only the perfective suffix *-e* occurs.

For verbs with singular A or S$_A$ arguments, there appears to be an aspectual distinction between punctual and durative. For punctual events – those that are instantaneous and occur only once (e.g. *split*) – the verb stem (which has Ø perfective marking) is followed directly by the A- or S$_A$-indexing suffix. For durative events – i.e. those that have just started but are incomplete (e.g. *carry*), or those that have occurred but still have an effect (e.g. *close*) – the inceptive suffix *-ang* follows the verb stem.

136 — 4 Verbal morphology

In the current inceptive tense, the A- or S_A-indexing suffixes for 1st person singular and nonsingular are the same as those we have seen for all the other tenses and aspects (*-èn* and *-m*). But here the suffix for 2nd and 3rd person singular is Ø (which means that in the punctual current, no overt suffix follows the verb stem for these persons). For 2nd and 3rd person dual, the suffix is *-nd*, and for 2nd and 3rd person plural (3+), it is normally *-i*.[29] The current inceptive verb endings are shown in Table 4.9.

Table 4.9: Current inceptive verb endings.

Perfective		INCEPTIVE	A- or Sa-indexing		Combined	
1SG		Ø/-**ang**	1SG	-èn	1SG	-èn-/angèn
2|3SG	Ø	Ø/-**ang**	2|3SG	Ø/-è	2|3SG	Ø/-ang/-angè
1DU	-**e**	Ø	1NSG	-m	1DU	-em
1PL	Ø	-**ang**			1PL	-angèm
2|3DU	-**e**	Ø	2|3DU	-nd	2|3DU	-end
2|3PL	Ø	-**ang**	2|3PL	-I	2|3PL	-angi

It is important to point out that in the inceptive tenses (including inceptive imperfective aspect, see section 4.9 below), the nonsingular A/S_A-indexing suffixes *-nd* and *-i* have functions different to those they have in remote and other current and recent tenses, both imperfective and perfective. Following the remote delimited marker *-m* (REM.DLT) (section 4.7.2) and the remote punctual marker *-ay* (REM.PUNC) (section 4.8.2 below), the *-nd* suffix indicates a 2nd or 3rd person nonsingular argument, either dual (2) or plural (3+). But in the inceptive tenses, it generally indicates only 2nd/3rd person dual (2|3DL). Following the *-t* nonsingular A/S_A-indexing suffix in the current and recent imperfective tenses, *-i* indicates a 2rd person plural argument (2PL). But in inceptive tenses, following *-ang* or *-tang*, it indicates a 2rd or 3rd person plural argument (2|3PL).

The following examples show current inceptive tense. Note that as already mentioned, this tense is frequently indicated by a lack of marking. This Ø-marking is shown in the examples in this section but not always in further examples.

(306) yèmo   ághè   ewanmo.   (NB5:2)
       yèmo   ághè   e-wanmo-Ø-Ø-Ø
       3SG.ERG dog   α.2|3NSGP-call-**PFV.ND-INC**-2|3SGA
       'He's just called the dogs.'

---

29 Note that there is one context where a 2|3PL A argument is indexed by the 2|3DU suffix *-nd*. (See example 323.)

4.8 Perfective tenses and aspects ── **137**

(307) *Yèmofem   ambum   yèsnend.*   (NB5:5)
Yèmofem   ambum   y-sènè-**e**-**Ø**-nd
3NSG.ERG   child   α.3SGP-kiss-**PFV.DU**-**INC**-2|3DUA
'They (2) just kissed the child.'

(308) *Yèndo   ángègh   márte   yumban.*   (NB10:54)
yèndo   ángègh   márèt-e   y-wumbè-**Ø**-**Ø**-èn
1SG.ERG   rope   tight-INST α.3SGP-tie-**PFV.ND**-**INC**-1SGA
'I just tied the rope tightly.'

(309) *nènamèn   nawatemb**ang**?*   (ND13)
nènamèn   n-awatembè-**Ø**-**ang**-**Ø**
why   α.ØP-run.away-**PFV.ND**-**INC**-2|3SGS$_A$
'Why did you run away?'

(310) *yèndfem   Sefor   mbilae   tè   yènès**mang**èm*
yèndfem   Sefor   mbilè-e   tè   y-n-sèmè-**Ø**-**ang**-m
1NSG.ERG   Sefor   axe-INST PRF   α.3SGP-VEN-hit-**PFV.ND**-**INC**-1NSGA
krotiyo.   (MD99(1))
krotiyo
to.death
'we killed Sefor with an axe.'

(311) *fèlismanam   yer**angi**.*   (PT:122)
fèlisman-am   y-werè-**Ø**-**ang**-i
policeman-ERG α.3SGP-hold-**PFV.ND**-**INC**-2|3PLA
'the policemen grabbed him.'

The examples below with the current inceptive tense illustrate that, as with the imperfective tenses, many morphophonemic changes occur with the perfective tenses. For instance, in nonduals where the perfective marker is Ø, if the verb stem ends in *r* or *o,* the initial *a* of the *-ang* inceptive suffix is deleted:

(312) *Nèmè   ár   nufar**ngi**?*   (DC20)
nèmè   ár   n-ufar-**Ø**-**ang**-i
what   person   α.ØP-arrive-**PFV.ND**-**INC**-2|3PLS$_A$
'What people just arrived?'

138 —— 4 Verbal morphology

(313) *Yèndfem ásáfogh yifongèm.* (NB1:27)
  yèndfem ásáfo-gh y-wifo-**Ø**-**ang**-m
  1NSG.ERG work-NOM α.3sGP-finish-**PFV.ND-INC**-1NSGA
  'We've just finished the work.'

Note that some verbs are intrinsically perfective, such as *winjo*/ which is glossed as 'see' but more accurately means something like 'perceive' or 'catch sight of'. Others are intrinsically imperfective, such as *fanda*/ which is also sometimes glossed as 'see' but more accurately means 'look at' or 'watch'. Other verbs have slightly different meanings, depending on whether they are imperfective or perfective – for example, *wèrè*/ 'hold, grab':

(314) a. *Yèmo wèrár yerete.* (NB5:6)
    yèmo wèrár y-werè-**ta**-e
    3SG.ERG wallaby α.3sGP-hold-**IPFV.ND**-2|3sGA
    'He held the wallaby.'
  b. *Yèmo wèrár yerè.* (NB5:6)
    yèmo wèrár y-werè-**Ø-Ø-Ø**
    3SG.ERG wallaby α.3sGP-hold-**PFV.ND-INC**-2|3sGA
    'He grabbed the wallaby.'

RECENT INCEPTIVE tense focuses on the starting point of activities or events that occurred in the not-so-distant past, and also those that will begin soon in the future. The recent inceptive is distinguished from the current inceptive by marking with the β set prefixes rather than the α set. The A- and S_A-indexing suffixes are the same as those for the current inceptive. The recent inceptive is frequently used for narrative past. Here are some examples:

(315) *Yèndo mbilè terangèn.* (NP11)
  yèndo mbilè t-werè-**Ø**-**ang**-èn
  1SG.ERG axe β.3sGP-hold-**PFV.ND-INC**-1SGA
  'I picked up the axe.'

(316) *Seforfè kokawend.* (MD99(1))
  Sefor-fè k-okawè-**e**-**Ø**-nd
  Sefor-COM β.ØP-fight-**PFV.DU-INC**-2|3DUS$_A$
  'He started fighting with Sefor.' (lit. 'with Sefor, the two started fighting.')

4.8 Perfective tenses and aspects — **139**

(317) *"Eso"*   *taramangèm.*   (YD11(3))
eso   ta-ramè-**Ø-ang**-m
thanks   β.2|3NSGP-give-**PFV.ND-INC**-1NSGA
'We gave them thanks.'[30]

(318) *Sèkw*   *kètè*   *kangsongè.*   (YD11(3))
sèkw   kètè   k-angso-**Ø-ang**-è
canoe   there   β.ØP-get.stuck-**PFV.ND-INC**-2|3SGS$_A$
'The canoe got stuck there.'

The following examples show another morphophonemic change: if a stem ends in *o*, this vowel is elided when followed by the dual perfective suffix *-e*:

(319) *Fá*   *nófnamend.*   (NB5:5)
fá   n-ófnamo-**e-Ø**-nd
3ABS   α.ØP-cross-**PFV.DU-INC**-2|3DUA
'They (2) just crossed.'

(320) *wèrár*   *tinjem.*   (MD93)
wèrár   t-winjo-**e-Ø**-m
wallaby   β.3SGP-see-**PFV.DU-INC**-1NSGA
'we (2) saw a wallaby.'

Compare the preceding example with the following one with Ø nondual perfective marking:

(321) *Yabun*   *bout*   *tinjongèm.*   (YD11(3))
yabun   bout   t-winjo-**Ø-ang**-m
big   boat   β.3SGP-see-**PFV.ND-INC**-1NSGA
'We (3+) saw a big boat.'

Note that as with transitive verbs in the imperfect tenses, dual perfective marking is used when the P argument has a dual referent as well as when the A argument does – for example:

---
30 Note that the verb \ramè/ can mean 'make' or 'give; depending on the context.

140 —— 4 Verbal morphology

(322) *tawinje* .... *áwumánghan.* .... (MD99(1))
ta-winjo-**e-Ø-Ø** .... áwumán-gh-an
β.2|3NSGP-see-**PFV.DU-INC**-2|3SGA wrestle-NOM-LOC2
'he saw the two wrestling.'

But unexpectedly, when a 2|3 plural A argument occurs with a dual P argument, it is indexed by what is normally the 2|3 dual suffix -*nd* instead of the 2|3 plural suffix -*i* – for example:

(323) *wèikor* *tawaranend.* .... (B&B:282)
wèikor ta-wa-ramè-**e-nd**
again β.3NSGP-APP-give-**PFV.DU**-2|3DUA
'they (3+) gave it again to them (2).'

In other contexts, this could also mean 'they (2) gave it again to them (3+) or 'they (2) gave it again to them (2)'. (This is also relevant to immediate imperatives [section 6.5.1]).

When used for the future, recent inceptive verbs are preceded by the preverbal future marker *so*:

(324) *Yau* *so* *kuwanongèm.* .... (YD11(3))
yau so k-uwano-**Ø-ang**-m
NEG FUT β.ØP-set.off-**PFV.ND-INC**-1NSGS$_A$
'We won't set off.'

(325) *(Fá)* *so* *kokáyend.* .... (B&B:60)
fá so k-okáyo-**e-Ø**-nd
3.ABS FUT β.ØP-stand.up-**PFV.DU-INC**-2|3DUS$_A$
'They (2) will stand up.'

(326) *Yèmon* *mèrès* *so* *tawaramangè.* .... (ND11)
yèmon mèrès so ta-wa-ramè-**Ø-ang**-è
3SG.ERG girl FUT β.2|3NSGP-APP-give-**PFV.ND-INC**-2|3SGA
'He'll give the girl to them.'

The recent inceptive tense is also used for imperatives (see section 6.5).

4.8.2 Remote punctual tense

The REMOTE PUNCTUAL tense indicates punctual events that occurred in the distant past. It is marked on the verb by the ɑ set prefixes and the suffix -*ay* (REM.PUNC), which directly follows the verb stem in the case of nondual arguments or follows the perfective marker -*e* in the case of duals. This is followed by one of the A- or S$_A$-indexing suffixes. These are the same as those for the imperfective remote delimited. The remote punctual verb endings are shown in Table 4.10. (Note that the stem final *è* is deleted when followed by the -*ay* perfective remote suffix, and that this suffix is the diphthong *ai* when it occurs at the end of the word.) Examples follow.

Table 4.10: Verb endings with perfective remote punctual suffixes.

Perfective		REM.PUNC	A- or S$_A$-indexing		Combined	
1/2|3SG	Ø	-**ay**	1SG	-*èn*	1SG	-*ayèn*
			2|3SG	Ø/-*ng*	2|3SG	-*ai/-ayèng*
1DU	-**e**	-**ay**	1NSG	-*m*	1DU	-*eayèm*
1PL	Ø				1PL	-*ayèm*
2|3DU	-**e**	-**ay**	2|3NSG	-*nd*	2|3DU	-*eayènd*
2|3PL	Ø				2|3PL	-*ayènd*

(327) *Sèlngwèl   yènramai.* (MD99(1))
sèlngwèl   yèn-ramè-Ø-**ay**-Ø
surprise   ɑ.1NSGP-give-PFV.ND-**REM.PUNC**-2\|3SGA
'It gave us a surprise.'

(328) *Yènd   a   afafè   álet* (MD93)
yènd   a   afè-afè   ále-t
1.ABS   and   father-COM   hunting-ALL
nuwaneayèm.
n-uwano-e-**ay**-m
ɑ.ØP-set.off-PFV.DU-**REM.PUNC**-1NSGS$_A$
'Me and father set off for hunting.'

(329) *fèyo   wagh   syèrnyayènd.* (ND13)
fèyo   wagh   s=y-rèny-Ø-**ay**-nd
then   singsing   PROX2=ɑ.3SGP-start-PFV.ND-**REM.PUNC**-2\|3NSGA
'then they (3+) started singing and dancing.'[31]

---

31 The proximal clitics (PROX) are described in section 5.4.1.

The following examples illustrate the same morphophonemic changes that occur in the inceptive tenses. In nonduals, where the perfective marker is Ø, if the verb stems ends in *r* or *o* the initial *a* of the remote suffix *-ay* is deleted, and again, the final *o* of a stem is deleted when followed by the dual perfective suffix *-e*:

(330) *Fá... nufaryènd.* (MK13(4))
      fá   n-ufar-Ø-**ay**-nd
      3.ABS α.ØP-arrive-PFV.ND-**REM.PUNC**-2|3NSGS$_A$
      'They arrived.'

(331) *ámb*   *ár*   *yinjeayèm.* (TE11(1))
      ámb   ár   y-winjo-e-**ay**-m
      another man α.3SGP-see-PFV.DU-**REM.PUNC**-1NSGA
      'we (2) saw another man.'

Again, compare the preceding example with the following one with the Ø nondual perfective suffix:

(332) *táf*   *yinjoyèm.* (MD99(1))
      táf   y-winjo-Ø-**ay**-m
      that.time α.3SGP-see-PFV.ND-**REM.PUNC**-1NSGA
      'at that time, we (3+) saw him.'

4.8.3 Remote inceptive tense

The remote punctual suffix *-ay* can also follow the inceptive suffix *-ang* to indicate the REMOTE INCEPTIVE tense. This emphasises the start of an event in the distant past. Unlike the remote punctual tense, β set rather than α set prefixes are used. Here, as in the inceptive tenses, the inceptive suffix *-ang* does not occur when there is a dual argument. With regard to verb endings, then, the remote inceptive tense would be indistinguishable from the remote punctual tense when there is a dual argument. However, it is questionable whether the remote inceptive tense actually occurs when there is a dual argument because these endings do not occur with β set prefixes in the data. (It appears that inceptive imperfective aspect is used instead –

4.8 Perfective tenses and aspects — **143**

see section 4.9.) Thus the remote inceptive suffixes combined with the A/S$_A$-indexing suffixes shown in Table 4.11 are only for verbs with nondual arguments.[32]

Table 4.11: Verb endings with remote inceptive suffixes.

1SG	*-angayèn*
2|3SG	*-angai/-angayèng*
1PL	*-angayèm*
2|3 PL	*-angayènd*

Some examples are:

(333)   *wèikor*   *tinjongai*   *mèrès*
           wèikor   t-winjo-Ø-**ang-ay**-Ø   mèrès
           again   β.3SGP-see-PFV.ND-**INC-REM.PUNC**-2\|3SGA   girl
           kètan.   (MD13(1))
           kètan
           there
           'Again, he saw the girl there.'

(334)   *mbilaro*   *tanangayèng.*   (MD99(2))
           mbilè-ro   t-wanè-Ø-**ang-ay**-ng
           axe-RSTR   β.3SGP-take-PFV.ND-**INC-REM.PUNC**-2\|3SGA
           'He took only an axe.'

(335)   *Árèm*   *táfátongayènd*   *wèrár.*   (DC37)
           ár-m   t-wáfáto-Ø-**ang-ay**-nd   wèrár
           man-ERG   β.3SGP-hide-PFV.ND-**INC-REM.PUNC**-2\|3NSGA   wallaby
           'The men hid the wallaby.'

(336)   *Rokár*   *kètorngai.*   (DC33)
           rokár   k-ètor-Ø-**ang-ay**-Ø
           thing   β.ØP-come.out-PFV.ND-**INC-REM.PUNC**-2\|3SGS$_A$
           'The thing came out.'

---

[32] Note that this is a revised analysis of that in Siegel (2015).

144 — 4 Verbal morphology

4.9 Inceptive imperfective aspect

The final type of tense/aspect in Nama is indicated by a combination of both perfective and imperfective verbal suffixes. (This exact phenomenon does not appear to be found in other languages – see section 8.2.2.) When markers for these two aspects are combined in Nama, imperfective as usual indicates that the event is not punctual, but once again the perfective focuses on inception rather than completion – i.e. on the beginning of an action or event rather than on the ending. The inceptive imperfective suffix is -tang (INC.IPFV), which appears to be a combination of the nondual imperfective suffix -ta and the nondual inceptive marker -ang. While both of these suffixes do not normally occur on verbs with dual arguments, the inceptive imperfective suffix -tang occurs on verbs with nondual arguments and on those with dual arguments as well, in which case it is followed by the dual perfective suffix -e. The A/S_A suffixes are the same as those for the inceptive (perfective current and recent) tenses. The set of inceptive imperfective verb endings combined with the A/S_A-indexing suffixes given in Table 4.12.

Table 4.12: Verb endings with inceptive imperfective suffixes.

1SG	-tangèn
2|3SG	-tangè
1DU	-tangem
1PL	-tangèm
2|3DU	-tangend
2|3 PL	-tangi

Note that similar to what occurs with the inceptive tenses, if a 2|3 plural A argument occurs with a dual P argument, the ending would be -tangend instead of -tangi.

As with the inceptive, the α set prefixes indicate an immediate "just now" or very immediate time frame whereas the β set prefixes indicate further back in time in narratives, or some time in the future in imperfective imperatives (see section 6.5.1). These are illustrated in the following examples. Example (340) also demonstrate that the same morphophonemic changes that apply to the addition of the imperfective suffix -ta also apply to the inceptive imperfective suffix -tang.

(337)   *mènamèn*   *fá*   *yèmotangèn.*   (B&B:149)
        mènamèn   fá   y-mo-**tang**-èn
        that's.why   3.ABS   α.3SGP-ask-**INC.IPFV**-1SGA
        'That's why I started asking him.'

4.9 Inceptive imperfective aspect —— **145**

(338) *Kafukèm*   *tèmotangè*   *"emofè*   (MD99(1))
    Kafuk-m   t-mo-**tang**-è   emo-afè
    Kafuk-ERG   β.3SGP-ask-**INC.IPFV**-2|3SGA   who-COM
    enèm"?
    e-nèm
    α.2|3NSGS$_P$-come.ND
    'Kafuk started asking her "who did you come with?"'

(339) *fèyo*   *kamotangend*   *"fèm*   *efe*
    fèyo   k-a-mo-**tang**-e-nd   fèm   efe
    then   β.ØP-REFL-ask-**INC.IPFV-PFV.DU**-2|3DUA   2.ABS   who.ABS
    nèm"?   (MD13(1))
    n-m
    α.2SGS$_P$-COP.ND
    'then they (2) started asking each other "who are you?".'

(340) *Yèmofem*   *fái*   *kwáutánangi.*   (FD13)
    yèmofem   fái   kw-wáutár-**tang**-i
    3NSG.ERG   OBL1   β.1SGP-help-**INC.IPFV**-2|3PLA
    'They should start helping me.'

The perfective remote punctual suffix -*ay* can also co-occur with the inceptive imperfective and the α set prefixes to indicate the start of a nonpunctual action further in the past (remote). This occurs most often when there is a dual argument. Thus, this marking may be what is used in dual contexts instead of the remote inceptive marking (section 4.8.3) which occurs only in nondual contexts. Some examples follow. (Note in example 341 that because the P argument is dual, the 2|3DU A-indexing suffix is used even though the A argument (*the young children*) is plural.)

(341) *Ambum*   *tèrtèram*
    ambum   tèrtèr-am
    child   small-ERG
    yènmotangeayènd.   (TE11(1))
    yèn-mo-**tang-e-ay**-nd
    α.1NSGP-ask-**INC.IPFV-PFV.DU-REM.PUNC**-2|3DUA
    'The young children started calling to us (2).'

146 —— 4 Verbal morphology

(342) *ámb   ndau   sifayè*
      ámb   ndau   sifayè
      other garden place
      *yèngwai**ndangeayèm.***   (TE11(1))
      y-ng-waing-**tang-e-ay**-m.
      α.3SGP-AND-pass-**INC.IPFV-PFV.DU-REM.PUNC**-1NSGA
      'we (2) went past another garden place.'

It also occurs rarely in nondual contexts, as in the following:

(343) *yènd   yènamè   nèngut**nangayèn**...*   (YD11(3))
      yènd   yènamè   n-ng-utár-**tang-ay**-n
      1.ABS   from.here   α.ØP-AND-shout-**INC.IPFV-REM.PUNC**-1SGA
      'from here I shouted...'

4.10 Summary of tense/aspect marking

The preceding sections have illustrated at least 11 distinct tense/aspect categories in Nama for transitive and A-aligned verbs. The 11 categories can be divided
into three groups with regard to perfectivity marking: imperfective, perfective
and inceptive imperfective (which has a combination of imperfective and perfective marking). Within each of these categories, there are three tense distinctions:
current, recent and remote. Two of the categories have further aspectual distinctions with 2nd and 3rd person A/S_A arguments: realis/irrealis in imperfective current
and durative/punctual in perfective current. A delimited/durative distinction also
occurs for all arguments in imperfective remote tense. Although verbal suffixes are
essential for marking tense, choice of one of the two prefix sets is also a necessary
component. These are summarised in Table 4.13.

In each of the three groups based on perfectivity marking, one set of suffixes
is used for marking both current and recent tense (bolded in Table 4.13). Thus,
each suffix set can have two different values. For these tenses, then, it is only the
choice of either the α set or the β set prefix that "sets the switch" at the appropriate
value – α for current or β for recent. On the other hand, the same two prefix sets
are also used for the remote tenses. In the case of the imperfective remote, the α
set is used for delimited and the β for durative, but the suffixes also differ (-*m* and
-*w* respectively). Therefore, as already discussed, it seems reasonable to conclude
that with regard to tense/aspect marking, the two sets have no intrinsic function of
their own.

Table 4.13: Prefix sets for tense/aspect categories (transitive and A-aligned intransitve verbs).

	α set	**β set**
IMPERFECTIVE	**current**	**recent**
	remote delimited	remote durative
PERFECTIVE	**current inceptive**	**recent inceptive**
	remote punctual	remote inceptive
INCEPTIVE	**current inceptive imperfective**	**recent inceptive imperfective**
IMPERFECTIVE	remote inceptive imperfective	

As is probably clear from the preceding account, the imperfective-perfective categories that underlie the tense/aspect system in Nama are based on morphological criteria. Nevertheless, as already mentioned, because of their inherent lexical semantics. some verbs occur almost exclusively with imperfective aspect: for example, \war/ 'hollow out'; \wafro/ 'do, make'; \rèsa/ 'carry'. Others are almost exclusively perfective: – for example, \winjo/ 'catch sight of, see'; \uwano/ 'set off'; \amnjo/ 'sit down'. On the other hand, some verbs are unexpectedly never imperfective – e.g. \nam/ 'shoot' – while others are unexpectedly never perfective – e.g. \wumbè/ 'tie up'. A large number of verbs, however, can be used in both categories, depending on the meaning – for example, \wauf/ 'blow, blow out' and \ramè/ 'make, give'.

4.11 P-Aligned intransitive verbs

As mentioned earlier, for P-aligned intransitive verbs, the single S argument is indexed by the same sets of prefixes as those for P arguments of transitive verbs (P-indexing prefixes). These are indicated by S_P in the gloss. A-indexing suffixes are absent. Like the A-aligned intransitives, an overt S argument NP itself is unmarked or consists of an absolutive pronoun. The P-aligned verbs are nearly all "stative intransitives", referred to as "positional verbs" by Evans (2014), whereas no A-aligned verbs are stative. Thus, lexical semantics appears to have conditioned split intransitivity (Creissels 2008: 149).

Examples follow with the verb stems \mor/ 'stay' (344a, 344b) and \mángo/ 'be piled up' (345):

148 — 4 Verbal morphology

(344) a. *Yènd* *tebolan* *sèfè* ***wèmor.*** (MD11(1))
       yènd   tebol-an   sèfè   **w-mor-Ø**
       1.ABS   table-LOC2 at.this.time **α.1SGSₚ-stay-PFV.ND**
       'At this time I was staying (sitting) at the table.'

    b. *sènonjo* *mèngotuan* *Mawai* *yèmor.* (MD99(1)
       sènonjo   mèngotu-an   Mawai   **y-mor-Ø**
       today   village-LOC1   Mawai   **α.3SGSₚ-stay-PFV.ND**
       'Today Mawai stays in the village.'

(345) *Wem* ***emángo.*** (NB7:89)
    wem   **e-mángo-Ø**
    yam   **α.2|3NSGSₚ-be.piled.up-PFV.ND**
    'The yams are piled up (in a heap).'

4.11.1 Differences between P-aligned intransitives and the other verb types

In addition to the absence of A- or S_A indexing suffixes, P-aligned intransitive verbs differ from transitive and A-aligned intransitive verbs, as already mentioned, in that they cannot be nominalised with the addition of the -*gh* suffix.

    P-aligned intransitive verbs also differ from transitive and A-aligned intransitive verbs in the marking of tense and aspect. First, P-aligned intransitives are always perfective, and take perfective marking (Ø nondual, -*e* dual) but never imperfective marking (-*ta* nondual). Thus, all P-aligned intransitives with a nondual S argument are Ø marked for perfectivity, as in the preceding examples. Those in current or recent tenses are most commonly treated as inceptive and also Ø marked. (In the examples that follow, Ø marking is not shown in the glosses for P-aligned intransitives.) Those with a dual argument are marked with the -*e* perfective dual suffix, but in contrast to transitive and A-aligned intransitive verbs, this follows a distinctive additional dual-marking suffix, -*ar* (DU.PA), as shown in (346a) and (346b):

(346) a. *sembyo* *ár* *kès* ***emorare.*** (PT:157)
       sembyo   ár   kès   **e-mor-ar-e**
       two   men just.there **α.2|3NSGSₚ-stay-DU.PA-PFV.DU**
       'two men are staying (sitting) just there.'

    b. *Wem* ***emángare.*** (NB7:45)
       wem   **e-mángo-ar-e**
       yam   **α.2|3NSGSₚ-be.in.a heap-DU.PA-PFV.DU**
       'The yams are in two heaps.'

4.11 P-Aligned intransitive verbs ⎯ **149**

P-aligned intransitives take both α and β set prefixes, similar to transitive and A-aligned intransitive verbs, as illustrated by examples (347a) and (347b), where they distinguish current and recent inceptive tense.

(347) a. *ndauan*   *so*   ***yènmor.***   (DC27)
      ndau-an   so   **yèn**-mor
      garden-LOC1   FUT   **α.1NSGS$_P$**-stay
      'we will stay in the garden.'

    b. *Jeffè*   *tè*   ***tènmor.***   (DC07)
      Jef-fè   tè   **tèn**-mor
      Jeff-COM   PRF   **β.1NSGS$_P$**-stay
      'we stayed with Jeff.'

Like the other two verb types, P-aligned intransitives can also be suffixed with the perfective inceptive marker *-ang* (which normally does not occur with dual). With stative verbs, it can emphasise the non-permanence as well as the inception of a state. Note that unlike with transitive and A-aligned intransitive verbs, the initial *a* in *-ang* is not deleted after P-aligned intransitive verb stems ending in *r* – e.g. *mor/* 'stay'.

(348) a. *Mer*   *njaran*   *yènmor**ang**.*   (DC15)
      mer   njar-an   yèn-mor-**ang**
      good   shade-LOC2   α.1NSGS$_P$-stay-**INC**
      'We're staying in good shade.'

    b. *ne*   *ghèrghèr*   *kètè*   *ekm**ang**.*   (ND13)
      ne   ghèrghèr   kètè   e-kèmè-**ang**
      shit   intestines   there   α.2|3NSGS$_P$-be.lying-**INC**
      'The guts were lying there.'

    c. *Yèndo*   *si*   *so*   *nafrotan*   *káye*   *ndèrnae*
      yèndo   si   so   n-afro-ta-èn   káye   ndèrnae
      1SG.ERG   story   FUT   α.2SGP-tell-IPFV.ND-1SG$_A$ yesterday   how
      yènd   *kwèmor**ang**.*   (MM13)
      yènd   kw-mor-**ang**
      1.ABS   β.1SGS$_P$-stay-**INC**
      'I'll tell you the story of how I stayed yesterday (i.e. about what happened to me).'

Although the inceptive suffix *-ang* does not normally occur when the subject of a P-aligned verb is dual, there is at least one exception. It still takes the dual-marking suffix *-ar*, but the *a* of the *-ar* suffix is deleted:

150 —— 4 Verbal morphology

(349) *yènd   yèmafè   yèntèmangre.*   (KS11)
yènd   yèmafè   yèn-tèm-**ang-ar-e**.
1.ABS   3SG.COM   α.1NSGS$_P$-be.united-**INC-DU.PA-PFV.DU**
'I have become united with him.' or 'We (2) have become united.'

In contrast to the other verb types, P-aligned intransitives are also marked differently in remote tense. For transitive and A-aligned intranstive verbs with imperfective aspect, a distinction is made in remote tense between delimited (marked
by -*m*) and durative (marked by *w*-) (see section 4.7.2). For those with perfective
aspect, remote tense is most often punctual (section 4.8.2). In contrast, P-aligned
verbs are never imperfective, and being predominantly sematically stative, they
would be incompatible with remote punctual. Instead, for these verbs remote tense
is marked with the same delimited-durative distinction used for transtive and
A-aligned intransitive verbs, but with different suffixes. For P-aligned verbs with a
nondual S$_p$ argument, these are -*èrmèn(g)* (REM.DLT.PA) for delimited and -*èrwèn(g)*
(REM.DUR.PA) for durative. Each one is preceded by the inceptive suffix -*ang*. (The
perfective nondual suffix is -*Ø*.) This is in contrast to the imperfective -*ta* + -*m* versus
-*ta* + -*w* in transitives and A-aligned intransitives (see examples 300 to 305 above).
However, similarly, the α set prefixes are used for delimited and the β set for durative, and the distinction is maintained between *m* in the suffix for delimited and *w*
for durative. The differences and similarities are shown in Table 4.14.

Table 4.14: Remote tense marking (nondual).

	transitive and A-aligned intransitive		P-aligned intransitive	
	imperfective	perfective	imperfective	perfective
delimited	-*ta*-**m**-A/S$_A$	—	—	-*Ø-ang-èrmèn(g)*
durative	-*ta*-**w**-A/S$_A$	—	—	-*Ø-ang-èrwèn(g)*
punctual	—	-*Ø-ay*-A/S$_A$	—	—

Some examples are:

(350) a.   *Sarghárèn   yèmorangèrmèn.*   (SK11)
Sarghár-n   y-mor-ang-**èrmèn**
Sarghar-LOC1   α.3SGS$_P$-stay-INC-**REM.DLT.PA**
'he stayed at Sarghar.'

  b.   *fèyo   kètè   támorangèrwèn.*   (MD11(2))
fèyo   kètè   tá-mor-ang-**èrwèn**
then   there   β.2|3NSGSP-stay-INC-**REM.DUR.PA**
'they were staying there.'

4.11 P-Aligned intransitive verbs — **151**

(351) a. *mer kèmegh yau yèkmangèrmèn.* (MD13(1))
mer kèmegh yau y-kèmè-ang-**èrmèn**
good sleeping NEG α.3sGSₚ-be.lying.down-INC-**REM.DLT.PA**
'he didn't sleep well.'

b. *kètè tèkmangèrwèng.* (DC34)
kètè t-kèmè-ang-**èrweng**
there β.3sGSP-be.lying.down-INC-**REM.DUR.PA**
'he was sleeping there.'

For most P-aligned intransitive verbs with a dual Sₚ argument, the remote suffixes are -*mèn* for delimited and -*wèn* for durative, both following the P-aligned intransitive dual suffix -*ar*:

(352) a. *Ámbiro sèmbár fètè ekmarmèn*
ámbiro sèmbár fètè e-kèmè-ar-**mèn**
one night VAL α.2|3NSGSₚ-be.lying.down-DU.PA-**REM.DLT.PA**
yènè mèngon. (MD13(1))
yènè mèngo-n
this house-LOC1
'One night they (2) actually slept in this house.'

b. *Kwambalisan tákmarwèn*
Kwambalisa-n tá-kèmè-ar-**wèn**
Kwambalisa-LOC1 β.2|3NSGSₚ-be.lying.down-DU.PA-**REM.DUR.PA**
ámbiro efogh soramè. (MD99(1))
ámbiro efogh soramè
one day later
'One day later, they (2) were sleeping at Kwambalisa.'

4.11.2 The copula and related forms

The copula in Nama is also a P-aligned intransitive verb. It has two forms, one for nondual Sₚ arguments and one for dual Sₚ arguments.

The nondual form of the copula is *m*/ or *mo*/. When there is no following tense or aspect marker, the two variants appear to be in free variation with *m*/ occurring most frequently. Another variant, *mon*/, also occurs occasionally. (Like for other P-aligned intransitives, Ø nondual perfective aspect marking and Ø current/recent perfective tense marking are not shown in glosses.)

(353) a. *Mèrès   wáiwái   **yèm.***   (NB3:42)
   mèrès   wáiwái   **y-m**
   girl   shy   α.3SG$_\text{SP}$-COP.ND
   'The girl is shy.'

   b. *Arufè   yamèndári   brubru   kètèf   **yèmo.***   (SK11(2))
   Arufè   yamèndári   brubru   kètè=f   **y-mo**
   Arufi   kundu.type   kundu   there=PROX1   α.3SG$_\text{SP}$-COP.ND
   'There is a "yamendari" kundu from Arufi.'

(354) a. *wèriafè   kár   **tèm.***   (PT:241)
   wèri-afè   kár   **t-m**
   drunkenness-COM   DUB   β.3SG$_\text{SP}$-COP.ND
   'He might have been drunk.'

   b. *tane   mèrès   tosè   enjne   **tèmo***
   tane   mèrès   tosè   enjne   **t-mo**
   1SG.GEN   girl   young.one   sickness   β.3SG$_\text{SP}$-COP.ND
   káye.   (MM13)
   káye
   yesterday
   'My little girl was sick yesterday.'

When the nondual form of the copula is followed by the inceptive suffix -*ang* (or the remote suffixes -*èrmèn(g)* or -*èrwèn(g)*), the \mo\ variant is always used (and, as with other verb classes, the initial *a* of the -*ang* suffix is deleted following *o*:

(355) *krúfèr   taim   **yèmong.***   (DC15)
   krúfèr   taim   **y-mo-ang**
   coldness   time   α.3SG$_\text{SP}$-COP.ND-INC
   'it's the cold time.'

(356) *mèrès   ámb   kèr   tè   **tèmong.***   (MD99(1))
   mèrès   ámb   kèr   tè   **t-mo-ang**
   girl   some   dead   PRF   β.3SG$_\text{SP}$-COP.ND-INC
   'one of his sisters has died.'

Although the nondual form of the copula takes the same remote tense suffixes as other verbs in this category (-*èrmèn(g)* and -*èrwèn(g)*), it differs in that the inceptive suffix -*ang* does not normally occur with the remote suffixes:

(357)   *kèr*   **yèmormèng.**   (MD99(1))
       kèr   **y-mo-èrmèng**
       dead   **α.3SGS$_\text{P}$-COP.ND-REM.DLT.PA**
       'he was dead.'

(358)   *mèrès*   *ghèrghèrsisafè*   **tèmorwèn.**   (MD13(1))
       mèrès   ghèrghèrsisè-afè   **t-mo-èrwèn**
       girl   sadness-COM   **β.3SGS$_\text{P}$-COP.ND-REM.DUR.PA**
       'the girl was sad.'

(359)   *fèyo*   *ghar*   *tèfnár*   *yèna*
       fèyo   ghar   tèfnár   yèna
       then   sorcery   practitioner   here
       emormèn.   (ND13)
       e-mo-èrmèn
       α.2|3NSGS$_\text{P}$-COP.ND-REM.DLT.PA
       'then sorcerers were here.'

The dual form of the copula is \r/. It is normally followed by the dual perfective suffix -*e* but unlike other P-aligned intransitives, without the dual suffix -*ar*:

(360)   *Sembyo*   *fèyotaro*   **ere.**   (MK13(2))
       sembyo   fèyotaro   **e-r-e**
       two   enough   **α.2|3NSGS$_\text{P}$-COP.DU-PFV.DU**
       'Two are enough.'

(361)   *femeliafayo*   *yènd*   **yènre.**   (MK13(2))
       femeli-afè-yo   yènd   **yèn-r-e**
       family-COM-EXCL   1.ABS   **α.1NSGS$_\text{P}$-COP.DU-PFV.DU**
       'We (2) are with family.'

According to consultants, the remote suffixes for P-aligned verbs with a dual argument, -*mèn* and -*wèn*, also occur with the dual form of the copula, but without the suffix -*ar* or the dual perfective marker -*e*. However, the only instance that occurs in the recorded data is the remote durative for 2nd and 3rd person:

154 —— 4 Verbal morphology

(362) *sembyo* *ár* *tawinjem* *sèkwan*
      sembyo ár ta-winjo-e-m sèkw-an
      two man α.2|3NSGP-see-PFV.DU-1NSGA canoe-LOC2

      tárwèn.   (YD11(3))
      tá-r-wèn
      α.2|3NSGSₚ-COP.DU-REM.DUR.PA
      'We saw two men were in the canoe.'

All inflected forms of the copula are shown in Table 4.15.

Table 4.15: Forms of the copula in Nama.

	Current	Recent	Inceptive-marked	Remote delimited	Remote durative
1SG	*wèm/wèmo*	*kwèm/kwèmo*	*wèmong*	*wèmormèn(g)*	*kwèmorwèn(g)*
2SG	*nèm/nèmo*	*kènèm/kènmo*	*nèmong*	*nèmormèn(g)*	*kènmorwèn(g)*
3SG	*yèm/yèmo*	*tèm/tèmo*	*yèmong*	*yèmormèn(g)*	*tèmorwèn(g)*
1PL	*yènèm/yènmo*	*tènèm/tènmo*	*yènmong*	*yènmormèn(g)*	*tènmorwèn(g)*
2|3PL	*em*	*tám/támo*	*emong*	*emormèn(g)*	*támorwèn(g)*
1DU	*yènre*	*tènre*	—	*yènèrmèn*	*tènèrwèn*
2|3DU	*ere*	*táre*	—	*enèrmèn*[33]	*tárwèn*

Note that applicative prefix *wa-* can also be used with the copula – for example:

(363) *yènè* *wem* ***kwam.***   (NB4:77)
      yènè wem kw-**wa**-m
      DEM yam β.1SGₚ-**APP**-COP.ND
      'This yam is for me.'

(364) *tárfár* *fifi* ***yènmare.***   (MK13(2))
      tárfár fifi yèn-**wa**-r-e
      plenty very α1NSGSₚ-**APP**-COP.DU-PFV.DU
      'it's really plenty for us two'

---

33 This is the form given by consultants, although one would expect *erèmèn*. However, neither occur in the recorded data.

(365) *fá* *fái* *yènamamèn* *áwèghafè* *wèi*   (FD13)
  fá fái yènamamèn áuwè-gh-afè wèi
  3.ABS really for.this.reason be.happy-NOM-COM now
  tam.
  t-**wa**-m
  β.3SGS$_P$-**APP**-COP.ND
  'for this reason there was really happiness now for them'

(366) *yènè* *mwighè* *kwamorwèng.*   (FD13)
  yènè mwighè kw-**wa**-mo-èrwèng
  DEM thought β.1SGS$_P$-**APP**-COP.ND-REM.DUR.PA
  'these thoughts were affecting me.'

Historically, the verb for 'come' is derived from the copula with the addition of the venitive deictic marker *n-*: *nèm(o)*/ 'come (nondual)' and *nèr*/ 'come (dual)'. Some examples are:

(367) *mbwito* *tènèm*   *eee* *tane* *kafkafan.* (MD11(1))
  mbwito t-**nèm**   eee tane kafkaf-an
  rat β.3SGS$_P$-**come.ND** PROL 1SG.GEN foot-LOC2
  'a rat came on my foot.'

(368) *afaf* *kès* *enre.*   (TE11(1))
  afè-af kès e-**nèr**-e
  father-PL just.now α.2|3NSGS$_P$-**come.DU**-PFV.DU
  '(our) parents are just coming.'

(369) *táf* *yènmormèn*   *Ghèráyan.*   (SK11(2))
  táf y-**nèmo**-èrmèn Ghèráyè-n
  then α.3SGS$_P$-**come.ND**-REM.DLT.PA Daraia-LOC1
  'then he came to Daraia.'

Similarly, the verb for 'go' appears to be derived from the copula with the addition of the andative deictic marker *ng-*: *ngèm(o)*/ 'go (nondual)', *ngèr*/ 'go (dual)':

(370) *wèikor* *yènngèm* *ámb* *mèngotuot.*   (YD11(3))
  wèikor yèn-**ngèm** ámb mèngotu-ot
  again α.1NSGS$_P$-**go.ND** another village-ALL
  'we went again to another village.'

156 — 4 Verbal morphology

(371) *tagase*　　　*fèmafayof*　　　*tángre?*　　　　　(DC29)
　　　ta-gase　　　fèmafè-yo=f　　　tá-**ngèr**-e?
　　　1SG-namesake　2SG.COM-EXCL=PROX1　β.2|3NSGS$_P$-**go.DU**-PFV.DU
　　　'did my namesake go with just you?'

(372) *yènd*　*mè*　*kwèngmorwèng.*　　　　　(TE11(2))
　　　yènd　mè　kw-**ngèmo**-èrwèng
　　　1.ABS　CONT　β.1SGS$_P$-**go.ND**-REM.DUR.PA
　　　'I kept going.'

4.12 Valency reduction: A-aligned intransitive

Most transitive verbs in Nama appear to be able to undergo valency decreasing operations to become intransitive verbs with one argument. The argument that is eliminated is what was in some cases the A argument of the transitive verb and in other cases, the P argument, or what Malchukov (2015: 96) refers to as either "subject removing" or "object removing". As we have seen, the argument that remains, the S argument, is indexed on the verb either in the same way as the A argument of a transitive verb (A-aligned intransitive), or as the P argument of a transitive verb (P-aligned intransitive). This section (4.12) looks at A-aligned intransitives and the following section (4.13) at P-aligned intransitives.

As already mentioned, the majority of transitive verb stems in Nama (220 out of the 392 recorded) have a corresponding A-aligned intransitive verb stem. This stem begins with a single vowel, most commonly *a-* or *á-*, that appears to be attached to the transitive verb stem, which always begins with a consonant. This seems to serve as a marker of intransitivity. In the analysis that follows, this vowel is glossed as as an INTRANSITIVE PREFIX (INTR). But whether or not it is a productive valency-reducing prefix will be discussed at the end of this section.

A-aligned intransitives with transitive counterparts are of several types with different functions, some corresponding to what have been traditionally called "voices". (See Malchukov 2015 for a typological overview of this "voice ambivalence".) These are illustrated in the following subsections.

4.12.1 Antipassive

In antipassive constructions, the A argument of the transitive verb becomes the S argument of the A-aligned intransitive; the P argument is omitted. These are "absolute" antipassives (Malchukov 2015: 98) rather than "proper" antipassives in which the P is demoted to an oblique (see Polinsky 2013). The following examples compare transitive and corresponding antipassive A-aligned intransitive verbs.

(373) a. *Yèmo   ngarènd   tèraye.*   (NB2:5)
        yèmo   ngarènd   t-raya-ta-e
        3SG.ERG   canoe   β.3SGP-paddle-IPFV.ND-2|3SGA
        'He paddled the canoe.'

    b. *Fá   nènarayat.*   *tèfefaf*   (YD11(3))
        fá   n-n-**a**-raya-Ø-t   tèfefaf
        3.ABS   α.ØP-VEN-**INTR**-paddle-IPFV.DU-2|3NSGS$_A$ 1NSG.ASS
        'They (2) paddled towards us.'

(374) a. *mer   kakayam   ke*   (TE11(1))
        mer   kakayam   ke
        good   bird.of.paradise sound
        tángyárwèm. . .
        tá-ng-yárè-Ø-w-m
        β.NSGP-AND-hear-IPFV.DU-REM.DUR-1NSGA
        'we heard (far away) the nice calls of the bird of paradise. . .'

    b. *yènd   ndenè   náyárayèn*   *nambyo*
        yènd   ndenè   n-**á**-yárè-Ø-ay-èn   nambyo
        1.ABS   like.this α.ØP-**INTR**-hear-PFV.ND-REM.PUNC-1SGS$_A$ three
        ghèrare   soramè   (ND13)
        ghèrare   soramè
        month   later
        'I heard this way three months later.'

Table 4.16 presents a list of other transitive verb stems with corresponding antipassive A-aligned intransitive stems.

4.12.2 Anticausative

In anticausative (or unaccusative) constructions, the P argument of the transitive verb becomes the S argument of the A-aligned intransitive; the A argument is omitted. The following examples compare transitive and corresponding anitcausative A-aligned intransitive verbs.

158 — 4 Verbal morphology

Table 4.16: Some transitives with corresponding antipassive intransitives.

Transitive		Antipassive intransitive	
far	'shape'	*afar*	'work at shaping'
faro	'write'	*afaro*	'be in the processes of writing'
kena	'refuse (something)'	*ákena*	'refuse'
limán	'pull'	*álimán*	'pull a bow to affix the string'
mèror	'quarrel with'	*ámèror*	'be quarrelsome'
mogh	'ask'	*ámogh*	'be asking'
rèny	'start (something)'	*árèny*	'start'
sèmè	'hit, fight with'	*ásèmè*	'be in the process of fighting'
wárámnja	'be jealous of'	*áwárámnja*	'be jealous'
wátárán	'complain about'	*áwátárán*	'complain'
werè	'hold (something)'	*awerè*	'be in the act of holding'

(375) a. *Yámrèsom*   *wèn*   *yèráfárngè.*   (NB2:52)
yá-mèrès-o-m   wèn   y-ráfár-Ø-ang-è
3SG.GEN-girl-SG-ERG   stick   α.3SGP-break-PFV.ND-INC-2|3SGA
'His sister just broke the stick.'

  b. *Wèn*   *náráfárngè.*   (NB2:52)
wèn   n-**á**-ráfár-Ø-ang-è
stick   α.ØP-**INTR**-break- PFV.ND-INC-2|3SGSA
'The stick just broke.'

(376) a. *Yèmofem*   *wìm*   *efwetat.*   (NB3:44)
yèmofem   wim   e-fwe-ta-t
3NSG.ERG   stone   α.2|3NSGP-roll-IPFV.ND-2|3NSGA
'They're rolling the stones.'

  b. *Wìm*   *nafwetat.*   (NB3:44)
wìm   n-**a**-fwe-ta-t
stone   α.ØP-**INTR**-roll-IPFV.ND-2|3NSGSA
'The stones are rolling.'

Table 4.17 presents a list of other transitive verb stems with corresponding anti-causative A-aligned intransitive stems.

4.12 Valency reduction: A-aligned intransitive — **159**

Table 4.17: Some transitives with corresponding anticausative intransitives.

Transitive		Anticausative intransitive	
fal	'warm something up'	*afal*	'become warmed up'
fam	'wet something'	*afam*	'get wet'
famngo	'smash, crack'	*afamngo*	'get smashed, crack'
fèli	'put inside (e.g. a basket)'	*afli*	'go inside'
fror	'spray'	*áfror*	'gush out'
káyo	'attach'	*ákáyo*	'become attached'
lawè	'put inside (e.g. a house)'	*èlawè*	'go inside'
manè	'close'	*amanè*	'become closed'
mar	'break off something'	*amar*	'break off'
mina	'increase something'	*èmina*	'increase'
ráf	'give birth to'	*áráf*	'be born'
rètè	'smash'	*árètè*	'become smashed'
sauèr	'open'	*ásauèr*	'become open'
tèno	'hide something'	*atèno*	'be hidden'
tiar	'break, shatter (something)'	*atiar*	'break, shatter'
tirár	'tear'	*átirár*	'become torn'
y	'plant'	*èy*	'become planted'

4.12.3 Reflexive/reciprocal

As shown in section 4.5.1, reflexive/reciprocal constructions occur when the A argument and P argument of a transitive verb would have had referents that were the same or that were simultaneously semantic agent and patient. In the examples that follow the single S argument in example b of each pair represents both of these arguments. (Here, for the sake of argument, the *-a/-á* suffix is glossed as INTR.)

(377) a. *Yèndo   min   yèsotan.*   (NB3:17)
           Yèndo   min   y-so-ta-èn
           1SG.ERG   nose   α.3SGP-scratch-IPFV.ND-1SGA
           'I'm scratching my nose.'

       b. *Fá   kámbi-kámbimèn   násote.*   (NB4:76)
           fá   kámbikámbi-mèn   n-á-so-ta-e
           3.ABS   ringworm-ORIG   α.ØP-**INTR**-scratch-IPFV.ND-2|3SGSA
           'She's scratching herself because of ringworm."'

160 —— 4 Verbal morphology

(378) a. *Yènd mè yènfande!* (DC20)
   yènd mè yèn-fanda-ta-e
   1.ABS CONT α.1NSGP-look.at-IPFV.ND-2|3SGA
   'It (a cassowary) is still looking at us!'

  b. *Fá nafande*   *glasan.* (NB9:19)
   fá n-**a**-fanda-ta-e   glas-an
   3ABS α.ØP-**INTR**-look.at-IPFV.ND-2|3SGS$_A$ mirror-LOC2
   'She's looking at herself in the mirror.'

(379) a. *Yèmofem yèmronat.*   *Les.* (NB7:39)
   yèmofem y-mèror-ta-t   Les
   3NSG.ERG α.3SGP-quarrel-IPFV.ND-2|3NSGA   Les
   'They're quarrelling with Les.'

  b. *wèri tèfnár snamronèt.* (PT:99)
   wèri tèfnár s=n-**a**-mèror-Ø-t
   drunkard PROX2=α.ØP-**INTR**-quarrel-IPFV.DU-2|3NSGSA
   'The two drunkards here are quarrelling with each other.'

4.12.4 Autocausative

In autocausative constructions,[34] the A argument of a transitive verb becomes the
S argument of the A-aligned intransitive, but semantically it is like the P argument
of the transitive verb – i.e. a patient, undergoing a change of state. The following
examples compare transitive and corresponding autocausative A-aligned intransitve verbs.

(380) a. *Yèndo sèrasèr ekmetan.* (NB2:12)
   yèndo sèrasèr e-kèmè-ta-èn
   1SG.ERG post α.2|3NSGP-lay.down-IPFV.ND-1SGA
   'I'm laying down the posts.'

  b. *Yènd nákmetan.* (NB2:74)
   yènd n-**á**-kèmè-ta-èn
   1.ABS α.ØP-**INTR**-lay.down-IPFV.ND-1SGS$_A$
   'I'm (in the process of) lying down.'

---

34 Autocausatives are defined by Parry (1998: 69) as reflexive constructions in which "the referent
represented by the subject combines the activity of actor and undergoes a change of state as a
patient".

(381) a. *Mèrsam*   *wiwi*   *kafwe*   (NB5:82)
           mèrès-am   wiwi   kafwe
           girl-ERG   mango   branch
           yefnoyènd.
           y-wefno-Ø-ay-nd
           α.3SGP-bend.down-PFV.ND-REM.PUNC-2|3NSGA
           'The girls bent down the branch of the mango tree.'
       b. *yène*   *yabun*   *ár*   (YD11(3))
           yène   yabun   ár
           DEM   big   man
           náwefnoyènd.
           n-**á**-wefno-Ø-ay-nd
           α.ØP-**INTR**-bend.down-PFV.ND-REM.PUNC-2|3NSGS$_A$
           'these big men bent down.'

4.12.5 Multiple functions

The apparent detransitivisation can result in different remaining arguments and
functions – for example, reflexive in (382b) and anticausative in (382c):

(382) a. *Yèmo*   *wèrár*   *yènfè.*   (NB5:26)
           yèmo   wèrár   y-nèfè-Ø-Ø
           child-SG-ERG   wallaby   α.3SGP-cut-PFV.ND-2|3SGA
           'He just cut the wallaby.'
       b. *Fá*   *fèt sìf*   *aweranghan*   *nánfè.*   (NB6:9)
           fá   fèt sìf   aweran-gh-an   n-**á**-nèfè-Ø-Ø
           3.ABS beard   shave-NOM-LOC2   α.ØP-**INTR**-cut-PFV.ND-2|3SGS$_A$
           'He cut himself shaving.'
       c. *Ángègh*   *nánfè.*   (NB5:69)
           ángègh   n-**á**-nèfè-Ø-Ø
           string   α.-**INTR**-cut- PFV.ND-2|3SGS$_A$
           'The string just got cut.'

This demonstrates that Nama does not have classes of verbs based on which of the
two transitive arguments remains or the resulting function of detransitivisation –
for example, one class resulting in antipassives (like *eat* in English) while another
in anticausatives (like *break*) (see Malchukov 2015: 110).

4.12.6 Discussion

From the preceding description, Nama appears to be similar to many other languages in having a marker on the transitive form of the verb to signal an intransitive construction with various functions. Kemmer (1993) presents examples from a wide range of languages (including Latin, German, Russian, Hungarian, Turkish, Guugu Yimidhirr, Bahasa Indonesia and Djola) with morphemes, termed "middle markers", that mark what she refers to as middle voice – namely anticausative, reflexive and reciprocal structures. Janic (2016) presents wide-ranging examples (from language families including Slavic, Cariban, Tacanan, South Caucasian, Pama-Nyungan, Chukotko-Kamchatkan and Manding) of morphemes that make a transitive verb intransitive and mark both reflexive and antipassive constructions.

However, while polyfunctional voice markers and voice ambivalence are common in languages, the use of one marker to indicate all of these functions with a wide range of verbs does not seem to be common. A case in point is the *-sja* suffix in Russian. It has reflexive, reciprocal, anticausative and antipassive functions, but these are restricted to verbs with certain semantics. For example, the antipassive meaning is found with only a few verbs of negative impact (Malchukov 2015: 113). In contrast, the apparent Nama intransitive prefix *a-/á-* has all these functions with a wide range of verbs with varying semantics. On the other hand, Janic (2013: 66–67, 2016: 167), referring to the work of Nedjalkov (2007), describes the *-tku/-tko* suffix in Chukchi (spoken in Siberia and a member of the Chukotko-Kamchatkan family), which appears to have similar polyfunctionality, and characterises this as being "typologically distinctive". It may be that Nama has this typologically distinctive characteristic as well.

However, there are two arguments against *a-/á-* being a productive detransitivising prefix. First, there are a large number of intransitive verbs that begin with *a-/á-* but do have have a corresponding transitive form (and are not used transitively). Some of these are listed in Table 4.18. Kemmer (1993: 22) refers to such verbs in other languages as "deponents", and argues that they actually have "middle" semantics . This subsumes the traditional notion of "subject affectedness" and along with the reflexive, is situated "intermediate in transitivity between one-participant and two-participant events" (Kemmer 1993: 3). However, it is difficult to see such middle semantics in most of these items in Nama – e.g. *afarman*/ 'drop down', *ála*/ 'hop' and *ásáfo*/ 'work'.

Second, there are many intransitive verbs that begin with vowels other than *a-/á-*. Some of these have transitive equivalents, such as *èmina*/ and *mina*/ 'increase' in Table 4.17. Many others have no transitive equivalents – such as the ones listed in Table 4.18 along with those beginning with *a-/á-*, such as *úrto* 'go up'. Again, it is difficult to see middle semantics (i.e. subject-affectedness) in most of these items – e.g. *ofing*/ 'whistle', *ómáng*/ 'sniff' and *úny*/ 'tell a lie'.

4.12 Valency reduction: A-aligned intransitive — **163**

Table 4.18: Some "deponents" in Nama.

afarman	'drop down'	*ásáfo*	'work'
afo	'get full'	*átán*	'jump'
afuran	'change'	*áurár*	'sleep close to fire'
ak	'remain'	*áwányo*	'deny'
anambè	'make repeated sound'	*eghno*	'slip'
ango	'return'	*ena*	'breathe'
armbo	'climb'	*enja*	'crawl'
awafer	'apologise'	*èghar*	'walk in water'
awambè	'scream, cry out'	*ófám*	'go over, cross'
ayo	'ripen'	*ófing*	'whistle'
áfála	'squat'	*ófmo*	'lie or fall down straight'
áker	'refuse'	*ómáng*	'sniff'
ála	'hop'	*umbè*	'become wrinkled'
ámyo	'accept, surrender'	*úny*	'tell a lie'
ánegha	'be eager, determined'	*úr*	'laugh'
ánjo	'dry out'	*úrto*	'go up, ascend'

In summary, all A-aligned intransitive verbs begin with a vowel, most commonly *a* or *á*, whereas all transitive verbs begin with a consonant. But at least 135 out of 370 recorded A-aligned intransitive verbs do not have transitive equivalents. Therefore, the initial vowel cannot be analysed as a detransitivising intransitive prefix because it marks all A-aligned intransitive verbs, not just those derived from transitives. One possible explanation is that historically, an initial vowel prefix became a marker of verbs with a single argument, particularly those indexed in the same way as the agent of a transitive verb – i.e. A-aligned intransitive verbs.[35] This prefix is not grammatically productive and currently serves to mark a lexical class of intransitive as opposed to transitive verbs.

Thus it is assumed that antipassive, anticausative and autocausative forms of transitive verbs, with their initial vowels, are lexicalised. However, it seems unlikely that every transitive verb that could be used reflexively or reciprocally would have a lexicalised reflexive/reciprocal form. As all such forms, when they occur, begin with either *a-* or *á-* and no other vowel, it is also assumed then that the *a-/á-* prefix is grammatically productive in this one context – that is, in deriving reflexive and reciprocal intransitives from transitives. In such constructions, the initial *a* or *á* is analysed as a distinct prefix, and glossed as reflexive/reciprocal (REFL), as in the examples in section 4.5.1 and the following:

---

[35] While historically there might have been some morphophonemic rules about the choice of vowel, currently none are apparent.

(383) a. *tane   mèrès yarawena.*   (B&B:234)
      tane   mèrès y-warawer-ta-Ø
      1SG.GEN girl   α.3SGP-take.care.of-IPFV.ND-2|3SGA.IRR
      'take care of my girl.'
   b. *kor   nèngawarawena.*   (DC07)
      kor   n-ng-**a**-warawer-ta-Ø
      careful α.ØP-AND-**REFL**-take.care.of-IPFV.ND-2|3SGSA.IRR
      'take care of yourself.'

(384) a. *ausam   ewinjoyènd.*   (SK11(2))
      ausè-m   e-winjo-Ø-ay-nd
      old.woman-ERG   α.2|3NSGP-see-PFV.ND-REM.PUNC-2|3NSGA
      'the old women saw them.'
   b. *nawinjeayènd. . .*   (MD13(1))
      n-**a**-winjo-e-ay-nd
      α.ØP-**REFL**-see-PFV.DU-REM.PUNC-2|3NSGA
      'they(2) saw each other. . .'

4.13 Valency reduction: P-aligned intransitive

In contrast to A-aligned intransitives, there is only a small number of P-aligned
intransitive verbs that appear to be derived from transitive verbs (21 out of 392).
However, the total number of P-aligned verbs is much smaller than that of A-aligned
verbs (35 versus 370). This means that nearly two thirds of P-aligned intransitives
are derived from transitives (as opposed to approximately half for A-aligned intransitives). Also, in contrast to A-aligned intransitives, the transitive verb stem usually
does not change. All P-aligned intransitive verbs with the same stem as transitive
verbs have stative functions.

4.13.1 Statives

The transitive verbs that can become P-aligned intransitives are all semantically
causative, and refer to a change of state which an agent (A argument) brings about
to an undergoer or patient (P argument). In most cases, it is the A argument that is
absent and the P argument that has become the S of the intransitive verb, which
then refers to the state, rather than a change of state. The following examples
compare the transitive and corresponding P-aligned intranstive verb.

4.13 Valency reduction: P-aligned intransitive — **165**

(385) a. *Árèm   sèsafne   yèsaurèt.*   (NB2:31)
ár-m   sèsafne   y-sauèr-Ø-t
man-ERG   door   α.3sGP-open-IPFV.DU-2|3NSGA
'The two men are opening the door.'

b. *Sèsafne   yèsauèr.*   (NB2:31)
sèsafne   y-sauèr
door   3sGS_P-be.open
'The door is open.'

(386) a. *Yosim   sáláme   ewámete.*   (NB2:45)
Yosi-m   sáláme   e-wámè-ta-e
Yoshie-ERG clothes   α.2|3NSGP-hang-IPFV.ND-2|3sGA
'Yoshie is hanging up the clothes.'

b. *Kwèng   yámang   wènan*   (NB2:45)
kwèng   y-wámè-ang   wèn-an
flying.fox α.3sGS_P-be.hanging-INC   tree-LOC2
'The flying fox is hanging in the tree.'

However, in some instances the A argument of the transitive can become the S of the P-aligned intransitive, as in the following:

(387) a. *karèf   fú   yèkmete.*   (MK13(1))
karèf   fú   y-kèmè-ta-e
earth.oven boundary   α.2|3NSGP-lay.down-IPFV.ND-2|3sGA
'She's laying down the earth oven boundary.'

b. *Fá   yèkèm.*   (NB2:11)
fá   y-kèmè
3.ABS   α.3sGS_P-be.lying.down
'She's lying down.' or 'She's sleeping.'

(388) a. *námb   tawakáyongè   wènan.*   (MD13(1))
námb   ta-wakáyo-ang-è   wèn-an
bow   β.2|3NSGP-stand.up-INC-2|3sGA   tree-LOC2
'he stood the bows up in the tree.'

b. *yènan   tane   ambum   syakái.*   (B&B:277)
yènan   tane   ambum   s=y-wakáy(o)
here-LOC1   1SG.GEN   child   PROX2=α.3sGS_P-be.standing
'my child is standing right here.'

166 —— 4 Verbal morphology

Some examples of transitive verb stems that can be used as P-aligned intransitives are given in Table 4.19.

Table 4.19: Some transitive and corresponding P-aligned intransitive verbs.

	Transitive meaning	Intransitive meaning
fayo	'put on top'	'be on top'
fèl	'put inside (basket, etc.)'	'be inside (basket, etc.)'
fiango	'spread out'	'be spread out'
káyo	'attach'	'be attached'
kèmè	'lay X down'	'be lying down'
kènyè	'bend'	'be bent'
lawè/nyawè	'put inside (house, garden, etc.)'	'be inside (house, garden, etc.)'
man	'close'	'be closed'
mègh	'lean X'	'be leaning'
sauèr	'open'	'be open'
sár	'put side by side'	'be side by side'
tèm	'put next to'	'be next to'
wafèli	'put in or on (plate, vehicle, etc)'	'be in or on (plate, vehicle, etc)'
wakáyo	'stand X up'	'be standing up'
wáfato	'hide X'	'be hiding'
wámè	'hang X'	'be hanging'
wármben	'put inside enclosed area'	'be inside enclosed area'
wumbè	'tie'	'be tied'
yáto	'do something until enough'	'be enough'
y	'plant'	'be planted, grow'

4.13.2 Transitive verbs with both P-aligned and A-aligned forms

A small number of transitive verbs have both A-aligned and P-aligned intransitive corresponding forms. For example, compare the A-aligned intranstive verbs in (389) and (390) below to the P-aligned ones in (387b) and (388b) above:

(389)   *Yènd*   *wèikor*   *ámb*   *mèngotuan*
    yènd   wèikor   ámb   mèngotu-an
    1.ABS   again   another   village-LOC1
    kákmangèm.   (YD11(3))
    k-ákèmè-Ø-ang-m
    β.ØP-lie.down-PFV.ND-INC-1NSGS$_A$
    'Again we laid down (i.e. went to sleep) in another village.'

4.13 Valency reduction: P-aligned intransitive — **167**

(390) *Ben so kokáyo.* (B&B:57)
 Ben so k-okáyo-Ø-Ø
 Ben FUT β.ØP-stand.up-PFV.ND-2|3SGS_A
 'Ben will stand up.'

Another example follows (a. transitive, b. P-aligned intransitive, c. A-aligned intransitive):

(391) a. *yènè wèrár yèngfayoyèn* (MK13(2))
 yènè wèrár y-ng-fayo-Ø-ay-èn
 DEM wallaby α.3SGP-AND-put.on.top-PFV.ND-REM.PUNC-1SGA
 wènan.
 wèn-an
 tree-LOC2
 'I put the wallaby up in the tree.'

 b. *mèrès stèfayongèrwèn* *wèn*
 mèrès s=t-fayo-ang-èrwèn wèn
 girl PROX2=β.3SGS_P-be.on.top-INC-REM.DUR.PA tree
 tukèn. (MD13(1))
 tuk-n
 top-LOC1
 'the girl was on the tree top.'

 c. *wiwi kafwean kafayongè.* (MB13)
 wiwi kafwe-an k-afayo-Ø-ang-è
 mango branch-LOC2 β.ØP-go.on.top-PFV.ND-INC-2|3SGS_A
 'it went onto the mango branch.'

These verb stems and others that have corresponding forms in all three verb classes are shown in Table 4.20.

Table 4.20: Some verbs with corresponding forms in all three verb classes.

	Transitive	P-aligned intransitive	A-aligned intransitive	
fayo	'put on top'	'be on top'	*afayo*	'go on top'
kèmè	'lay X down'	'be lying down'	*ákèmè*	'lie down'
man	'close'	'be closed'	*aman*	'become closed'
nyawè	'put inside'	'be inside'	*ènyawè*	'go inside'
sauèr	'open'	'be open'	*ásauèr*	'become open'
tèm	'put next to'	'be next to'	*átèm*	'go alongside'
wakáyo	'stand X up'	'be standing'	*ókáyo*	'stand up'

168 —— 4 Verbal morphology

4.13.3 Distinguishing P-aligned intransitives from transitives

Besides the lack of A-indexing suffixes, P-aligned intransitives can be distinguished from their transitive counterparts in four ways.

First, for some verbs, the final vowel of the stem (*è*, or *o* following *y*) is deleted in the current and remote tenses. This is illustrated in examples (387b) and (388b) above, and in the following, contrasting a transitive and corresponding P-aligned intranstive verb:

(392) a. *Yèmo*   *karèf*   *mbwe* *njamkean*
     yèmo   karèf   mbwe njamke-an
     3SG.ERG earth.oven stone food-LOC
     efayo.   (NB2:64)
     e-**fayo**-Ø-Ø
     α.2|3NSGS_P-**put.on.top**-PFV.ND-2|3SGA
     'She just put the earth oven stones on the food.'

   b. *Ambum*   *kitárè-kitárayan* *efai.*[36]   (NB2:93)
     ambum   kitárèkitárayè-an e-**fay(o)**-Ø
     child   platform-LOC2   α.2|3NSGS_P-**be.on.top**
     'The child is on top of the platform.'

Second, as already mentioned (section 4.11.1), in contrast to transitives (and A-aligned intransitives), the P-aligned counterparts can never occur with the -*gh* nominalising suffix (section 4.1).

Third, as also mentioned in section 4.11.1, P-aligned intransitives have special dual suffixes, as shown in example (346) and the following, which contrasts a transitive and corresponding P-aligned intransitive verb:

(393) a. *Yèmofem* *mèngon*   *yènyawèt.*   (NB2:97)
     yèmofem mèngo-èn   y-nyawè-Ø-t
     3NSG.ERG house-LOC1 α.3SGP-put.inside-IPFV.DU-2|3NSGA
     'They (2) put it in the house.'

   b. *Fâ* *mèngon*   *enyaware.*   (NB2:45)
     fá   mèngo-èn   e-nyawè-**ar-e**
     3.ABS house-LOC1 α.2|3NSGS_P-be.inside-**DU.PA-PFV.DU**
     'They (2) are inside the house.'

---

[36] As already noted, it is an orthographic convention in Nama not to end a word in a semivowel, \<y\> or \<w\>.

Fourth, again mentioned in section 4.11.1, P-aligned intransitives have distinctive suffixes for the remote tenses, as illustrated in examples (350) and (351) and the following, again comparing transitive and corresponding P-aligned intranstive verbs:

(394) a. *námb*   *kasèn*   *yènè*   (MD99)2)
        námb   kasèn   yènè
        bamboo small.bamboo DEM
        eitamèng.
        e-y-ta-**m**-ng
        α.2|3NSGP-plant-IPFV.ND-**REM.DLT**-2|3SGA
        'he planted this bamboo.'
    b. *námb*   *eyangèrmèn.*   (NB6:26)
        námb   e-y-ang-**èrmèn**
        bamboo α.2|3NSGS_P-be.planted-INC-**REM.DLT.PA**
        'the bamboo was planted.'

(395) a. *kende*   *tawumbetawèm*   *wem.* (MK13(4))
        kend-e   ta-wumbè-ta-**w**-m   wem
        bark-INST β.2|3NSGP-tie-IPFV.ND-**REM.DUR**-1NSGA yam
        'we were tying the yams with bark.'
    b. *ángègh* *tumbangèrwèn.*   (NB6:27)
        ángègh t-wumbè-ang-**èrwèn**
        rope   β.3SGS_P-be.tied-INC-**REM.DUR.PA**
        'the rope was tied.'

Finally, here is an example of a single utterance with both types of remote suffixes:

(396) *ADC-m* *enyawetamèng*   *diburè* *mèngot*
      ADC-m   e-nyawè-ta-**m**-ng   diburè mèngo-t
      ADC   α.2|3NSGP-put.inside-IPFV.ND-**REM.DLT**-2|3SGA prison house-ALL
      kètè   *tanyawangèrwèn.*   (MD99(1))
      kètè   ta-nyawè-ang-**èrwèn**
      there   β.2|3NSGS_P-be.inside-INC-**REM.DUR.PA**
      'The ADC [Assistant District Commisioner] put them inside the prison (and) they stayed inside there.'

5 Additional word classes and phrase structure

This chapter looks at word classes and subclasses in addition to nominals and verbs, and at the structure of the two major phrase types, the nominal phrase (NP) and the verb phrase (VP). The NP is the word or group of words that can make up an argument of the verb. Its head is always a nominal, as defined in chapter 3. The VP is headed by a verb, as defined in chapter 4.

5.1 Nominal phrase: Subclasses of nominals

Nominals are divided into subclasses by both their function and morphological behaviour, especially the degree to which they take case suffixes.

5.1.1 Nouns

The subclass of nouns can take the ergative suffix *-(a)m* and a wide range of other case marking suffixes, depending on semantic constraints. As in other languages, nouns in Nama are generally the names of particular objects or sets of objects, living entities, places, actions, qualities and ideas. In addition, words that appear to function as adjectives do in other languages are also classified as nouns in Nama because of their morphological and syntactic behaviour. To illustrate this, we will look at two words normally translated as meaning 'big', as in the following:

(397)   **Ndimbal**   *mbormboram*   *ewutfoi*
      ndimbal   mbormbor-am   e-wutfo-Ø-ay-Ø
      big   wind-ERG   α.2|3NSGP-disrupt-PFV.ND-REM.PUNC-2|3SGA
      sauan.   (Mark 4:37)
      sau-an
      sea-LOC2
      'A big storm came up [lit. 'suddenly disrupted them'] on the sea.'

(398)   **yabun**   *bout*   *tinjongèm.*   (YD11(3))
      yabun   bout   t-winjo-Ø-ang-m
      big   boat   β.3SGP-see-PFV.ND-INC-1NSGA
      'we saw a big boat.'

However, both these words can take a wide range of case markers, including the ergative:

https://doi.org/10.1515/9783111077017-005

5.1 Nominal phrase: Subclasses of nominals — **171**

(399)   *áuwègh*   **ndimbalam**   *yènèrse.*   (B&B:160)
        áuwè-gh   **ndimbal-am** yèn-rèsa-ta-e
        be.happy-NOM **big-ERG**   α.1NSGP-carry-IPFV.ND-2|3SGA
        'we are carried away by great happiness.'

(400)   *nófnamoyèm*   *wanje* **yabunan.**   (YD11(3))
        n-ófnamo-Ø-ay-m   wanje   **yabun-an**
        α.ØP-cross.over-PFV.ND-REM.PUNC-1NSGA river   **big-LOC2**
        'we crossed over a big river.'

(401)   **yabunofnar** *njamke* ... *ekèm.*   (PT:62)
        yabun-ofnar njamke ... e-kèmè
        big-PRIV   food   α.2|3NSGSₚ-be.lying.down
        'they're lying down without plenty of food.'

(402)   **Yabunmèn** *kimb* *noron*   *rèsaghofnar*   (NB7:90)
        yabun-mèn kimb n-oro-Ø-èn   rèsa-gh-ofnar
        big-ORIG   pig   α.ØP-leave-PFV.ND1SGSₐ carry-NOM-PRIV
        tèmorwèn
        t-mo-èrwèn
        β.3SGSₚ-COP.ND-REM.DUR.PA
        'Because of its big size, I just left the pig; it wasn't carried.'

A better translation in (402), then, might be 'largeness' rather than 'big'.

With regard to function, nouns generally can modify other nouns as a noun adjunct – for example:

(403)   *yáne*   **njamke** **mèngo** *tèndomè*
        yáne   **njamke** **mèngo** tèndo-mè
        3SG.GEN **food**   **house** side-PERL
        enènfetamèm ...   (MD99(1))
        e-n-nèfè-ta-m-m
        α.2|3NSGP-VEN-cut-IPFV.ND-REM.DLT-1NSGA
        'we cut them alongside his kitchen...'

172 —— 5 Additional word classes and phrase structure

(404) *yènd* *syènèm*   **mèngotu** **ár.**   (B&B:111)
yènd s=yèn-m   **mèngotu** **ár**
1.ABS PROX1-α.1NSGS_P-COP.ND **village** **person**
'We here are village people.'[37]

The analysis here, then, is that what appear to be adjectives in some contexts are actually noun adjuncts.

There are many other clear examples of a word functioning as both a prototypical noun and prototypical adjective:

(405) a. ***krufèr*** *taim* *yèmong.*   (DC15)
krufèr taim y-mo-ang
cold time α.3SGS_P-COP.ND-INC
'it's the cold time.'

b. ***Krúfram*** *wèfinde.*   (NB10:25)
krúfèr-am w-fing-ta-e
cold-ERG α.1SGP-affect-IPFV.ND-2|3SGA
'I became cold.' (lit. 'cold affected me.')

(406) a. *Wanje* **kènjkam** *yèm.*   (NB6:80)
wanje **kènjkam** y-m
river **deep** α.3SGS_P-COP.ND
'The river is deep.'

b. *Wagif* *wanje* **kènjkamamè** *narende.*   (NB6:80)
wagif wanje **kènjkam-amè** n-areng-ta-e
fish river **deep-PERL** α.ØP-move.around-IPFV.ND-2|3SGA
'The fish is swimming [lit. moving around] along the deep part of the river.'

(407) a. *Fèm* **sakèr** *táre.*   (PC/MD: 1.5.19)
fèm **sakèr** tá-r-e
2.ABS **tired** β.2|3NSGS_P-COP.DU-PFV.DU
'You (2) are tired.'

b. ***Sakram*** *áligh-mèn* *tè* *wifo.*   (YD11(3))
sakèr-am áligh-mèn tè w-wifo-Ø-Ø
tired-ERG walking-ORIG PRF α.1SGP-finish-PFV.ND-2|3SGA
'I got tired from walking.' (lit. 'tiredness finished me'.)

---

37 The proximal clitics (PROX) are described in section 5.4.1 below.

5.1 Nominal phrase: Subclasses of nominals ── **173**

(408) a. *tane* *mèrès* *tosè* ***enjne*** *tèmo* *káye.* (MM13)
　　　 tane mèrès tosè **enjne** t-mo káye
　　　 1SG.GEN girl young.one **sick** β.3SGSₚ-COP.ND yesterday
　　　 'my small daughter was sick yesterday'.

　 b. *nèmamèn* *kèr* *tèmong* – ***enjnemèn*** *o*
　　　 nèmamèn kèr t-mo-ang **enjne-mèn** o
　　　 why dead β.3SGSₚ-COP.ND-INC **sick-ORIG** or
　　　 ngimèn. (MD99(1))
　　　 ngi-mèn
　　　 murder-ORIG
　　　 'why she died – from sickness or from murder.'

Another good example is *kèr*, meaning 'dead' or 'death', which can be used as both a head noun and a noun adjunct:

(409) a. *mèrès* *ámb* ***kèr*** *tè* *tèmong.* (MD99)1))
　　　 mèrès ámb **kèr** tè t-mo-ang
　　　 girl some **dead** PRF β.3SGSₚ-COP.ND-INC
　　　 'one of our sisters is dead/has died.'

　 b. *mènamèn* ***kèr*** *kètè* *tèngmorwèng.* (ND11)
　　　 mènamèn **kèr** kètè t-ng-mo-èrwèng
　　　 because **death** there β.3SGSₚ-AND-COP.ND-REM.DUR.PA
　　　 'because death was going to stay there.'

　 c. *yènd* *yáne* ***kèr*** *siam* *sèlngwèl*
　　　 yènd yáne **kèr** si-am sèlngwèl
　　　 1.ABS 3SG.GEN **death** word-ERG surprise
　　　 yènramai. (MD99(1))
　　　 yèn-ramè-Ø-ay-Ø
　　　 α.1NSGP-give-PFV.ND-REM.PUNC-2|3SGA
　　　 'word of her death surprised us.'

Following a noun with a human referent, *kèr* is translated as 'dead' or 'late', but its meaning is closer to 'dead person' and it can take the plural marking suffix, and other case suffixes:

174 — 5 Additional word classes and phrase structure

(410) a. *Sefor* **kèr-am** *áuyè*
Sefor **kèr-am** áuyè
Sefor **dead.person-ERG** cassowary
yèsmetamèng... (MD99(1))
y-sèmè-ta-m-ng
α.3SGP-kill-IPFV.ND-REM.DLT-2|3SGA
'The late Sefor killed a cassowary...'

b. *áki* **kèrafene** *yam* *ndernae*
áki **kèr-af-e-ne** yam ndernae
grandfather **dead.person-PL-DAT-GEN** custom how
tèmorwèn. (B&B:148)
t-mo-èrwèn
β.3SGS$_P$-COP.ND-REM.DUR.PA
'how the ancestors' custom was.'

So with regard to both morphological and syntactic behaviour, there is no criterion for distinguishing a subclass of adjectives from the subclass of nouns.

Nevertheless, while we have just seen that some nouns function like adjectives without any change in form, other nouns are marked with the comitative suffix *-afè* when they have an adjective-like function, as shown in chapter 3 (examples 52–54). Additional examples are:

(411) *ár* **sotafè** *tè* *náyáto.* (B&B:210)
ár **sot-afè** tè n-áyáto-Ø-Ø
man **bone-COM** PRF α.ØP-end.up-PFV.ND-2|3S$_A$
'the man ended up strong.'

(412) *Fèm* *so* *kamndè* **sú-afè.** (Luke 1.31)
Fèm so k-amndè-Ø-Ø **sú-afè**
2.ABS FUT β.ØP-become-PFV.ND-2|3S$_A$ **pregnancy-COM**
'You will become pregnant.'

However, nouns themselves have several subclasses, First there is the subclass of nouns that can take the nonsingular suffix *-af/-f.* Semantically, this subclass includes nouns with human referents (especially relations – i.e. terms for relatives and friends), and some animals that are important sources of food. This subclass has its own small subclass that can take the singular suffix *-o.* However, as noted before, the use of these suffixes is relatively rare, and the exact distribution of this subclass and sub-subclass is not clear.

Second, a more clearly delineated noun sub-subclass is comprised of terms for relationships that take pronominal possessive prefixing (section 3.7). A subset of these terms have bound forms that only occur with the pronominal prefix.

Third, personal names form a subclass, as evidenced by their unique behaviour with regard to the following case marking suffixes: ergative (section 3.3.1), comitative (section 3.3.5), privative (3.3.6), ablative (3.3.8) and purposive (3.3.11). The subclass of personal names differs from other nouns in that they are not preceded by determiners or modified by other nominals. (This is also true of place names.)

The fourth noun subclass is nominalised verbs (section 4.1). Like other nouns, they can function like adjectives and modify other nouns – e.g. *waramnjagh ár* 'jealous man', *únyègh ár* 'liar' (lit. 'lying person'), *ásáfogh tèfnár* 'hard worker' (lit. 'working practitioner'). Other examples are:

(413) …***ámnjogh*** *sifayan.*   (MD11(1))
       ámnjo-gh sifayè-n
       sit-NOM place-LOC1
       '…on the sitting place.'

(414) ***watamegh*** *mèngo* *yènwároi.*   (FD13)
       watame-gh mèngo y-n-wáro-Ø-ay-Ø
       teach-NOM house α.3SGP-VEN-set.on.fire-PFV.ND-REM.PUNC-2|3SGA
       'he set the school house on fire.' (lit. 'teaching house')

(415) *Si* ***waitogh*** *tèfnáram*   *burarè* *yèna*
       si **waito-gh** tèfnár-am   burarè yèna
       story **tell-NOM** practitioner-ERG flute here
       fyaufete.   (DC34)
       f=y-waufè-ta-e
       PROX2-α.3SGP-blow-IPFV.ND-2|3SGA
       'The story teller is just blowing the flute here.'

These are distinguished not only by the *-gh* nominalising suffix, but also by their ability to occur in inchoative constructions with *wafngo*/ and *ufngo*/ (section 4.1).

Two other subclasses pattern with nominalised verbs: action nominals and stative nominals. The action nominals include words such as *anu* 'bathing', *ále* 'hunting', *fèn* 'laughing', *fès* 'cooking', *ngi* 'killing', *susi* 'fishing', *wagh* 'singsing (singing and dancing)' and *ye* 'crying'. Like nominalised verbs, these occur in inchoative constructions:

176 — 5 Additional word classes and phrase structure

(416) **álet**   *nuɸngo ...*   (DC25)
       ále-t   n-ufngo-Ø-Ø
       hunting-ALL   α.ØP-start-PFV.ND-2|3S$_A$
       'start hunting ...'

(417) **fès**   *engwaɸngoyènd.*   (TE11(2))
       fès   e-ng-wafngo-Ø-ay-ng
       cooking   α.2|3NSGP-AND-start-PFV.ND-REM.PUNC-2|3SGA
       'she started cooking them.'

(418) *fèyo*   **wagh**   *kuɸngongi.*   (ND13)
       fèyo   **wagh**   k-ufngo-Ø-ang-i
       then   **singsing**   β.ØP-start-PFV.ND-INC-2|3PLS$_A$
       'then they started singing and dancing.'

The stative nominals are *mat* 'knowing', *maɸina* 'not knowing', *mènde* 'wanting, liking' and *mèndeɸnar* 'not wanting, not liking'.

     Finally, there is the subclass of temporal and orientational nouns. These can act as adverbs without any additional marking, unlike other nouns with temporal meaning, such as the times of the day – e.g. *akw-an* 'in the morning'. They also frequently appear to modify other nouns, putting them in a time frame or indicating their position or orientation. When they do so, they follow the noun they are modifying. The temporal and orientational nouns include: *soramè* 'later (time)', *kènjún* 'inside (location)', and *tambèn* 'side (position)' or 'origin'.

(419) **soramè**   *amam*   *njamke*
       soramè   amè-am   njamke
       later   mother-ERG   food
       ewaɸlitam ...   (MD93)
       e-wafèli-ta-m-Ø
       α.2|3NSGP-put.in-IPFV.ND-REM.DLT-2|3SGA
       'later mother served the food ...'

(420) *fèyo*   *yènè*   **soramè**   *naɸleayèm*
       fèyo   yènè   **soramè**   n-afèl-e-ay-m
       then   DEM   **later**   α.ØP-go.short.distance-PFV.DU-REM.PUNC-1NSGS$_A$
       ndauot.   (TE11(1))
       ndau-ot
       garden-ALL
       'then after this we (2) went a short distance to the garden.'

5.1 Nominal phrase: Subclasses of nominals ⎯ **177**

(421) *náyárayèn* *nambyo* *ghèrare* **soramè.** (TE13)
n-áyárè-Ø-ay-èn nambyo ghèrare **soramè**
α.ØP-listen-PFV.ND-REM.PUNC-1SGS_A three month **later**
'I found out three months later.'

(422) *Yosi* *mèngo* **kènjún** *yèm.* (NB4:59)
Yosi mèngo **kènjún** y-m
Yoshie house **inside** α.3SGS_P-COP.ND
'Yoshie is inside the house.'

(423) *Ár* *enyárayèng* *yawar*
Ár e-n-yárè-Ø-ay-ng yawar
person α.2|3NSGP-VEN-hear-PFV.ND-REM.PUNC-2|3SGA black.palm
kènjún. (MD99(2))
kènjún
inside
'He heard people inside the black palm.'

(424) *fèyo* *kèwewongè* *tuk* **tambèn.** (MD13(1))
fèyo k-èwewo-Ø-ang-è tuk **tambèn**
then β.ØP-look.up-PFV.ND-INC-2|3SGS_A top **side**
'then he looked up to the top side.'

(425) *nurtotamèm* *yènè* *sèkw* **tambèn.** (YD11(3))
n-urto-ta-m-m yènè sèkw **tambèn**
α.ØP-get.out-IPFV.ND-REM.DLT-1NSGS_A DEM canoe **origin**
'we got out from that canoe.'

(426) *Ghèráyè* **tambèn** *yènè* *wagh.* (SK11(1))
Ghèráyè **tambèn** yènè wagh
Daraia **origin** DEM singsing
'this singsing from Daraia'

While it may be tempting to analyse these forms as postpositions, two factors lead to their analysis as a subclass of nouns: First they can take case-marking suffixes (although rarely), as in examples (427)–(429) below, and second, they can be modified by other case-marked nouns or pronouns, as in (430):

178 —— 5 Additional word classes and phrase structure

(427) *yènè   mè   tánlaweta . . .*   **soramamèn.**   (PT:285)
   yènè   mè   tá-n-lawè-ta-Ø   **soramè-mèn**
   DEM   still   α.2|3NSGP-VEN-put.inside-IPFV.ND-IRR   **later-ORIG**
   'still put these inside . . . for later'

(428) *fèyo   fètè   yanai*   *nu*
   fèyo   fètè   y-wanè-Ø-ay-Ø   nu
   then   VAL   α.3SGP-take-PFV.ND-REM.PUNC-2|3SGA   water
   kènjúnmè.   (YD11(2))
   kènjún-mè
   inside-PERL
   'then it really took her along inside of the water.'

(429) *ta   **tambèneta**   árkè   si   so   náyáretati . . .*   (ND13)
   ta   **tambèn-e-ta**   árkè   si   so   n-áyárè-ta-t-i . . .
   1SG.DAT   **side-DAT-ABL**   that   story   FUT   α.ØP-hear-IPFV.ND-2|3NSGS$_A$-2PL
   'from my side, the story that you all will listen to . . .'

(430) ***tèfène   kènjún***   *kèrtè   tè   nufar.*   (MD99(1))
   tèfène   kènjún   kèrtè   tè   n-ufar
   3NSG.GEN   inside   trouble   PRF   α.ØP-arrive
   'inside our [family] trouble has already arrived.'

5.1.2 Pronouns

A second subclass of nominals are pronouns, introduced in chapter 3. These are
defined as words that can function by themselves as a nominal phrase, referring
to participants in the discourse (e.g. *yènd* 'I', *fèm* 'you') or to someone or something
already mentioned in the discourse or familiar to the participants (e.g. *fá* 'he, she,
they'). Similar to personal and place names, they are not preceded by determiners
or modified by other nominals.

There are three categories of pronouns: ergative and absolutive (shown in
Table 3.4), and dative (Table 3.5). Interrogative and relative pronouns corresponding to these are described in sections 6.4 and 7.3. Dative pronouns plus case-marking suffixes form independent pronoun sets for the following cases: genitive
(section 3.3.4), privative (3.3.6), purposive (3.3.11), associative (3.3.12) and originative (3.3.13). A comitative pronoun set incorporates ergative and absolutive pronouns (3.3.5).

5.1.3 Demonstrative

There is only a single demonstrative in Nama *yènè*, which can mean 'this', 'that', 'these' or 'those'. Thus, it does not reflect the two-way spatial distinction in the language that is indicated by spatial adverbs (section 5.3.6). It is the most frequent word in the recorded data.

(431)	*Nèmamèn*	**yènè**	*yáf*	*yèm?*	(DC9)
	nèmè-mèn	**yènè**	yáf	y-m	
	what-ORIG	**DEM**	basket	α.3SGS$_P$-COP.ND	

'What's this basket made of?'

(432)	**yènè**	*akwan*	*wèi*	*kuwanongèm.*	(YD11(3))
	yènè	akw-an	wèi	k-uwano-Ø-ang-m	
	DEM	morning-LOC2	now	β.ØP-set.off-PFV.ND-INC-1NSGS$_A$	

'that morning now we set off.'

(433)	*fèyo*	*tánetawèm*		**yènè**	*wagif.*	(TE11(2))
	fèyo	tá-ne-ta-w-m		**yènè**	wagif	
	then	α.2|3NSG-eat-IPFV-ND-REM.DUR-1NSGA	**DEM**	fish	

'then we ate these fish.'

(434)	**yènè**	*ár*	*nèngufaryènd.*	(MD99(1))
	yènè	ár	n-ng-ufar-Ø-ay-nd	
	DEM	person	α.ØP-AND-arrive-PFV.ND-REM.PUNC-2|3NSGS$_A$	

'those people arrived.'

Again, the demonstrative normally precedes the nominal it is modifying, including a noun with preceding modifiers:

(435)	**yènè**	*árofnar*	*sifayan*...	(NB10:29)
	yènè	ár-ofnar	sifayè-n	
	DEM	person-PRIV	place-LOC1	

'in this place without people...'

180 —— 5 Additional word classes and phrase structure

In addition, *yènè* can function as a demonstrative pronoun:

(436) **yènè** tè *yèfande.* (PT:575)
   yènè tè y-fand-ta-e
   DEM PRF α.3sGP-look.at-IPFV.ND-2|3sGA
   'you already looked at this.'

(437) *árèm* **yènè** *tánèwitat* (MD13(1))
   ár-m **yènè** tá-n-wi-ta-t
   man-ERG **DEM** β.3NsGP-VEN-throw-IPFV.ND-2|3NsGA
   'the men threw those (in this direction).'

In the few occasions when *yènè* follows rather than precedes the nominal it is modifying, it appears to be used as a focus marker, as in the following:

(438) *ghakèr* **yènè** *nuwanoi.* (PT:700)
   ghakèr **yènè** n-uwano-Ø-ay-Ø
   boy **DEM** α.ØP-set.off-PFF.ND-REM.PUNC-2|3sGS$_A$
   'the boy, he set off.'

(439) *námb* *kasèn* **yènè**
   námb kasèn **yènè**
   big.bamboo small.bamboo **DEM**
   eitamèng. (MD99(2))
   e-y-ta-m-ng
   α.3NsGP-plant-IPFV.ND-REM.DLT-2|3sGA
   'big bamboo, small bamboo, he planted them.'

The demonstrative can be appended by the restrictive suffix *-ro* – e.g.:

(440) **yènaro** *sif* *tèmo* *ndau* *ásáfoghmèn.* (TE11(1))
   yènè-ro si=f t-mo ndau ásáfo-gh-mèn.
   DEM-RSTR si=PROX2 β.3sGSP-COP.ND garden work-NOM-ORIG
   'this was only just a story about garden work.'

As shown in section 3.6.1, it appears that historically case suffixes were appended to the demonstrative to form adverbs and adverbial subordinating conjunctions. But example (485) in section 5.3.1 is an indication that the process may still be productive.

5.1.4 Quantifiers

Quantifiers generally function to indicate the quantity of a noun. They differ from nouns and pronouns morphologically in that they normally do not take case-marking morphology. However, like other nominals, they can be the head of a nominal phrase.

The most frequently used quantifier is *ámb* which generally means 'some' or 'any' as in these examples:

(441) **Ámb** *ár* *álet* *fengèm.* (NB7:68)
   Ámb ár ále-t f=e-ngèm
   some man hunting-ALL PROX2=α.2|3NSGSP-go.ND
   'Some men just went hunting.'

(442) ***ámb*** *misi-misi-afè* *nuwanoyèm*
   ámb misimisi-afè n-uwano-Ø-ay-m
   some missionary-COM α.ØP-set.off-PFV.ND-REM.PUNC-1NSGS$_A$
   Kiungat. (YD11(3))
   Kiungè-t
   Kiunga-ALL
   'we set off with some pastors to Kiunga.'

(443) ***ámb*** *árèm* *yau* *wáutánam.* (MK13(4))
   ámb ár-m yau w-wáutár-ta-m-Ø
   any person-ERG NEG α.1SGP-help-IPFV.ND-REM.DLT-2|3SGA
   'no one helped me.'

(444) *fainár* ***ámb*** SO *eitati?* (DC24)
   fainár **ámb** SO e-y-ta-t-i
   pineapple **any** FUT α.2|3NSGP-plant-2|3NSGS$_A$-2PL
   'will you plant any pineapple?'

By an extension of meaning to 'more of the same' *ámb* can also mean '(an)other' or 'next':

(445) *kuwanongèm* ***ámb*** *mèngotuot.* (YD11(3))
   k-uwano-Ø-ang-m **ámb** mèngotu-ot
   β.ØP-set.off-PFV.ND-INC-1NSGS$_A$ **other** village-ALL
   'we set off for another village.'

(446) *wèikor* **ámb** *akwan*
      wèikor **ámb** akw-an
      again **next** morning-LOC2
      nènangoi.   (MD13(1))
      n-n-ango-Ø-ay-Ø
      α.ØP-VEN-return-PFV-ND-REM.PUNC-2|3SGS$_A$
      'he returned again the next morning.'

The combination of *ámb* and *yau* (NEG), meaning 'no other', can follow a nominal meaning 'nothing but':

(447) *wem* **ámb** **yau** *so* *eitati?*   (DC24)
      wem **ámb** **yau** so e-y-ta-t-i
      yam **some** **NEG** FUT α.2|3NSGP-plant-IPFV.ND-2|3NSGA-2PL
      'will you all plant nothing but yams?'

The combination can also have the adverbial meaning of 'never':

(448) *Yèmo* *ndènè* *rokár* **ámb** **yau**
      yèmo ndènè rokár **ámb** **yau**
      3SG.ERG like.this thing **some** **NEG**
      yinjoi   *ghèrsayan.*   (MD13(1))
      y-winjo-Ø-ay-Ø   ghèrsè-yan
      α.2|3SGP-see-PFV.ND-REM.PUNC-2|3SGA life-LOC2
      'He never saw something like this in his life.'

With regard to position in the NP, *ámb* usually occurs before the nominal it modifies; however, it can also follow, as in (444) above and the following:

(449) *susi*   **ámb** *yèfenan.*   (MK13(1))
      susi   **ámb** y-fer-ta-èn
      fishing.line **some** α.3SGP-unroll-IPFV.ND-1SGA
      'I unrolled some fishing line.'

Ámb can also function as an indefinite pronoun, and be the lone constituent of an absolutive pronominal argument:

5.1 Nominal phrase: Subclasses of nominals —— **183**

(450) **ámb** *ngákmetawèt.* (ND13)
ámb ng-ákèmè-ta-w-t
some AND-lie.down-IPFV.ND-REM.DUR-2|3NSGS_A
'some were going to sleep.'

(451) *... mato* **ámb** *tangfande.* (PT;192)
mato **ámb** ta-ng-fanda-ta-e
possibly **other** β.NSGP-AND-look.at-IPFV.ND-2|3SGA
'... maybe you're looking at the others.'

There are no examples in the data of this quantifier taking any case marker, but one with the restrictive marker:

(452) **ámbro** *ghakèr tawaufrongè...* (YD11(3))
ámb-ro ghakèr ta-waufro-Ø-ang-è
other-RSTR boy β.NSGP-tell-PFV.ND-INC-2|3SGA
'he told the other guys...'

The other common quantifier is *tárfár* 'many, plenty of'. This occurs frequently both before and after the noun it modifies:

(453) *yèmo* **tárfár** *wagif táreindawèng.* (TE11(2))
yèmo **tárfár** wagif tá-reny-ta-w-ng
3SG.ERG **many** fish β.NSGP-catch-IPFV.ND-REM.DUR-2|3SGA
'She was catching many fish.'

(454) *engwaindamèm* **tárfár** *yúnjèf.* (FD13)
e-ng-waing-ta-m-m **tárfár** yúnjèf
α.2|3NSGP-AND-pass-IPFV.ND-REM.DLT-1NSGA **many** year
'we passed many years.'

(455) *fá mè yámbya tèmorangèrwèn yúnjèf*
fá mè yámbya t-mor-ang-èrwèn yúnjèf
3.ABS CONT 3SG.alone β.3SGS_P-stay-INC-REM.DUR.PA year
tárfár. (SK11(2))
tárfár
many
'he kept staying alone many years.'

184 —— 5 Additional word classes and phrase structure

(456) *satèf* **tárfár** *náyátoyènd.* (NP11)
satèf **tárfár** n-áyáto-Ø-ay-nd
heap **many** ɑ.ØP-end.up-PFV.ND-REM.PUNC-2|3NSGS_A
'it ended up that there were many heaps [of sago].'

Tárfár can also be the lone head of an NP and act as an absolutive pronominal argument:

(457) *tawinjon* **tárfár.** (MK13(1))
ta-winjo-Ø-èn **tárfár**
β.2|3NSGP-see-IPFV.ND--1SGA **many**
'I just saw many.'

It appears that the only nominal morphology this quantifier takes is the exclusive marker *-yo* – e.g.:

(458) ... *yèmorangèrmèn* *yúnjèf* **tárfáryo.** (SK11(1))
y-mor-ang-èrmèn yúnjèf **tárfár-yo**
ɑ.3SGS_P-stay-INC-REM.DLT.PA year **many-EXCL**
'... he stayed many years.'

The numeral *nambyo* 'three' with the restrictive suffix *-ro* can also function as a quantifier, meaning 'a few':

(459) **nambyoro** *ár* *fengèm* *álet.* (NB7:68)
nambyo-ro ár f-e-ngèm ále-t
three-RSTR man PROX2-ɑ.2|3NSGS_P-go.ND hunting-ALL
'A few men went hunting.'

(460) *Fèmo* **nambyoro** *waram.* (NB6:84)
fèmo **nambyo-ro** w-wa-ramè-Ø-Ø
2SG.ERG **three-RSTR** ɑ.1SGP-APP-give-PFV.ND-2|3SGA
'You just gave me only a few.'

The word *mèinyotyo* 'all' often appears to function as a quantifier (as in 461 below), but it is actually a noun, as it frequently takes case-marking morphology (as in 462 and 463).

5.1 Nominal phrase: Subclasses of nominals —— **185**

(461) **mèinyotyo** *ár* *korayangi*
 mèinyotyo ár k-oray-Ø-ang-i
 all person β.ØP-say-PFV.ND-INC-2|3PLS$_A$
 nangotam. (ND13)
 n-ango-ta-m
 α.ØP-return-IPFV.ND-1NSGS$_A$
 'All the people said let's go back.'

(462) *nène* *tawaramwèm*   **mèinyotyoe.** (MK13(4))
 nène ta-wa-ramè-Ø-w-m   **mèinyotyo-e**
 food β.2|3NSGP-APP-give-IPFV.DU-REM.DUR-1NSGA   **all-DAT**
 'We (2) were giving food to all.'

(463) *fès* **mèinyotyoyam** *yènè* *syanai*
 fès **mèinyotyo-yam** yènè s=y-wanè-Ø-ay-Ø
 fire **all-ERG** DEM PROX1=α.3SGP-take-PFV.ND-REM.PUNC-2|3SGA
 wèn . . . (ND13)
 wèn
 tree
 'all the fire got the tree . . . .'

Note that *mèinyotyo* can mean either 'all' or 'nearly all', as in the following:

(464) **Mèinyotyo** *ár* *álet* *fengèm* *wènde*
 mèinyotyo ár ále-t f=e-ngèm wènde
 nearly.all man hunting-ALL PROX2-α.2|3NSGS$_P$-go.ND but
 sembyo *ár* *námnjend.* (NB7:68)
 sembyo ár n-ámnjo-e-nd
 two man α.ØP-remain-PFV.DU-2|3DUS$_A$
 'Nearly all the men went hunting but two men stayed behind.'[38]

---

38 Note that in Nama the word for 'all' can mean 'nearly all' but not 'most', unlike in Nen (Evans 2017: 587). For example, it could be used for 27 or 28 out of 30, but not for 20 out of 30.

186 —— 5 Additional word classes and phrase structure

(465) sembyo ámbiro fèyo kèmègh kètè **mèinyotyo**
sembyo ámbiro fèyo kèmègh kètè **mèinyotyo**
two one then sleeping there **all**
nefoyènd. (ND13)
n-efo-Ø-ay-nd
α.ØP-finish-PFV.ND-REM.PUNC-2|3NSGS_A
'One or two were sleeping there; all (the rest) had finished.'

Numerals form a subclass of quantifiers. These numerals, unlike the numbers used in ceremonial yam counting (section 1.1.3), are used to quantify nouns in everyday discourse. There were originally only three numerals in Nama: *ámbiro* 'one', *sembyo/sómbyo* 'two' and *nambyo* 'three'. Combinations of these were used for higher numbers, e.g. *sembyo sembyo* 'four', but nowadays. numerals borrowed from English are used. Some examples are:

(466) **ámbiro** safèt soramè yènè ár
ámbiro safèt soramè yènè ár
one week after DEM man
nèngufaryènd. (MD99(1))
n-ng-ufar-Ø-ay-nd
α.ØP-AND-arrive-PFV.ND-REM.PUNC-2|3NSGS_A
'after one week, that man arrived.'

(467) olman Dárfi-m **sembyo** yúnjèf tawife. (MD99(1))
olman Dárfi-m **sembyo** yúnjèf ta-wifo-e-Ø
old.man Darfi-ERG **two** year β.2|3NSGP-finish-PFV.DU-2|3sGA
'Old man Darfi finished two years (in prison).'

(468) **nambyo** efogh yènmorarmèn. (MK13(4))
nambyo efogh yèn-mor-ar-mèn
three day α.1NSGS_P-stay-DU.PA-REM.DLT.PA
'we (2) stayed three days.'

Numerals almost always precede the noun they modify, but there are some exceptions – e.g.:

(469) *Mèngotu* **ámbiro** *yafromènd.*   (MD13(1))
       mèngotu **ámbiro** y-wafro-Ø-m-nd
       village **one** α.3sGP-make-IPFV.DU-REM.DLT-2|3NSGA
       'They (2) made one village.'

Numerals can also function as heads of a NP:

(470) **sembyo** *enwaneayèn.*   (NP11)
       sembyo e-n-wanè-e-ay-èn
       two α.2|3NSGP-VEN-take-PFV.DU-REM.PUNC-1SGA
       'I brought the two.'

Numerals can be followed by the restrictive suffix as in examples (459–460) above.
And there is one instance in the data where a numeral appears to take a case marker:

(471) *fá* *mat* *ere* **ámbiroero.**   (B&B:283)
       fá mat e-r-e **ámbiro-e-ro**
       3.ABS knowing α.2|3NSGS_P-COP.DU-PFV.DU **one-INST-RSTR**
       'they (2) know [it should be] only one at a time.'

Note that the numerals all begin with *(C)Vmbi or (C)Vmby*. As shown in section 3.6.2,
the addition of -*notmè* to this beginning gives a form meaning 'X number of times':
ámbinotmè 'one time, once', *sembinotmè* 'two times, twice' and *nambinotmè* 'three
times'. The ending -*mbinotmè*, as in the numerals, also occurs with the quantifier
tárfár: *tárfármbinotmè* 'many times'.

5.1.5 Temporal nominals

Like the temporal and orientational subclass of nouns (section 5.1.1), temporal nominals
can have adverbial functions without any additional marking. However, they differ in
that they do not take case-marking suffixes and they precede rather than follow other
nominals they are modifying. The temporal nominals include *sènonjo* 'today, at this
time', *káye* 'tomorrow, yesterday', *nambèt* 'day after tomorrow, day before yesterday, a
few days ago', *tándái* 'a long time ago, very early', *fronde* 'first'. Examples follow:

(472) **Sènonjo** *ámb* *si* *so* *yaitotan.*   (MD13(1))
       sènonjo ámb si so y-waito-ta-èn
       today some story FUT α.3sGP-tell-IPFV.ND-1SGA
       'Today I'll tell you a story.'

188 —— 5 Additional word classes and phrase structure

(473) **Sènonjo** *sèmbár yaufi onjo* *fyèkmang.* (NB2:31)
    sènonjo sèmbár yaufi onjo f=y-kèmè-ang
    today night bad dream PROX2=α.3SGP-be.lying.down-INC
    'Last night he had a bad dream.'

(474) **káye** *akwan* *so* *nèngangom*
    káye akw-an so n-ng-ango-Ø-m
    tomorrow morning-LOC2 FUT α.ØP-AND-return-PFV.ND-1NSGS_A
    mèngotuot. (MK13(1))
    mèngotu-ot
    village-ALL
    'tomorrow morning we'll return to the village.'

(475) **káye** *ámb* *tè* *táwifárèt* *yau?* (DC16)
    káye ámb tè tá-wifár-Ø-t yau
    yesterday some PRF β.2|3NSGP-chase-IPFV.DU-2DUA NEG
    'yesterday you (2) chased some, didn't you?'

(476) **Nambèt** *yafrotamèn.* (DC13)
    nambèt y-wafro-ta-m-èn
    day.before.yesterday α.3SGP-do-IPFV.ND-REM.DLT-1SGA
    'I did it the day before yesterday.'

(477) **Yènè** *syèm* *latu* **tándái**
    yènè s=y-m latu **tándái**
    DEM PROX2=α.3SGP-COP.ND flute **long.ago**
    stamorwèn. (DC37)
    s=ta-mo-èrwèn
    PROX2=β.2|3NSGP-COP.ND-REM.DUR.PA
    'This here sacred flute existed here long ago.'

(478) **tándái** *akwan* *nènangoi.* (MD13(1))
    tándái akw-an n-n-ango-Ø-ay-Ø
    early morning-LOC2 α.ØP-VEN-return-PFV.ND-REM.PUNC-2|3SGS_A
    'he returned early morning'

5.1 Nominal phrase: Subclasses of nominals — **189**

(479)   ***Fronde***   *Les*   *yárafè*   *so*
        fronde   Les   yá-rafè   so
        first   Les   3SG-father FUT
        kèntorngè.   (B&B:198)
        k-n-ètor-Ø-ang-è
        β.ØP-VEN-come.out-PFV.ND-INC-2|3SGS$_A$
        'First Les'his father will come out.'

(480)   ***Fronde***   *ár*   *yènd*   *wèm.*   (KS11)
        fronde   ár   yènd   w-m
        first   man   1.ABS   α.1SGSP-COP.ND
        'I'm the first man.'

Like other nominals, the temporal nominals can take exclusive and restrictive suffixes and act as a heads of a NP:

(481)   *Njam*   *yègh*   *ewafngoi?*   *Yau*
        njam   y-gh   e-wafngo-Ø-ay-Ø?   Yau
        when   plant-NOM   α.2|3NSGP-start-PFV.ND-REM.PUNC-2|3SGA   NEG
        nambètyo.   (DC12)
        nambèt-yo
        few.days.ago-EXCL
        'When did you (just) start planting them? No, a few days ago.'

(482)   ***Tándáiyo***   *nènguwanend.*   (DC26)
        tándái-yo   n-ng-uwano-e-nd
        early-EXCL   α.ØP-VEN-set.off-PFV.DU-2|3DUS$_A$
        'They (2) set off very early.'

(483)   *Bulu*   *Matán*   ***frondeyo***   *námnjoi.*   (SK11(2))
        Bulu   Matán   **fronde-yo**   n-ámnjo-Ø-ay-Ø
        Bulu   Matan   **first-EXCL**   α.ØP-stay-PFV.ND-REM.PUNC-2|3SGS$_A$
        'Bulu Matan settled [here] first.'

(484)   *efe*   ***fronde***   *kembnean*   *yèm?*   (DC18)
        efe   **fronde**   kembne-an   y-m
        who.ABS   **first**   game-LOC2   α.3SGS$_P$-COP.ND
        'who is the first in the game?

5.2 Nominal phrase structure

As mentioned above, a nominal phrase is headed by a nominal. The word order can be flexible except that in a NP that is an ergative argument, the final nominal is the head – either an ergative pronoun or a noun marked by the ergative suffix, as in examples (397) and (399).

As also mentioned, within a nominal phrase, a noun can modify another noun. with the preferred order being the modifying or adjunct noun occurring before the noun it is modifying. The preferred order for other nominals is also to precede the noun they are modifying or to precede the adjunct noun if there is one. The demonstrative normally precedes a quantifier and other constituents. Example (435) illustrates this ordering.

5.3 Verb phrase: Other word classes

The head of a verb phrase is always a verb. Chapter 4 describes the three subclasses of verbs: transitive, A-aligned intransitive and P-aligned intransitive. This section introduces other word classes found in the verb phrase. The most common of these are the periphrastic tense, aspect and modality markers (TAM markers). All these markers serve to provide further information about the action, event or state described by the verb and they all have the same syntactic behaviour: occurring as separate words and directly before the verb, with only a few exceptions. Also described in this section are quasi-modals and adverbs. The TAM markers and quasi-modals are listed in Table 5.1.

5.3.1 Tense marker

Nama has only one tense marker *so,* which indicates future tense (FUT), as shown in Chapter 4. While other tenses are indicated with bound verb morphology, this marker is the only way to specify the future. Many examples with the future marker have already been given above. Here are two more:

(485)   *nèmè*   *yènae*   *so*   *yafrote?*   (FD11(1))
        nèmè   yènè-e   so   y-wafro-ta-e
        what   DEM-INST   **FUT**   ɑ.3SGP-do-IPFV.ND-2|3SGA
        'what will you do with this?'

5.3 Verb phrase: Other word classes ■ **191**

Table 5.1: TAM markers and quasi-modals in Nama.

marker	type	name	abbreviation
so	tense marker	future	FUT
tè	aspect marker	perfect	PRF
fèf / fèfè	aspect marker	inchoate 1	INCH1
tèf / tèfè	aspect marker	inchoate 2	INCH2
mè	aspect marker	continuative	CONT
sèno	modal	possibility	POSB
mètè	modal	potential 1	POT1
wè	modal	potential 2	POT2
kár	modal	dubitative	DUB
siya / sái	modal	intentional	INT
tèu	modal	permissive	PERM
miya	modality + aspect marker	obligative continuative	OBL.CONT
yèf	modal	evidential	EVID
yèta	modal	attemptive	ATT
fèyotaro	quasi-modal	abilitative	ABIL
fiya / fái	quasi-modal	obligatative 1	OBL1
táf	quasi-modal	obligatative 2	OBL2
mato	quasi-modal	investigative	INV
fètè	quasi-modal	validative	VAL

(486) *Ben **so** kokáyo.*   (B&B:57)
(=390) Ben **so** k-okáyo-Ø-Ø
       Ben **FUT** β.ØP-stand up-PFV.ND-2|3SGS$_A$
       'Ben will stand up.'

The marker *so* is also used in counterfactual sentences (section 7.2.5) to refer to events that did not actually occur – for example:

(487) *Yau njam ausam   terangai*
       yau njam ausè-m   t-werè-Ø-ang-ay-Ø
       NEG if   old.woman-ERG β.3SGP-hold-PFV.ND-INC-REM.PUNC-2|3SGA

       *sèkw **so** káráfnau   **so***
       sèkw **so** k-áráfár-ta-w-Ø   **so**
       canoe **FUT** β.ØP-break-IPFV.ND-REM.DUR-2|3SGS$_A$ **FUT**

       kolitotawèm.   (YD11(3))
       k-olito-ta-w-m
       β.ØP-sink-IPFV.ND-REM.DUR-1NSGS$_A$
       'If the old lady hadn't held it, the canoe would have broken, we would have sunk.'

5.3.2 Aspect markers

Nama has several aspect markers:

tè perfect (PRF)

The perfect marker *tè* indicates that an action or event is complete or that a state existed in the past. It is often translated as 'already':

(488) *fès*   ***tè***   *yètayamènd*...   (YD11(3))
       fès   **tè**   y-taya-Ø-m-nd
       fire   **PRF**   α.3sGP-light.a.fire-IPFV.DU-REM.DLT-2|3NSGA
       'they (2) already lit a fire . . .

(489) *Ben*   *áuyamèn*   *káuwetau*...
       Ben   áuyè-mèn   k-áuwè-ta-w-Ø
       Ben   cassowary-ORIG   β.ØP-be.happy-IPFV.ND-REM.DUR-2|3sGS$_A$
       ákim   ***tè***   *tènamnde.*   (KK11)
       áki-m   **tè**   t-nam-ta-e
       grandfather-ERG   **PRF**   β.3sGP-shoot-IPFV.ND-2|3sGA
       'Ben was happy about the cassowary . . . that grandpa had shot.'

(490) *káye*   *ámb*   ***tè***   *táwifárèt*   *yau?*   (DC16)
       káye   ámb   **tè**   tá-wifár-Ø-t   yau?
       yesterday   other   **PRF**   β.3sGP-chase-IPFV.DU-2|3NSGA   NEG
       'yesterday you (2) chased others [in hunting], didn't you?'

In almost all examples, *tè* occurs directly before the verb, but there is a handful of exceptions in the data – for example:

(491) *mèrès*   ***tè***   *ghèrghèrsisam*   *yifo.*   (MD13(1))
       mèrès   **tè**   ghèrghèrsisè-am   y-wifo-Ø-Ø
       girl   **PRF**   feeling.sorry-ERG   α.3sGP-finish-PFV.ND-2|3sGA
       'The girl was already feeling sorry.'

fèf/fèfè inchoate 1 (INCH1)

The inchoate marker *fèf* or *fèfè* indicates that an action, event or state has just occurred or has just begun, but not necessarily that it has finished. The meaning of this marker differs according to the tense/aspect morphology of the verb it modifies.

5.3 Verb phrase: Other word classes — **193**

With the imperfective current tense, it is similar to the immediate perfect (or "hot news" perfect) in English:

(492)   *biskar*   *wèn*   *fèsot*   **fèf**
        biskar   wèn   fès-ot   **fèf**
        cassava plant cooking-PURP **INCH1**
        enèrsèt.   (DC29)
        e-n-rèsa-Ø-t
        a.2|3NSGP-VEN-carry-IPFV.DU-2|3NSGA
        'you (2) have just carried cassava plants for cooking.'

(493)   *Yèmo*   **fèfè**   *yènfete.*   (NB8:34)
        yèmo   **fèfè**   y-nèfè-ta-e
        3SG.ERG **INCH1** a.3SGP-cut-IPFV.ND-2|3SGA
        'He has just cut it.'

(494)   **fèfè**   *ewawaitotan . . .*   (ND13)
        fèfè   e-wa-waito-ta-èn
        INCH1 a.2|3NSGP-APP-tell-IPFV.ND-1SGA
        'I've just told this for you all . . .'

With the perfective current inceptive tense, it can mean that the action or event was started but not accomplished – giving the meaning 'almost' or 'nearly':

(495)   *Fá*   **fèfè**   *nuyang*   *ásta.*   (NB7:93)
        fá   **fèfè**   n-uy-Ø-ang-Ø   ás-ta
        1.ABS **INCH1** a.ØP-fall-PFV.ND-INC-2|3SGS$_A$ coconut-ABL
        'He nearly fell from the coconut tree.'

(496)   *Yèmo*   **fèf**   *yènfè.*   (NB8:34)
        Yèmo   **fèf**   y-nèfè-Ø-Ø
        3SG.ERG **INCH1** a.ØP-cut-PFV.ND-2|3SGA
        'He's almost cut it.'

Compare (496) with (493) above.

With remote tenses, *fèf/fèfè* also implies that the action or event has not yet finished or was not accomplished:

194 —— 5 Additional word classes and phrase structure

(497) *Yènd **fèfè** násáfotamèn.* (NB5:46)
yènd **fèfè** n-ásáfo-ta-m-èn
1.ABS **INCH1** ɑ..ØP-work-IPFV.ND-REM.DLT-1SG$_A$
'I worked (but didn't finish).'

(498) *Yèmo **fèfè** yènfetam.* (NB8:34)
yèmo **fèfè** y-nèfè-ta-m-Ø
3SG.ERG **INCH1** ɑ..3SGP-cut-IPFV.ND-REM.DLT-2|3SGA
'He cut it (but may not have finished).'

(Compare this to *Yèmo tè yènfetam.* 'He cut it (and finished).')

(499) *namgèt **fèfè** yafngoyèn.* (TE11(2))
nam-gh-t **fèfè** y-wafngo-Ø-ay-èn
shoot-NOM-ALL **INCH1** ɑ.3SGP-start-PFV.ND-REM.PUNC-1SGA
'I started to shoot it (but was unsuccessful).'

(500) *kambanam yènd **fèfè** wèrnam.* (YD11(3))
(=8) kamban-am yènd **fèfè** w-rár-ta-m-Ø
snake-ERG 1.ABS **INCH1** ɑ.1SGP-bite-IPFV.ND-REM.DLT-2|3SGA
'the snake nearly bit me.'

Note that this aspect marker is most likely the origin of proximal clitic *f* (see section 5.4.1).

tèf/tèfè inchoate 2 (INCH2)
The marker *tèf/tèfè* has functions very similar to those of *fèf/fèfè* and the differences between the two are not well understood. It can have the immediate perfect meaning, as in the following:

(501) *sáyáf **tèf** naflitat; kès*
sáyáf **tèf** n-afèli-ta-t kès
prawn **INCH2** ɑ.ØP-go.inside-IPFV.ND-2|3NSGS$_A$ just.there
eflitan. (DC36)
e-fèli-ta-èn
ɑ.2|3NSGP-put.inside-IPFV.ND-1SGA
'the prawns are already just going in [the basket]; there, I put them in.'

5.3 Verb phrase: Other word classes —— **195**

(502) *Kimbam ndau **tèfè** yasarmbote.*   (NB4:48)
      kimb-am ndau **tèfè** y-wasarmbo-ta-e
      pig-ERG garden **INCH2** α.3SGP-spoil-IPFV.ND-2|3SGA
      'The pig has just spoiled the garden,'

(503) *wagif **tèfè** enèrnyèt.*   (TE11(2))
      wagif **tèfè** e-n-rèny-Ø-t
      fish **INCH2** α.2|3NSGP-catch-IPFV.DU-2|3NSGA
      'you (2) have just caught the fish'

With the perfective current inceptive tense and remote tenses, it can also indicate that the action or event was started but not accomplished or finished:

(504) *Yèmo náifae mèrès **tèfè** yaghnai.*   (NB4:48)
      yèmo náifè-e mèrès **tèfè** y-waghan-Ø-ay-Ø
      3SG.ERG knife-INST girl **INCH2** α.3SGP-kill-PFV.ND-REM.PUNC-2|3SGA
      'He nearly killed the girl with the knife.'

(505) *Yèmo **tèfè** tènfetau.*   (NB8:34)
      yèmo **tèfè** t-nèfè-ta-w-Ø
      3SG.ERG **INCH2** β.3SGP-cut-IPFV.ND-REM-DUR-2|3SGA
      'He was cutting it (but didn't finish).'

mè continuative (CONT)

The continuative marker *mè* indicates a durative or continuing action, event or state, with the meaning 'still' or 'keep on':

(506) *yènd **mè** yènfande!*   (DC20)
(=378) yènd **mè** yèn-fanda-ta-e
      1.ABS **CONT** α.1NSGP-look.at-IPFV.ND-2|3SGA
      'it's still looking at us!'

(507) *Nèmamèn kètè **mè** nokawend?*   (PT:583)
      nèmamèn kètè **mè** n-okawè-e-nd
      why there **CONT** α.ØP-fight-PFV.DU-2|3DUS$_A$
      'Why are they (2) still fighting there?'

196 —— 5 Additional word classes and phrase structure

(508) *minde* *wèri nu* ***mè*** *tánetau.* (PT:738)
minde   wèri nu   **mè**   tá-ne-ta-w-Ø
excessive   alcohol   **CONT**   β.2|3NSGP-drink-IPFV.ND-REM.DUR-2|3SGA
'he kept on drinking too much alcohol.'

Although *mè* generally occurs directly before the verb, there are several instances where it does not – for example:

(509) *nuam*   *túmèn*   *fá*   ***mè***   *fètè*
nu-am   túmèn   fá   **mè**   fètè
water-ERG   always   3.ABS   **CONT**   VAL
tèlálau . . .   MD13(1))
t-lála-ta-w-Ø
β.3SGP-reflect-IPFV.ND-REM.DUR-2|3SGA
'The water kept on really reflecting her . . .'

(510) *yènd* ***mè*** *arengègh*   *ár*   *wèm.* (DC27)
yènd   **mè**   areng-gh   ár   w-m
1.ABS   **CONT**   roam.around-NOM   man   α.1SGSP-COP.ND
'I'm still a roamer.'

5.3.3 Modals (modality markers)

Nama also has several modality markers (modals). Although they most frequently occur before the verbs they modify, some occasionally occur in other locations.

sèno possibility (POSB)
The modal *sèno* indicates possibility, and is most often translated as 'maybe':

(511) *Kemnangè*   *nuan*   *sèno*
k-emán-Ø-ang-è   nu-an   sèno
β.ØP-think-PFV.ND-INC-2|3SGS_A   water-LOC2   **POSB**
yughár.   (MD13(1))
y-wughár
α.3SGSP-be.submerged
'He thought maybe she is submerged in the water.'

5.3 Verb phrase: Other word classes — **197**

(512)   *yènè*   *áryo*   **sèno**   *fyèm.*   (PT:495)
       yènè   ár-yo   **sèno**   f-y-m
       DEM   man-EXCL   **POSB**   PROX1=α.3SGS_P-COP.ND
       'maybe it's just this man.'

(513)   *mband*   *kènjún*   *sèno*   *yèm*   *mèrès.*   (MD13(1))
       mband   kènjún   sèno   y-m   mèrès
       ground   inside   **POSB**   α.3SGS_P-COP.ND   girl
       'maybe the girl is in the ground.'

In one elicited example, *sèno* does not occur directly before the verb:

(514)   **Sèno**   *Benèm*   *fyèráfne.*   (NB2:79)
       sèno   Ben-m   f=y-ráfár-ta-e
       POSB   Ben-ERG   PROX1=3SGP-break-IPFV.ND-2|3SGA
       'Maybe Ben just broke it.'

mètè potential 1 (**POT1**)

The modal *mètè* indicates the potential to occur, and is translated as 'might':

(515)   *tárfár*   **mètè**   *káyátongi.*   (MK13(2))
       tárfár   **mètè**   k-áyáto-Ø-ang-i
       plenty   **POT1**   β.ØP-end.up-PFV.ND-INC-2|3PLS$_A$
       'they might end up as plenty.'

(516)   *yènè*   *fène-fènen*   **mètè**   *kilawangè.*   (PT:517)
       yènè   fènefène-n   **mètè**   k-ilawè-Ø-ang-è
       DEM   opening-LOC1   **POT1**   β.ØP-go.inside-PFV.ND-INC-2|3SGS$_A$
       'it might go inside this opening.'

(517)   *Mer*   *efogh*   *o*   *mer*   *sèite*   *o*   *nèmè*   *efogh*
       mer   efogh   o   mer   sèite   o   nèmè   efogh
       good   day   or   good   afternoon   or   what   time
       mètè   *yèngèm*   *feyot.*   (FD13)
       mètè   y-ngèm   feyot
       POT1   α.3SGS_P-go.ND   2SG.PURP
       'Good day or good afternoon or whatever time might go for you.'

198 —— 5 Additional word classes and phrase structure

This modal is used in apprehensive constructions:

(518) *Kor,* ***mètè*** *tètiar.* (NB10:7)
kor **mètè** t-tiar-Ø-Ø
careful **POT1** α.3sGP-break-PFV.ND-2|3sGA
'Careful, you might break it.'

(519) *Kor,* *fèm* ***mètè*** *kuyè.* (NB10:7)
kor fèm **mètè** k-uy-Ø-è
careful 2SG.ABS **POT1** β.ØP-fall.down.PFV.ND-2|3sGS$_A$
'Careful, you might fall down.'

Mètè occasionally appears at the beginning of the sentence:

(520) ***Mètè*** *yau* *osferean* *em.* (PT:84)
mètè yau osfere-an e-m
POT1 NEG hospital-LOC2 α.2|3NSGS$_P$-COP.ND
'It might be that they're not in hospital.'

wè potential 2 (POT2)

The modal *wè* has a meaning similar to *mètè* 'might' but indicates a weaker potential. (This modal has been described by consultants but does not appear in the naturalistic recordings.) Irrealis aspect (section 4.7.1) is used with 2nd or 3rd person singular arguments in imperfective current and recent tense, as in (521).

(521) *Fá* ***wè*** *násáfota.* (NB5:57)
fá **wè** n-ásáfo-ta-Ø
3.ABS **POT2** α.ØP-work-IPFV.ND-2|3sGS$_A$.IRR
'He might work.'

(522) *Gasiyam* ***wè*** *tanangè.* (NB2:39)
gasiyè-am **wè** t-wanè-Ø-ang-è
crocodile-ERG **POT2** β.3sGP-take-PFV.ND-2|3sGA
'A crocodile might take (someone).'

kár dubitative (DUB)

The modal *kár* expresses uncertainty or doubt that the proposition is true or has actually occurred or will occur:

5.3 Verb phrase: Other word classes — **199**

(523) *yabun kiwál **kár** nènuwano.* (ND13)
yabun kiwál **kár** n-n-uwano-Ø-Ø
big west.wind **DUB** α.ØP-VEN-set.off-PFV.ND-2|3SGS_A
'a big westerly wind might have come up.'

(524) *kemnangèn* *kor* ***kár***
k-emán-Ø-ang-èn kor **kár**
β.ØP-think-PFV.ND-INC-1SGSA actually **DUB**
nárotan. (MK13(2))
n-áro-ta-èn
α.ØP-go.hunting-IPFV.ND-1SGS_A
'I thought actually whether or not to go hunting.'

(525) *...nde yènd yau Mátamèn **kár** wèm.* (TE13)
nde yènd yau Mátè-mèn **kár** w-m
that 1.ABS NEG Mata-ORIG **DUB** α.1SGS_P-COP.ND
'...that I might not be from Mata.'

siya/sái intentional (INT)

The modal *siya* or *sái* indicates intention to accomplish an action or event or a goal to be reached.[39]

(526) tane mèrès **siya** tanangè fene
tane mèrès **siya** t-wanè-Ø-ang-è fene
1SG.GEN girl **INT** β.3SGP-take-PFV.ND-INC-2|3SGA 2SG.GEN
ambumom. (B&B:32)
ambum-o-m
child-SG-ERG
'your son intends to take my girl.'

(527) *nèmè mwighè **sái** tètiarngayèn?* (FD12:71)
nèmè mwighè **sái** t-tiar-Ø-ang-ay-èn
what mind **INT** β.3SGP-break-PFV.ND-INC-REM.PUNC-1SGA
'what will I intend to decide?'

The goal-oriented meaning of *sái/siya* can be seen in its use with the stative nominal *mènde* 'wanting, liking':

---

39 Some consultants say that *siya* is Nama and *sái* is Namat (Mibini).

(528) *Fèm* *yènè* *wiwi* *mènde* **siya** *nèm?* (NB7:79)
     fèm yènè wiwi mènde **siya** n-m?
     2.ABS DEM mango liking **INT** α.2SGS$_P$-COP.ND
     'Do you like this mango?'

Note that the modal *siya/sái* also occurs in subordinate adverbial clauses where it has a purposive or hypothetical meaning (section 7.2.5).

tèu permissive (PERM)

Although the modal *tèu* is labelled "permissive" it more often expresses general approval for the action or event of the verb, meaning something like 'it's OK that...'. Some examples are:

(529) *Tane* *náifè* **tèu** *tèngwan.* (NB7:78)
     Tane náifè **tèu** t-ng-wanè-Ø-Ø
     1SG.ERG knife **PERM** β.3SGP-AND-take-PFV.ND-2|3SGA
     'You can take my knife.'

(530) *Ros* **tèu** *yèngèm* *skulot.* (NB8:28)
     Ros **tèu** y-ngèm skul-ot
     Rose **PERM** α.3SGS$_P$-go.ND school-PURP
     'Rose can go to school.'

(531) *yèndo* *yènè* *yús* **tèu** *tèngènfetawèn.* (FD13)
     yèndo yènè yús **tèu** t-ng-nèfè-ta-w-èn
     1SG.ERG DEM grass **PERM** β.3SGP-AND-cut-IPFV.ND-REM.DUR-1SGA
     'it was OK that I was cutting this grass.'

miya obligative continuative (OBL.CONT)

The marker *miya* (alternatively *mèa*) combines modal and aspectual functions, meaning 'must or have to keep on doing something'. Note the the use of irrealis aspect in (532).

(532) *Fá* **miya** *násáfota.* (NB5:57)
     fá **miya** n-ásáfo-ta-Ø
     3.ABS **OBL.CONT** α.ØP-work-IPFV.ND-2|3SGS$_A$.IRR
     'He must keep on working.'

5.3 Verb phrase: Other word classes — **201**

(533) Tane   amaf   tèmndangèn   yèna   **miya**
      tane   amaf   t-mèndè-Ø-ang-èn   yèna   **miya**
      1SG.GEN wife   β.3SGP-say.to-PFV.ND-INC-1SGA here   **OBL.CONT**
      nèmor.   (MK13(1))
      n-mor
      α.2SGS_P-stay
      'I said to my wife you must keep on staying here.'

yèf evidential (EVID)
This modal indicates that the speaker is witnessing or has witnessed the action, event or state and implies its continuation. (Note that this modal has very limited use in the data.)

(534) Ár   **yèf**   násáfote.   (NB7:91)
      ár   **yèf**   n-ásáfo-ta-e
      man   **EVID**   α.ØP-work-IPFV.ND-2|3SGS_A
      'The man is working.'

(535) Yèna   **yèf**   yèm   ánde   mèngo
      yèna   **yèf**   y-m   ánde   mèngo
      here   **EVID**   α.3SGS_P-COP.ND where   house
      yèrametan.   (NB7:91)
      y-ramè-ta-èn
      α.3SGP-make-IPFV.ND-1SGA
      'It's here where I'm making the house.'

(536) ambum   sèmta   **yèf**   enwakái.   (B&B:252)
      ambum   sèm-ta   **yèf**   e-n-wakáy(o)
      child   back-ABL **EVID**   α.2|3NSGS_P-VEN-be.standing
      'children are standing at the back.'

yèta attemptive (ATT)
Nama has two different preverbal markers that can be translated as 'try': the modal yèta, described here, and the quasi-modal mato, described in section 5.3.5 below. The modal yèta refers to having tried or attempted to do or accomplish something, normally unsuccessfully:

202 — 5 Additional word classes and phrase structure

(537)   *wèkeye*   **yèta**   *karafnawèm*   *witma.*   (YD11(3))
        wèkeye   **yèta**   k-áráfár-ta-w-m   witma
        quickly   **ATT**   β.ØP-steer-IPFV.ND-REM.DUR-1NSGS$_A$ side
        'quickly we tried to steer (the canoe) to the side.'

(538)   *nu*   *kènjúnmè*   **yèta**   *ronjagh*   *tafngongè.*   (MD13(1))
        nu   kènjún-mè   **yèta**   ronja-gh   t-wafngo-Ø-ang-è
        water   inside-PERL   **ATT**   search-NOM   β.3SGP-start-PFV.ND-INC-2|3SGA
        'he tried to start searching in the water.'

(539)   *Wèrár*   **yèta**   *tèlaitan.*   (NB6:82)
        wèrár   **yèta**   t-lai-ta-èn
        walaby   **ATT**   β.3SGP-shoot-IPFV.ND-1SGA
        'I tried to shoot the wallaby.'

More frequently than other modals, *yèta* is found in positions other than directly preceding the verb – e.g.:

(540)   **yèta**   *yènamè*   *tètanau*   *yau.*   (MD13(1))
        yèta   yènamè   t-tar-ta-w-Ø   yau
        ATT   along.here   β.3SGP-dig-IPFV.ND-REM.DUR-2|3SGA   NEG
        'he tried unsuccessfully to dig along here.'

5.3.4 Combinations of TAM markers

Some TAM markers can be combined, as shown in the following examples:

(541)   *yène*   *si*   **mè**   **so**   *yafarotam*
        yène   si   **mè**   **so**   y-wafaro-ta-m
        DEM   story   **CONT**   **FUT**   α.3SGP-remember-IPFV.ND-1NSGA
        yèndfem.   (ND13)
        yèndfem
        1NSG.ERG
        'we will keep on remembering this story.'

(542)   *Káye*   *yènd*   **kár**   **so**   *Darut*   *wèngèm.*   (NB6:20)
        káye   yènd   **kár**   **so**   Daru-t   w-ngèm
        tomorrow   1.ABS   **DUB**   **FUT**   Daru-ALL   α.1SGS$_P$-go.ND
        'Tomorrow I might go to Daru.'

(543) *mèinyotyo ...* **mè** **kár** *nèretat.* (ND13)
mèinyotyo **mè** **kár** n-ère-ta-t
all **CONT** **DUB** α.ØP-be.awake-IPFV.ND-2|3NSGS$_A$
'they all ... might still be awake.'

(544) *kèma nde* **tè** **kár** *nawer.* (DC21)
kèma nde **tè** **kár** n-awerè-Ø-Ø
idea that **PRF** **DUB** α.ØP-get.married-PFV.ND-2|3SGS$_A$
'he thought that you might already be married.'

(545) **sèno** **tè** *enwanangi.* (PT:97)
sèno **tè** e-n-wanè-Ø-ang-i
POSB **PRF** α.2|3NSGP-VEN-take-PFV.ND-INC-2|3PLA
'maybe they already took them.'

5.3.5 Quasi-modals

There are some words that appear to function as modals when they occur before
the verb or a TAM marker, but in other contexts function as members of other word
classes.

fèyotaro abilitative (ABIL)
As a modal, *fèyotaro* has an abilitative function – i.e. meaning 'can, able to':

(546) **fèyotaro** *tawinjon* *tárfár.* (MK13(1))
fèyotaro ta-winjo-Ø-èn tárfár
ABIL β.2|3NSGP-see-PFV.ND-1SGA plenty
'I could see plenty.'

(547) *yákae* **fèyotaro** *so yúyote*
yákè-e **fèyotaro** so y-wúyo-ta-e
digging.stick-INST **ABIL** FUT α.3SGP-lift-IPFV.ND-2|3SGA
wem. (B&B:217)
wem
yam
'with the digging stick you can dig up [lift out] a yam.'

(548) mèrèn mèinyotyo **fèyotaro** so yásinde.   (B&B:223)
      mèrèn mèinyotyo **fèyotaro** so y-wásing-ta-e
      family all   **ABIL**   FUT α.3SGP-take.care.of-IPFV.ND-2|3SGA
      'all the family will be able to take care of it.'

However, the word *fèyotaro* more generally means 'enough' or 'capable', as in the following:

(549) sembyo **fèyotaro** ere.   (MK13(2))
      sembyo **fèyotaro** e-r-e
      two   **enough**   α.2|3NSGS$_P$-COP.DU-PFV.DU
      'two are enough.'

(550) etarmèm   ndènd   kètanotyo
      e-tar-Ø-m-m   ndènd   kètanotyo
      α.2|3NSGP-dig-IPFV.DU-REM.DLT-1NSGA   worm   until
      fèyotaroan ewinjeayèm.   (TE11(2))
      fèyotaro-an e-winjo-e-ay-m
      enough-LOC2 α.2|3NSGP-see-PFV.DU-REM.PUNC-1NSGA
      'we (2) dug worms until we saw there were enough.'

(551) tane   ambum **fèyotaro** ár   tè   namndè.   (B&B:201)
      tane   ambum **fèyotaro** ár   te   n-amndè-Ø-Ø
      1SG.GEN child **capable**   man PRF α.ØP-become-PFV.ND-2|3SGS$_A$
      'my child has become a capable man.'

fiya/fái obligative 1 (OBL1)

Two different adverbs, *fiya/fai* and *táf*, also appear to function as modals with an obligative meaning. When used as a modal, *fiya* and its alternate form *fái*, have the meaning of 'should' or 'have to'. In imperfective current and recent tenses, irrealis aspect is again used with with 2nd and 3rd person singular arguments, as in (552).

(552) Fèm   **fiya**   násáfota.   (NB5:57)
      fèm   **fiya**   n-ásáfo-ta-Ø
      2.ABS **OBL1** α.ØP-work-IPFV.ND-2|3SGS$_A$.IRR
      'You should work.'

5.3 Verb phrase: Other word classes — **205**

(553) *fèyo   yèmofem   **fái**   kwáutanangi.*   (FD13)
fèyo   yèmofem   **fái**   kw-wáutár-tang-i
then   3NSG.ERG   **OBL1**   β.1SGP-help-INC.IPFV-2|3PLA
'then they should have helped me.'

(554) *yèndo   yau   **fiya**   tankafetawèn...*   (FD13)
yèndo   yau   **fiya**   ta-n-kafè-ta-w-èn
1SG.ERG   NEG   **OBL1**   β.2|3NSGP-pick.up-IPFV.ND-REM.DUR-1SGA
'I shouldn't have been picking them up...'

More frequently, *fái* is an adverb meaning 'really' or acting as an emphatic discourse marker:

(555) *yáki   yènè   fyèm   **fái**   mèngon.*   (PT:688)
yá-ki   yènè   f=y-m   **fái**   mèngo-n
3SG-grandfather   DEM   PROX1=α.3SGS$_P$-COP.ND   **EMPH**   house-LOC1
'It was really his grandfather in the house.'

(556) *ngangongi   **fái**   waghèt.*   (MK13(4))
ng-ango-Ø-ang-i   **fái**   wagh-t
AND-return-PFV.ND-INC-2|3PLS$_A$ **EMPH**   singsing-ALL
'they went right back to the singsing.'

(557) *fá   **fái**   yènamamèn   áuwèghafè   wèi*
fá   **fái**   yènamamèn   áuwè-gh-afè   wèi
3.ABS   **really**   for.this.reason   be.happy-NOM-COM   now
tam.   (FD13)
t-wa-m
β.3SGS$_P$-APP-COP.ND
'for this reason there was really happiness now for them.'

In story-telling, sentences frequently end in *fái* or the phrase *fái kès yèm* (literally 'it's really just there') for emphasis (see section 5.3.6). Examples are:

(558) *yifoyènd   yènè   ninyè   **fái**.*   (ND13)
y-wifo-Ø-ay-nd   yènè   ninyè   **fái**
α.3SGP-finish-PFV.ND-REM.PUNC-2|3NSGA   DEM   witch   **EMPH**
'they really finished (off) that witch.'

206 —— 5 Additional word classes and phrase structure

(559) ndenaro   mèf   kufrotau
ndenè-ro   mè=f   k-ufro-ta-w-Ø
like.this-RSTR CONT=PROX1 β.ØP-do-IPFV.ND-REM.DUR-2|3SGS_A
fái   kès   yèm.   (MD13(1))
fái   kès   y-m
EMPH   just.there α.3SGS_P-COP.ND
'he kept on doing only this, that's really it.'

táf obligative 2 (OBL2)

Adverb táf 'then, at that time' can also act as an obligative modal, with stronger
force than fiya/fái. Again, note the use of irrealis aspect in the following examples:

(560) Fá   táf   nasafota.   (NB7:10)
fá   táf   n-ásáfo-ta-Ø
3.ABS OBL2   α.ØP-work-IPFV.ND-2|3SGS_A.IRR
'He must work.'

(561) Ás mère   táf   yèrameta.   (DC19)
ás mèr-e   táf   y-ramè-ta-Ø
coconut.cream-INST OBL2   α.3SGP-make-IPFV.ND-2|3SGA.IRR
'You should make it with coconut cream.'

(562) wagif yaufi yau táf eneta.   (DC07)
wagif yaufi yau táf e-ne-ta-Ø
fish   bad   NEG OBL2 α.2|3NSGP-eat-IPFV.ND-2|3SGA.IRR
'you mustn't eat bad fish.'

Táf used as a modal can co-occur with other TAM markers – for example:

(563) Fá   mètè táf so   yènèm.   (NB6:20)
fá   mètè táf so   y-nèm
3.ABS POT1 OBL2 FUT α.3SGS_P-come.ND
'He might have to come (in the future).'

Here are examples of *táf* functioning as an adverb meaning 'then':

(564) *kènangotawèt*   *yènè*   *waghta*
       k-n-ango-ta-w-t   yènè   wagh-ta
       β.ØP-VEN-return-IPFV.ND-REM.DUR-2|3NSGS$_A$ DEM   singsing-ABL
       táf.   (ND11)
       táf
       then
       'they were coming back from the singsing at that time.'

(565) *fès*   *njam*   *náwároi,*   *fá*   ***táf***   *kèmègh*
       fès   njam   n-áwáro-ay-Ø   fá   **táf**   kèmègh
       fire   when   α.ØP-burn-REM.PUNC-2|3SGS$_A$ 3.ABS   **then**   sleeping
       kèr.   (ND13)
       kèr
       dead
       'when the fire started burning, she was then fast asleep.'

mato investigative (INV)

When used as a modal, *mato* means 'try out' or 'investigate if something can be done'. This differs from the modal *yèta* (section 5.3.3) that can also be translated as 'try' but refers attempting something. *Mato* occurs most often with imperatives and hortatives, as in example (292) above and (568) below. Once again, imperfective verbs with 2|3SG arguments have irrealis aspect, as in (566) and (567).

(566) *sèkw*   ***mato***   *tèfanda.*   (MK13(1))
       sèkw   **mato**   t-fanda-ta-Ø
       canoe   **INV**   β.3SGP-look.at-IPFV.ND-2|3SGA.IRR
       'try and look at the canoe.'

(567) *Fèyotaro*   ***mato***   *tèmota?*   (DC20)
       fèyotaro   **mato**   t-mo-ta-Ø
       ABIL   **INV**   β.3SGP-ask-IPFV.ND-2|3SGA.IRR
       'Can you try and ask him?'

(568) *yènè*   ***mato***   *tafandangem.*   (PT:292)
       yènè   **mato**   ta-fanda-tang-e-m
       DEM   **INV**   β.2|3NSGP-look.at-INC.IPFV-PFV.DU-1NSGA
       'let's try to look at these.'

208 —— 5 Additional word classes and phrase structure

In other contexts, *mato* appears to be an adverb meaning 'possibly':

(569) **mato** *sómbyo* *si* *yèna* *ere.* (PT:427)
    mato sómbyo si yèna e-re
    possibly two word here α.2|3NSGS$_P$-COP.DU
    'Possibly two words are here.'

(570) *kemnangè*   **mato** *mband* *kènjún* *sèno*
    k-emán-Ø-ang-è   **mato** mband kènjún sèno
    β.ØP-think-PFV.ND-INC-2|3SGS$_A$ **possibly** ground inside POSB
    yèm. (MD13(1))
    y-m
    α.3SGS$_P$-COP.ND
    'he thought possibly it might be in the ground.'

fètè validative (VAL)

As a modal, *fètè* signals that a current action, event or state is true, right or certain. It could also be analysed as an adverb meaning 'truly' or 'rightly' but it most frequently occurs before a verb or another modal and its meaning varies according to the tense of the verb. Some examples:

(571) *fèyo* *fèrfèram* *wèi* *fá* **fètè**
    fèyo fèrfèr-am wèi fá **fètè**
    then fear-ERG now 3.ABS **VAL**
    yifoi. (ND13)
    y-wifo-Ø-ay-Ø
    α.3SGP-finish-PFV.ND-REM.PUNC-2|3SGA
    'then she became really afraid.' (lit. . . .'fear really finished her.')

(572) *yárefè* *nèkw* **fètè** *naramayèng.* (MD99(2))
    yá-refè nèkw **fètè** n-aramè-Ø-ay-ng
    3SG-father angry **VAL** α.ØP-do-PFV.ND-REM.PUNC-2|3SGA
    'his father really got angry.'

(573) *kèr* **fètè** *tèm.* (MD13(1))
    kèr **fètè** t-m
    dead **VAL** β.3SGS$_P$-COP.ND
    'he really was dead.'

5.3 Verb phrase: Other word classes —— **209**

Referring to the hypothetical future, *fètè* signals that an action, event or state would be good or right (implying mild obligation). In this context the irrealis construction is used with 2nd or 3rd person singular arguments, as in (574).

(574) *Fá* **fètè** *násáfota.* (NB5:57)
      Fá **fètè** n-ásáfo-ta-Ø
      3.ABS **VAL** α.ØP-work-IPFV.ND-2|3SGS$_A$.IRR
      'It would be good if he works.'

(575) *yènd* *ndenè* **fètè** *nafandan.* (TE13)
      yènd ndenè **fètè** n-a-fanda-ta-èn
      1.ABS like.this **VAL** α.ØP-REFL-look.at-IPFV.ND-1SGS$_A$
      'It would be good if I look at myself this way.'

When *fètè* occurs in positions other than before a verb, especially at the end of a sentence, it is used for emphasis.

(576) *yèrametan* *mènamèn* **fètè** *ár* *mèrnam*
      y-ramè-ta-èn mènamèn **fètè** ár mèrèn-am
      α.3sGP-make-IPFV.ND-1SGA because **EMPH** person family-ERG
      yènetat. (DC11)
      y-ne-ta-t
      α.3sGP-drink-IPFV.ND-2|3NSGA
      'I'm making it really because family members drink it.'

(577) *ámbiro* *fá* **fètè.** (PT:599)
      ámbiro fá **fètè**
      one 3.ABS **EMPH**
      'it's just one.'

In this way it is like *fái*, with which it can co-occur:

(578) *eso* *fèfe* *ewaramangèm* *ndenè*
      eso fèfe e-wa-ramè-Ø-ang-m ndenè
      thanks INCH1 α.2|3NSGP-APP-give-PFV.ND-INC-1NSGA like.this
      fètè **fái** *kès* *yèm.* (B&B:145)
      fètè **fái** kès y-m
      EMPH **EMPH** just.there α.3sGS$_P$-COP.ND
      'we're just giving thanks to you all, just really like this.'

5.3.6 Adverbs (ADV)

We have seen in chapter 3 that nominals with the instrumental suffix -*e* can function as adverbs of manner in a verb phrase (examples 34 and 35), and nominals with the locative suffix -*an* can function as adverbs of location and time (examples 80 and 81). Some additional examples are:

(579) *wèi*   *kètè*   **tekofnare**
      wèi   kètè   **tek-ofnar-e**
      now there **long.time-PRIV-INST**
      nafrendamènd.   (MD99(1))
      n-áfreng-ta-m-nd
      a.3sGP-get.ready-IPFV.ND-REM.DLT-2|3NSGS$_A$
      'now they got ready there immediately.'

(580) **kèghátan**   *yèndo*   *tákman*   *tanjo*
      kèghát-an   yèndo   tá-kèmè-Ø-èn   tanjo
      bush-LOC2   1SG.ERG   β.2|3NSGP-lay.down-PFV.ND-1SGA   1SG.own
      yáf.   (YD11(3))
      yáf
      basket
      'I laid down my own baskets in the bush.'

(581) **sèmbár-sèmbáran**   *wèkeye*   *wèi*   *nènangend.*   (DC26)
      sèmbársèmbár-an   wèkeye   wèi   n-n-ango-e-nd
      early.dawn-LOC2   quickly   now   a.ØP-VEN-return-PFV.DU-2|3DUS$_A$
      'they (2) returned quickly in the early dawn.'

And earlier in this chapter, we have seen that temporal and orientational nouns (examples 419–426) and temporal nominals (472–480) can also function as adverbs.

However, Nama has a small set of words that function only as adverbs in a verb phrase. Some adverbs of manner, such as *wèkeye* 'quickly' in example (537) and (581) above, appear to be derived from nominals because they end with instrumental -*e*. However, equivalent nominals without -*e* do not occur. Another example is *áfráye* 'slowly, quietly':

(582) *Yènd*   **áfráye**   *so*   *wèngèm.*   (NB6:55)
      yènd   **áfráye**   so   w-ngèm
      1.ABS   **slowly**   FUT   a.1sGP-go.ND
      'I'll go slowly.'

5.3 Verb phrase: Other word classes ——— **211**

Other adverbs of manner include *amro* 'regrettably', *minde* 'intensively, exceedingly' and *yènèmès* 'suddenly, by surprise':

(583) *Yènd*   ***amro***   *Darut*   *wèngmormèn*
       yènd   **amro**   Daru-t   w-ngmo-èrmèn
       1.ABS **regrettably** Daru-ALL α.1SG$_P$-go.ND-REM.DLT.PA
       dingian.   (NB8:37)
       dingi-an
       dinghy-LOC2
       'I regrettably went to Daru in a dinghy.'

(584) *kètè*   ***minde***   *fèrèk*   *ngasurngè.*   (MD99(2))
       kètè   **minde**   fèrèk   ng-asur-Ø-ang-è
       there   **exceedingly** blood AND-spill-PFV.ND-INC-2|3SG$_A$
       'there blood poured out exceedingly.'

(585) *mbwito*   ***yènèmès***   *kèntone*
       mbwito   **yènèmès**   k-n-ètor-ta-e
       rat   **suddenly**   β.ØP-VEN-come.out-IPFV.ND-2|3SG$_A$
       mámèminde.   (MD11(1))
       mámèminde
       very.exceedingly
       'a rat suddenly came out very quickly.'

Locational adverbs include *kaka* 'near, close':

(586) *fèyo*   *fá*   *nènaflai*   ***kaka***
       fèyo   fá   n-n-afèl-Ø-ay-Ø   **kaka**
       then   3.ABS α.ØP-VEN-go.short.distance-PFV.ND-REM.PUNC-2|3SG$_A$ **near**
       kámnjongè.   (TE11(1))
       k-ámnjo-Ø-ang-è
       β.ØP-sit-PFV.ND-INC-2|3SG$_A$
       'then she came near and sat down.'

Note that its opposite *nayu* 'far' is a actually a nominal which functions as an adverb, as it takes nominal morphology (as in example 558 below). A more accurate meaning would be 'long distance'.

(587) tane   mèngo   syèm   **nayu**   yèm.   (MD13(1))
     tane   mèngo   s=y-m   **nayu**   y-m
     3SG.GEN house   PROX2=α.3SGS_P-COP.ND   **long.distance** α.3SGS_P-COP.ND
     'my house, it's far away.'

(588) nderè   **nayu**   Sarghár yèm?   **nayuofnar**
     nderè   **long.distance**   Sarghár y-m   **nayu-ofnar**
     how   **long.distance**   Sarghar α.3SGS_P-COP.ND   **long.distance-PRIV**
     yèm.   (DC26)
     y-m
     α.3SGS_P-COP.ND
     'how far is Sarghar? it's not far.'

The basic spatial adverbs are *yèna* 'here' (proximal) and *kètè* 'there' (distal).[40] The adverb *yèna* 'here' is clearly related to the demonstrative *yènè* (section 5.1.3).

(589) **yèna**   so   wèmorang.   (MM13)
     yèna   so   w-mor-ang
     here   FUT   α.1SGS_P-stay-INC
     'I will stay here.'

(590) ne   ghèrghèr   **kètè**   ekmang   kèsou.   (ND13)
     ne   ghèrghèr   **kètè**   e-kèmè-ang   kèsou
     shit intestines **there** α.2|3NSGS_P-be.lying.down-INC over.there
     'the guts are lying over there.'

Other spatial adverbs, such as *yènan* 'here, in here' and *kètan* 'there, at that place' are mentioned in section 3.6.1 (see examples 87, 123 and 198–200). It is clear that *kétan* and related forms such *kètanmè* 'along there' are derived from *kètè* 'there' plus case suffixes. It is not clear whether *yènan* and related forms such *yènanmè* 'along here' are derived from *yèna* 'here' or from the demonstrative *yènè*, as indicated in 3.6.1. So it could be argued *kètè* and *yèna* are actually nominals with adverbial functions, as they occur with case suffixes. However, here they are considered to be adverbs, and the result of historical word formation rather than productive affixation.

Closely associated with the spatial adverbs is the immediacy adverb *kès*. It is used to indicate the nearness of an entity, action, event or state in space or time. It also implies that the listener should be aware of the entity, action, event or state

---

40 Unlike other Yam languages, such as Kamnzo (Döhler 2018:105), there is no medial.

because of its proximity or prominence. It is glossed in this grammar as 'just here', 'just there', 'just now' or 'just then', depending on the context.

(591)   *sáyáf*   *tèf*   *naflitat,*   **kès**
        sáyáf   tèf   n-afèli-ta-t   **kès**
        prawn   INCH2   α.ØP-go.inside-IPFV.ND-2|3NSGA   **just.here**
        eflitan.   (DC36)
        e-fèli-ta-èn
        α.2|3NSGP-put.inside-IPFV.ND-1SGA
        'the prawns are going inside, I'm putting them inside just here.'

(592)   *Sargháràn*   **kès**   *yèmo.*   (ND13)
        Sarghár-n   **kès**   y-mo
        Sarghar-LOC1   **just.there**   α.3SGS_P-COP.ND
        'she's just there in Sarghar.'

(593)   *afaf*   **kès**   *enre.*   (TE11(1))
        afè-af   **kès**   e-nèr-e
        father-PL   **just.now**   α.2|3NSGS_P-come.DU-PFV.DU
        'our parents are just coming.'

(594)   *yènè*   *murè-murè*   **kès**   *táneghan.*   (MM13)
        yènè   murèmurè   **kès**   tá-negh-ta-èn
        DEM   medicine   **just.then**   β.2|3NSGP-cook-IPFV.ND-1SGA
        'I cooked the medicine just then.'

The expression *kès yèm*, literally 'it's just there', has come to be commonly used as a discourse marker meaning 'that's it' or 'just so' or 'just right'. (See section 6.8.) An example is:

(595)   *sìf*   *so*   *nákèrnat*   **kès**   *yèm.*   (DC19)
        sìf   so   n-ákèrár-ta-t   **kès**   y-m
        hair   FUT   α.ØP-burn-IPFV.ND-2|3NSGS_A   **just.there**   α.3SGS_P-COP.ND
        'the hair will burn just so!'

This usage is very common in coversation and story-telling, making the immediacy adverb the sixth most frequent word in the recorded data.

     Temporal adverbs include *wèi* 'now', *túmèn* 'always' and *yènau* 'before':

(596) *Ánde* **wèi** *Tinè* *yèm?* (DC24)
ánde **wèi** Tinè y-m
where **now** Tina α.3sGSₚ-COP.ND
'Where is Tina now?'

(597) *Yènè* *mèrès* **túmèn** *nararmbete.* (NB8:41)
Yènè mèrès **túmèn** n-ararmbè-ta-e
DEM girl **always** α.ØP-talk.too.much-IPFV.ND-2|3sGSₐ
'This girl always talks too much.'

(598) *Yèmo* **yènau** *tèfandawèng.* (NB8:25)
yèmo **yènau** t-fanda-ta-w-ng
3sG.ERG **before** β.2|3sGP-look.at-IPFV.ND-REM.DUR-2|3sGA
'She was watching him before.'

Another common adverb is *ndenè* 'like this/that; this/that way', which corresponds to *olsem* in Tok Pisin. Some examples are:

(599) mètare **ndenè** endmè tèngmorwèn. (PT:699)
mètar-e **ndenè** end-mè t-ngèmo-èrwèn
quiet-INST **like.this** road-PERL β.3sGSₚ-go-REM.DUR.PA
'quietly like this he went along the road.'

(600) yèmo **ndenè** rokár ámb yau
(=448) yèmo **ndenè** rokár ámb yau
3sG.ERG **like.this** thing any NEG
yinjoi *ghèrsayan.* (MD13(1))
y-winjo-Ø-ay-Ø ghèrsè-yan
α.3sGP-see-PFV.ND-REM.PUNC-2|3sGA life-LOC2
'He never saw something like this in his life.'

(601) sènonjo yènamamèn yènè yam **ndenè**
sènonjo yènamamèn yènè yam **ndenè**
today that's.why DEM custom **this.way**
mè fètè yakái. (ND11)
mè fètè y-wakáy(o)
CONT really α.3sGSₚ-be.standing
'today that's why this custom is still really standing this way.'

5.4 Verb phrase clitics

Three clitics occur as part of the verb phrase.

5.4.1 Proximal clitics (PROX)

Nama has two commonly used proximal clitics, one temporal and one locational. They are not required grammatically, but are used stylistically, especially in story telling, to give extra immediacy to an utterance.

Temporal proximal clitic f (PROX1)
The temporal proximal clitic f (PROX1) emphasises that the activity or event just happened or is about to happen or that a state has just started or is about to start. It also seems to have an emphatic function in some contexts. It is most commonly a proclitic on the verb:

(602)   *yèndo*   *yènándmè*   *fengwifnan*
       yèndo   yènándmè   **f**=e-ng-wifár-ta-èn
       1SG.ERG   along.here   **PROX1**=ɑ.2|3NSGP-AND-chase-IPFV.ND-1SGA
       wagif.   (ND13)
       wagif
       fish
       'I just chased the fish along here.'

(603)   *yèndon*   *kèma*   *fyèfandan*
       yèndon   kèma   **f**=y-fanda-ta-èn
       1SG.ERG   idea   **PROX1**=ɑ.3SGP-look.at-IPFV.ND-1SGA
       ausè   *kár*   *yèm.*   (PT:24)
       ausè   kár   y-m
       old.woman   DUB   ɑ.3SGSₚ-COP.ND
       'I thought I just saw there might be an old woman.'

(604)   *Endene*   *ambum?*   *Pitoene*   *yènè*   *fyèm.*   (DC14)
       endene   ambum?   Pitè-o-e-ne   yènè   **f**=y-m
       whose   child   Peter-SG-DAT-GEN   DEM   **PROX1**=ɑ.3SGSₚ-COP.ND
       'Whose child?' 'It's just Peter's.'

216 —— 5 Additional word classes and phrase structure

But if the verb begins with a consonant, *f* can be an enclitic to a TAM marker or other word preceding the verb and ending in a vowel:

(605)   *mer*   *basiau*   *sènonjo*   *sof*   *yènet.*   (DC19)
            mer   basiau   sènonjo   so=**f**   y-ne-Ø-t
            good   soup   today   FUT=**PROX1**   α.3SGP-eat-IPFV.DU-2|3NSGA
            'you (2) will just eat good soup today.'

(606)   *yènèf*   *yèráfnan.*   (DC13)
            yènè=**f**   y-ráfár-ta-èn
            DEM-**PROX1**   α.3SGP-finish.making-IPFV.ND-1SGA
            'I'm just finishing this.'

(607)   *soramè*   *fáf*   *kènèwaitotawèt.*   (SK13(3))
            soramè   fá=**f**   k-n-èwaito-ta-w-t
            later   3.ABS=**PROX1**   β.ØP-follow-IPFV.ND-REM.DUR-2|3NSGSA
            'later they just followed.'

Note that the aspect marker *tèf,* as in example (501) above, might actually be analysed as *tè=f.*

A possible origin for the temporal proximal clitic is the inchoate aspect marker *fèf/fèfè* (described in section 5.3.2).

Locational proximal clitic s (PROX2)

The locational proximal clitic *s* (PROX2) emphasises a closeness to the location of the activity, event or state being described. It also seems to have an evidential or emphatic function in some contexts. Unlike the temporal proximal clitic, it is always a proclitic on the verb:

(608)   *Ghèrayan*   *fá*   *snufaryèng.*   (SK11(2))
            Ghèrayè-n   fá   **s**=n-ufar-Ø-ay-ng
            Daraia-LOC1   3.ABS   **PROX2**=α.ØP-arrive-PFV.ND-REM.PUNC-2|3SGS_A
            'He arrived here in Daraia.'

(609)   *tot*   *mèrès*   *ghèrare*   *syawewote.*   (PT:159)
            tot   mèrès   ghèrare   **s**=y-wawewo-ta-e
            young   girl   moon   **PROX2**=α.3SGP-look.up.at-IPFV.ND-2|3SGA
            'the young girl is here looking up at the moon.'

(610) *kètándmè   wagif   wifárègh*
      kètándmè   wagif   wifár-gh
      along.there fish   chase-NOM
      senwafŋgoi.   (ND13)
      s=e-n-wafŋgo-Ø-ay-Ø
      PROX2=α.2|3NSGP-VEN-start-PFV.ND-REM.PUNC-2|3SGA
      'she started chasing the fish right there.'

It occurs most frequently with the copula to have the meaning 'be right here/there':

(611) *felisman*   *syèm.*   (PT:347)
      felisman   s=y-m
      policeman **PROX2**=α.3SGS$_P$-COP.ND
      'a policeman is right here.'

(612) *Yènè*   *rokár*   *syèm.*   *burarè.*   (DC34)
      yènè   rokár   s=y-m   burarè
      DEM   thing   **PROX2**=α.3SGS$_P$-COP.ND   flute
      'this thing right here is a flute.'

(613) *yèyáretamènd*   *fèyo*   *kètándmè*
      y-yárè-ta-m-nd   fèyo   kètándmè
      α.3SGP-hear-IPFV.ND-REM.DLT-2|3NSGA   then   from.where
      syèmormèn.   (ND13)
      s-y-mo-èrmèn
      PROX2=α.3SGS$_P$-COP.ND-REM.DLT.PA
      'they heard her then from right where she was.'

A likely origin for *s=* is the immediacy adverb *kès* 'just here/there', as shown in examples 591–594).

5.4.2 Mirative proclitic (MIR)

There is a proclitic *r=* that is sometimes added to the plural form of the nondual copula *em/emo* and the plural form of the verb 'go' *engèm*. Its usage is rare, and its exact meaning is unclear, but it appears to indicate an unusally large number or distance, or something that is extraordinary in another way. Thus it is labelled as mirative (MIR). (An alternative analyis of *r=* is given at the end of section 6.9.)

218 —— 5 Additional word classes and phrase structure

(614) *Ár   kèghátan   rem.*   (NB10:42)
ár   kèghát-an   r=e-m
man   bush-LOC2   **MIR**=ɑ.2|3NSGS_P-COP.ND
'Very many men are in the bush.'

(615) *tárfár   tè   náyátongi,*   *fèyotaro*
tárfár   tè   n-áyáto-Ø-ang-i   fèyotaro
plenty   PRF   ɑ.ØP-end.up-PFV.ND-INC-2|3PLS_A   enough
rem.   (MK13(1))
r=e-m
MIR=ɑ.2|3NSGS_P-COP.ND
'they ended up to be plenty, more than enough.'

(616) *súmb   kès   rengèm.*   (KS11)
súmb   kès   r=e-ngèm
pool   just.there   **MIR**=ɑ.2|3NSGS_P-go.ND
'the pools are far away.'

(617) *yau   tèfene   wèrár   yène   mèndefè   remo.*   (MK13(2))
yau   tèfene   wèrár   yène   mènde-fè   r=e-mo
NEG   1NSG.GEN   wallaby   DEM   liking-COM   **MIR**=-ɑ.2|3NSGS_P-COP.ND
'our wallabies (unexpectedly) didn't taste good.'

5.5 Verb phrase structure

The verb phrase (VP) is made up of an obligatory inflected verb and optional TAM
markers and adverbs (including nominals functioning as adverbs). These optional
constituents normally occur before the verb, but there are exceptions, especially
for nominal constituents, as in the following:

(618) *nèngangotam*   *mèngot.*   (MD13(1))
n-ng-ango-ta-m-Ø   mèngo-t
ɑ.ØP-AND-return-IPFV.ND-REM.DLT-2|3SGS_A   house-ALL
'he returned home.'

(619)   *sèkw*   *namghetam*   *yabun*

           sèkw   n-amghè-ta-m-Ø   yabun

           canoe   α.ØP-go.alongside-IPFV.ND-REM.DLT-2|3SGS$_A$ big

           wènfaf.   (YD13(3))

           wèn-faf

           tree-ASS

           'the canoe went alongside a big place with trees.'

(620)   *efe*   *fronde*   *nufaryèng*   *Ghèrayè*

           efe   fronde   n-ufar-Ø-ay-ng   Ghèrayè

           who.ABS   first   α.ØP-arrive-PFV.ND-REM.PUNC-3SGS$_A$ Daraia

           mèngotuan . . .   (SK11(2))

           mèngotu-an

           village-LOC1

           'whoever first arrived at Daraia village . . .'

6 Simple sentences

This chapter describes the structure of simple sentences in Nama (those with only one clause) and the four types of simple sentences: declarative, interrogative, imperative and exclamative. It also covers minor sentences (those without a verb phrase), the use of discourse markers and morphological discord in simple sentences.

6.1 Simple sentence structure

With the exception of minor sentences (section 6.6), the basic simple sentence in Nama is made up of a single clause with an obligatory verb phrase (which, by definition, contains an inflected verb). As outlined in chapter 1, a sentence with a transitive verb has two underlying core arguments, the A argument and the P argument, and one or both of these can be overtly specified by a nominal phrase. A sentence with an intransitive verb has one underlying core argument, the S argument, which may or may not be overtly specified by a nominal phrase.

Examples in earlier chapters have shown that it is very common, especially in conversation and story telling, not to specify the arguments for both transitive and intransitive verbs, as in these additional examples:

(621) *sèno tè   enwanangi.*   (PT:97)
  sèno tè   e-n-wanè-Ø-ang-i
  POSB PRF a.2|3NSGP-VEN-take-PFV.ND-INC-2|3PLA
  'maybe they already brought them.'

(622) *wèikor tándái   akwan   nènangoi.*   (MD13(1))
  wèikor tándái   akw-an   n-n-ango-Ø-ay-Ø
  again   very.early morning-LOC2 a.ØP-VEN-return-PFV.ND-REM.PUNC-2|3SGS$_A$
  'in the early morning he returned again.'

As mentioned in section 1.4.2, when arguments are overtly specified, the basic order of constituents is APV for a sentence with a transitive verb and SV for a sentence with an intransitive verb – e.g.:

(623) *ágham   wèrár   yifnamènd.*   (MD93)
  ághè-am wèrár   y-wifár-ta-m-nd
  dog-ERG   wallaby a.3SGP-chase-IPFV.ND-REM.DLT-2|3NSGA
  'the dogs chased the wallaby.'

https://doi.org/10.1515/9783111077017-006

(624) *Yènè árom yènd wawèrsote.* (NB4:41)
yènè ár-o-m yènd w-wawèrso-ta-e
DEM man-SG-ERG 1.ABS α.1SGP-trouble-IPFV.ND-2|3SGA
'That man is troubling me.'

(625) *nu nèfnoi.* (SK11(2))
nu n-èfno-Ø-ay-Ø
water α.ØP-spurt.out-PFV.ND-REM.PUNC-2|3SGS_A
'water spurted out.'

However, there are many exceptions with transitive verbs (but not intransitive verbs) showing different ordering – for example:

(626) *yènè fèlismanam fyárametat*
yènè fèlisman-am f=y-wa-ramè-ta-t
DEM policeman-ERG PROX1=α.3SGP-APP-give-IPFV.ND-2|3NSGA
rokár-rokár. (PT:357)
rokár~PL
things
'That policemen gave him things.'

(627) *fèyo yènfetam afam.* (MD93)
fèyo y-nèfè-ta-m-Ø afè-am
then α.3SGP-cut-IPFV.ND-REM.DLT-2|3SGA father-ERG
'then father cut it.'

(628) *wèikor tinjongai mèrès kètan.* (MD13(1))
wèikor t-winjo-Ø-ang-ay-Ø mèrès kètan
again β.3SGP-see-PFV.ND-INC-REM.PUNC-2|3SGA girl there
'again he saw the girl there.'

PAV ordering generally occurs in a particular type of construction, described in the following subsection (6.1.1). There is also phrasal coordination of nominal phrases, as described in section 6.1.2.

6.1.1 P-focussed constructions

In Nama, there is a distinct construction for simple transitive sentences concerning the experiencing of particular physiological processes (such as sweating), sensations

222 ——— 6 Simple sentences

(such as feeling hungry) and emotions (such as being angry). In such constructions, the experiencer is referenced by the P argument and the stimulus – a process, sensation or emotion (or its cause) – by the A argument. Here, for reasons given below, these are called P-focussed constructions.[41] They occur most frequently with a small set of verbs, including \fing/ 'affect'; \ramè/ 'make, give' (but in this context meaning 'affect') and \wifo/ 'finish'. In such constructions, there is often no overt P argument – for example:

(629) *yeam   wèfinde.*   (B&B:77)
　　　ye-am　　 w-fing-ta-e
　　　crying-ERG　α.1SGP-affect-IPFV.ND-2|3SGA
　　　'I feel like crying.' (lit. 'crying is affecting me.')

(630) *Nuam   wèramete.*   (NB1:65)
　　　nu-am　　 w-ramè-ta-e
　　　water-ERG　α.1SGP-affect-IPFV.ND-2|3SGA
　　　'I'm thirsty.' (lit. 'water is affecting me.')

(631) *nakwam   yifoi*   *minde.*   (MD99(2))
　　　nakw-am　　y-wifo-Ø-ay-Ø   minde
　　　anger-ERG　α.3SGP-finish-PFV.ND-REM.PUNC-2|3SGA   really
　　　'he got really angry.' (lit. 'anger really finished him.')

When there is an overt P argument in this construction, it almost always precedes rather than follows the A argument. This appears to signal that the affectedness of the semantic patient is more important than any action or force of the semantic agent, and therefore the focus is on the P argument, rather than on the A argument. Some examples follow:

(632) *Fá   siogham   yèfinde.*   (NB1:26)
　　　fá　　 siogh-am　　y-fing-ta-e
　　　3.ABS　 hunger-ERG　α.3SGP-affect-IPFV.ND-2|3SGA
　　　'He is hungry.'

(633) *Fá   sakram   efingwè.*   (NB1:26)
　　　fá　　 sakèr-am　　e-fing-Ø-wè-Ø
　　　3.ABS　 tiredness-ERG　α.2|3NSGP-affect-IPFV.DU-DA-2|3SGA
　　　'They (2) are tired.'

---

41 They are called "experiencer object constructions" by Pawley, Gi, Majnep and Kias (2000:154).

6.1 Simple sentence structure — **223**

(634) *Yènd   neam   wèfinde.*   (NB1:65)
yènd   ne-am   w-fing-ta-e
1.ABS   shit-ERG   α.1SGP-affect-IPFV.ND-2|3SGA
'I have to shit.'

(635) *tane   tikèf   áuwègham   tèramete.*   (MM13)
tane   tikèf   áuwè-gh-am   t-ramè-ta-e
1SG-GEN   heart   be.happy-NOM-ERG   β.3SGP-affect-IPFV.ND-2|3SGA
'my heart was happy.'

(636) *mèrès   ghèrghèrsisam   yifoi.*   (MD13(1))
mèrès   ghèrghèrsisè-am   y-wifo-Ø-ay-Ø
girl   feeling.sorry-ERG   α.3SGP-finish-PFV.ND-REM.PUNC-2|3SGA
'the girl was feeling sorry for him.'

P-focussed constructions – with animate P arguments preceding inanimate A arguments – also occur with other verbs:

(637) *Fá   njamkeam   tè   yèremngo.*   (NB8:79)
fá   njamke-am   tè   y-remngo-Ø-Ø
3.ABS   food-ERG   PRF   α.3SGP-satisfy-PFV.ND-2|3SGA
'He was satified with the food.' (lit. 'The food satisfied him.')

(638) *Yènd   wemam   wafo.*   (NB2:69)
yènd   wem-am   w-wafo-Ø-Ø
1.ABS   yam-ERG   α.1SGP-choke-PFV.ND-2|3SGA
'I'm choking on the yam.' (lit. 'The yam choked me.')

(639) *Yu   yèsambnetam   sokreream.*   (NB4:53)
yu   y-sambnè-ta-m-Ø   sokrere-am
place   α.3SGP-shake-IPFV.ND-REM.DLT-3SGA   earthquake-ERG
'There was an earthquake.' (lit. 'An earthquake shook the place.')

A subset of P-focussed constructions involve a small subclass of transitive verbs: agentless transitive verbs. As the name indicates, these verbs never have a specified overt A argument, but they take the verb morphology (A-indexing suffixes) of a 2nd or 3rd person singular agent. Semantically, agentless intransitives are like intransitives with the P argument having the role of subject. Agentless transitive verbs include \fèrnè/ 'float', \ramngo/ 'die', \wawaling/ 'forget', \wawena/ 'cough' , \wáwálimán/ 'go in a large group' and \yo/ 'travel'. Some examples:

224 —— 6 Simple sentences

(640) *Yènd* **wèfèrnete.** (NB3:18)
yènd w-fèrnè-ta-e
1.ABS α.1sGP-float-IPFV.ND-2|3sGA
'I'm floating.'

(641) *Yènd náifè rèsaghèt tè* **kwawalinde.** (NB5:43)
yènd náifè rèsa-gh-t tè kw-wawaling-ta-e
1.ABS knife carry-NOM-ALL PRF β.1sGP-forget-IPFV.ND-2|3sGA
'I forgot to bring the knife.'

(642) *Fá* **engwáwálimndam**
fá e-ng-wáwáliman-ta-m-Ø
3.ABS α.2|3NSGP-go.in.large.group-IPFV.ND-REM.DLT-2|3sGA
Darut. (PC/MD:1.5.19)
Daru-t
Daru-ALL
'They went to Daru in a large group.'

(643) *kuwanongèm* **yènyote**
k-uwano-Ø-ang-m yèn-yo-ta-e
β.ØP-set.off-PFV.ND-INC-1NSGS$_A$ α.1NSGP-travel-IPFV.ND-2|3sGA
yènyote. . . (YD11(3))
yèn-yo-ta-e
α.1NSGP-travel-IPFV.ND-2|3sGA
'we set off and we travelled and travelled.'

6.1.2 Phrasal coordination

As a result of coordination, a simple sentence in Nama can have an argument with
more than one nominal phrase. The conjunctive coordinating conjunction is *a* 'and':

(644) *yènd* **a** *tanjo amaf susiot nuwaneayèm.* (TE11(2))
yènd a tanjo amaf susi-ot n-uwano-e-ay-m
1.ABS **and** 1sG.own wife fishing-ALL α.ØP-set.off-PFV.DU-REM.PUNC-1NSGS$_A$
'me and my wife, we went fishing.'

6.1 Simple sentence structure — **225**

(645) *námb* **a** *besi* *ewaneayèm.*   (MD93)
 námb **a** besi e-wanè-e-ay-m
 bow **and** arrow α.2|3NSGP-take-PFV.DU-REM.PUNC-1NSGA
 'we (2) took bows and arrows.'

The alternative coordinating conjunction is *o* 'or':

(646) *kètè* *ausè-ausè* **o** *efero* *támorwèn*
 kètè ausèausè **o** efe-ro tá-mo-èrwèn
 there old.women **or** who-RSTR β.2|3NSGSp-COP.ND-REM.DUR.PA
 waghan.   (ND13)
 wagh-an
 singsing-LOC2
 'old women or whoever were there at the singsing.'

(647) *nènamèn* *kèr* *tèmong* –   *enjnemèn* **o**
(=408b) nènamèn kèr t-mo-ang   enjne-mèn **o**
 why dead β.2|3NSGSp-COP.ND-INC sickness-ORIG **or**
 ngimèn.   (MD99(1))
 ngi-mèn
 killing-ORIG
 'why she died – from sickness or from murder.'

Note that structurally, Nama often uses nominal morphology in place of what is done with phrasal coordination – for example:

(648) *Tonifem* *Benfè* *wèrar* *yènfèt.*   (NB10:5)
 Toni-f-e-m Ben-fè wèrar y-nèfè-Ø-t
 Tony-NSG-DAT-ERG Ben-COM wallaby α.3SGP-CUT-IPFV.DU-2|3NSGA
 'Tony and Ben are cutting the wallaby.' (lit. 'Tony with Ben. . .')

Here the A argument is interpreted as being dual, just as the argument with phrasal coordination in (644). This is an example of an "inclusory construction" (Lichtenberk 2000), as described in detail for Komnzo (Döhler 2018: 276–293).

6.1.3 Multi-verb construction

Two inflected verbs may occur together in a single clause without any coordinating conjunction to describe a single event. The first verb is generally the P-aligned verb meaning 'go'; the second verb can be preceded by an overt P argument – for example:

(649) *fá   yèngèm   mbákw enèsmete.*   (MK13(1))
fá   y-ngèm   mbákw e-n-sèmè-ta-e
3.ABS α.3SGSₚ-go.ND bark   α.2|3NSGP-VEN-strike-IPFV.ND-2|3SGA
'she's going to remove the bark (from the trees).'

(650) *yènd   wèngmormèng   fainár   tèfrárèn.*   (TE11(1))
yènd   w-ngèmo-èrmèng   fainár   t-frár-Ø-èn
1.ABS α.1SGSₚ-go.ND-REM.DUR.PA pineapple α.3SGP-pick-PFV.ND-1SGA
'I went and picked a pineapple.'

6.2 Declarative sentences

Declarative sentences, or statements, are the most common type of sentence in Nama. The vast majority of examples given so far are declarative sentences.

6.3 Negative sentences

Negative sentences in Nama contain the negative marker *yau*. Several examples occur in preceding chapters, such as (520), (525), (554) and (600) in chapter 5. Some additional examples are:

(651) *afè-amafem   yau kwáutanawèt.*   (FD13)
afè-amè-af-e-m   **yau** kw-wáutár-ta-w-t
father-mother-PL-DAT-ERG   **NEG** β.1SGP-help-IPFV.ND-REM.DUR-2|3NSGA
'the parents weren't helping me.'

(652) *tèfene   mèrèn kènjún mwighè yau*
tèfene   mèrèn kènjún mwighè **yau**
3NSG.GEN family inside thought **NEG**
kafrendau.   (MD99(1))
k-afreng-ta-w-Ø
β.ØP-be.proper-IPFV.ND-REM.DUR-2|3SGSₐ
'inside our family this way of thinking was not proper.'

(653) *yèndfem* *Nèmè* *si* **yau** *so* *nèmdetam.*   (B&B:114)
yèndfem Nèmè si **yau** so n-mèndè-ta-m
1NSG.ERG Nama language **NEG** FUT α.2SGP-talk.to-IPFV.ND-1NSGS$_A$
'we won't talk in Nama language to you.'

In a negative sentence, the negative marker does not always precede the verb and its modals – for example:

(654) *yènd* **yau** *wènde* *nèkw* *kwèmorwèn.*   (FD13)
yènd **yau** wènde nèkw kw-mo-èrwèn
1.ABS **NEG** but anger β.1SGS$_P$-COP.ND-REM.DUR.PA
'but I wasn't angry.'

(655) *yèndo* *yènè* *ghèrsè* *kènjún* **yau** *yènè* *rokár*
yèndo yènè ghèrsè kènjún **yau** yènè rokár
1SG.ERG DEM life inside **NEG** DEM thing
yinjoyèn.   (MD13(1))
y-winjo-Ø-ay-èn
α.3SGP-see-PFV.ND-REM.PUNC-1SGA
'in my life, I haven't seen this thing.'

6.4 Interrogative sentences

Interrogative sentences are of two types: yes/no questions and question word (or wh) questions.

6.4.1 Yes/no questions

Yes/no questions are generally indicated by rising intonation rather than any morphological marking or change of word order – for example:

(656) *yèna* *so* *yènkmare?*   (TE11(1)
yèna so yèn-kèmè-ar-e
here FUT α.1NSGS$_P$-be.lying.down-DU.PA-PFV.DU
'will we (2) sleep here?'

228 — 6 Simple sentences

(657) *Nu   yèkw   nefo?*   (DC11)
nu   yèkw   n-efo-Ø-Ø
water   hole   α.ØP-finish-PFV.ND-2|3SGS$_A$
'Is the water well finished?'

(658) *Marim   ámb   so   kèmaramangè?*   (DC13)
Mari-m   ámb   so   kèn-wa-ramè-Ø-ang-è
Murry-ERG   something   FUT   β.2SGP-APP-give-PFV.ND-INC-2|3sgA
'Will Murry give something to you?'

Yes/no questions can a be negative:

(659) *Yárafè   yau   syèm?*   (DC14)
yá-rafè   yau   s=y-m
3SG-father   NEG   PROX2=α.3SGS$_P$-COP.ND
'Isn't his father here?'

(660) *Wèt   yèmo   yau   nèmáutane   Robatèm?*   (DC12)
wèt   yèmo   yau   nèn-wáutár-ta-e   Robat-m
so   3SG.ERG   NEG   α.2SGP-help-IPFV.ND-2|3SGA   Robert-ERG
'So Robert isn't helping you?'

Yes/no questions can also end with *o yau* 'or not', or just with *o*:

(661) *mè   enetat*   *o   yau?*   (PT:245)
mè   e-ne-ta-t   o   yau
CONT   α.2|3NSGP-eat-IPFV.ND-2|3NSGA   or   NEG
'are they still eating them or not?'

(662) *Fèm   tè   násáfote*   *o   yau?*   (DC27)
fèm   tè   n-ásáfo-ta-e   o   yau
2.ABS   PRF   α.ØP-work-IPFV.ND-2|3SGS$_A$ or   NEG
'Have you finished working or not?'

(663) *wem   ámb   yau   so   eitati*   *o?* (DC24)
wem   ámb   yau   so   e-y-ta-t-i   o
yam   some   NEG   FUT   α.2|3NSGP-plant-IPFV.ND-2|3NSGA-2PL   or
'will you all plant more yams or. . .?'

6.4 Interrogative sentences —— **229**

(664) *yúnjèfan*   *wèi*   *ndenè*   *tè*   *nafalati*
      yúnjèf-an   wèi   ndenè   tè   n-afala-ta-t-i
      season-LOC2   now   like.this PRF   α.ØP-warm.up-IPFV.ND-2|3NSGS$_A$-2PL
      táf   *o?*   (DC15)
      táf   o
      this.time **or**
      'is it the season now that you're already warmed up, or. . .?'

The discourse particle *si* can also indicate a yes-no question (see section 6.8).

6.4.2 Question word questions

Interrogative sentences with question words are divided into two types, depending
of the type of question words: interrogative pronouns or interrogative adverbs.

Interrogative pronouns
Interrogative pronouns are question words that take the place of nominals. The
first three correspond to the three types of pronouns described in chapter 3 and in
section 5.1.2: *emo* 'who (ergative)', *efe* 'who/whom (absolutive)' and *ende* 'to whom,
for whom (dative)'.

(665) **emo**   *wèláite?*   (MD13(1))
      emo   w-lái-ta-e
      who.ERG α.1SGP-shoot.at-IPFV.ND-2|3SGA
      'who was shooting at me?'

(666) **efe**   *fronde*   *kembnean*   *yèm?*   (DC18)
(=484) **efe**   fronde   kembne-an   y-m
       who.ABS first   game-LOC2   α.3SGS$_P$-COP.ND
       'who was first in the game?'

(667) *Marim*   **ende**   *náifè*   *yaram?*   (NB2:43)
      Mari-m   **ende**   náifè   y-wa-ramè-Ø-Ø
      Murry-ERG   **who.DAT** knife   α.3SGP-APP-give-PFV.ND-2|3SGA
      'Who did Murry give the knife to?'

Like other pronouns, these question words (except for *efe*) can take nominal morphology for different functions and even plural marking – for example:

230 — 6 Simple sentences

(668) ***emo-fè*** *fenèm.* (MD99(1))
emo-fè f=e-nèm
who.ERG-COM PROX1=α.2SGSₚ-come.ND
'who did you come with?'

(669) ***Endeot*** *yèramete?* (DC13)
ende-ot y-ramè-ta-e
who.DAT-PURP α.3SGP-give-IPFV.ND-2|3SGA
'Who are you making it for?'

(670) ***Endene*** *wem em?* (DC12)
ende-ne wem e-m
who.DAT-GEN yam α.2|3NSGSₚ-COP.ND
'Whose yams are they?'

(671) ***Endefene*** *njamke?*[42] (DC26)
ende-f-e-ne njamke
who.DAT-PL-DAT-GEN food
'Who all's food?'

There are also two additional interrogative pronouns taking the place of inanimate nominals and modifying nominals: *nèmè* 'what' and *árkè* 'which':

(672) ***Nèmè*** *yafrote?* (DC13)
nèmè y-wafro-ta-e
what α.3SGP-do-IPFV.ND-2|3SGA
'What are you doing?'

(673) ***Árkè*** *kafusi mènde nèm?* (NB3:21)
árkè kafusi mènde n-m
which cup liking α.2SGSₚ-COP.ND
'Which cup do you like?'

The interrogative pronoun *nèmè* 'what' can take nominal morphology to express different meanings, including forming adverbial question words for 'why' and 'when':

---

42 This is actually a minor sentence (see section 6.6)

(674) **Nèmamèn** *fèmo* *skul* *yèwáro?* (FD13)
nèmè-mèn fèmo skul y-wáro-Ø-Ø
what-ORIG 2SG.ERG school α.3SGP-set.on.fire-PFV.ND-2|3SGA
'Why did you set the school on fire?'

(675) **Nèmamèn** *yènè* *yáf* *em?* (DC10)
nèmè-mèn yènè yáf e-m
what-ORIG DEM basket α.2|3NSGS$_P$-COP.ND
'what are these baskets made of?'

(676) **Nèmat** *yènè* *yu* *yèrtete?* (DC24)
nèmè-t yènè yu y-rètè-ta-e
what-ALL DEM place α.3SGP-clear-IPFV.ND-2|3SGA
'Why are you clearing this place?'

(677) **Nèmayan** *so* *engèsmete?* (DC20)
nèmè-yan so e-ng-sèmè-ta-e
what-LOC2 FUT α.2|3NSGP-AND-kill-IPFV.ND-2|3SGA
'When will you kill them?'

Again, it is not clear whether these forms are the result of historical or productive affixation.

The interrogative pronouns *árke* 'which' can also have limited nominal morphology:

(678) **Árkamè** *enèrse?* (DC21)
árkè-mè e-n-rèsa-ta-e
which-PERL α.2|3NSGP-VEN-carry-IPFV.ND-2|3SGA
'Along where did you carry them?'

Note that some of the interrogative pronouns can also function as indefinite pronouns. These are *emo* 'someone, whoever', *efe* 'someone, whoever/whomever' and *nème* 'whatever'. An example can be seen in (646); another is:

232 — 6 Simple sentences

(679) mer   efogh   o   mer   sèite   o   **nèmè**   efogh
       mer   efogh   o   mer   sèite   o   **nèmè**   efogh
       good day   or   good afternoon or   **what**   time
       mètè yèngèm   feyot.   (FD13)
       mètè y-ngèm   feyot
       POT1 α.3sGS$_P$-go.ND 2SG.PURP
       'good day or good afternoon or whatever time might suit you.'

Further examples are given in section 7.4.3, where interrogative pronouns are used to introduce clausal complements.

In addition, *emo yau* and *efe yau* both mean 'no one, nobody':

(680) yènd **emo yau** so   kwitrongè.   (TE13)
       yènd **emo yau** so   kw-witro-Ø-ang-è
       1.ABS **who NEG** FUR β.1SGP-move-PFV.ND-INC-2|3SGA
       'no one will move me.'

(681) **efe   yau** nutne.   (ND13)
       efe   yau n-utár-ta-e
       who   NEG α.ØP-call.out-IPFV.ND-2|3SGS$_A$
       'nobody shouted.'

The expressions *emo ámb yau* and *efe ámb yau* mean 'no one/nobody else' and *nèmè ámb yau* 'nothing else':

(682) **emo   ámb   yau** ngaram-ngaram tèrametau.   (Mark 5:4)
       emo   ámb   yau ngaramngaram   t-ramè-ta-w-Ø
       who   other NEG calmness   β.3SGP-give-IPFV.ND-REM.DUR-2|3SGA
       'no one else could subdue him.'

(683) nu   kènjún **nèmè ámb   yau** yinjoi.   (MD13(1))
       nu   kènjún **nèmè ámb   yau** y-winjo-Ø-ay-Ø
       water inside **what other NEG** α.3SGP-see-PFV.ND-REM.PUNC-2|3SGA
       'he saw nothing else in the water.'

Interrogative adverbs

Interrogative adverbs are question words that take the place of adverbs indicating for example, time and manner. They include *njam* 'when', *ánde* 'where', *nderè* 'how', *ndernae* 'how (in what way)' and *njenamb* 'how many'. Some examples:

(684) **Njam** *yègh* *ewafngoi?* (DC12)
njam y-gh e-wafngo-Ø-ay-Ø
when plant-NOM α.2|3NSGP-start-PFV.ND-REM.PUNC-2|3SGA
'When did you start planting them?'

(685) **Ánde** *kakanè* *yèm?* (DC26)
ánde kakanè y-m
where big.brother α.3SGSP-COP.ND
'Where is big brother?

(686) **Nderè** *teke* *so* *eramete?* (DC10)
nderè tek-e so e-ramè-ta-e
how long.time-INST FUT α.2|3NSGP-make-IPFV.ND-2|3SGA
'How much time will it take to make them?'

(687) **Nderè** *yabun* *yèm?* (NB3:23)
nderè yabun y-m
how big α.3SGSP-COP.ND
'How big is it?

(688) **Ndernae** *yèna* *wèi* *kufanat?* (PT:592)
ndernae *yèna* *wèi* k-ufar-ta-t
how here now β.ØP-arrive-IPFV.ND-2|3NSGSA
'How did they arrive here now?"

(689) **Njènamb** *efogh* *nèngèm?* (DC07)
njènamb efogh n-ngèm
how.many day α.2SGSP-go.ND
'How many days are you going (for)?'

(690) **Nderèmènè** *nu* *yèm?*[43] (NB3:24)
nderèmènè nu y-m
how.much water α.3SGSP-COP.ND
'How much water is there?'

---

43 This form, a combination of *nderè* 'how' and *mènè* 'thing', does not occur in the naturalistic recordings.

234 —— 6 Simple sentences

There is also a comparatively rare alternative for *ánde* 'where' – that is *ándan* (which could be *ándè* plus the locative case marker *-n* or *-an*):

(691)   *Fèm*   **ándan**   *nuwano?*   (DC11)
           fèm   **ándan**   n-uwano-Ø-Ø
           2.ABS   **where**   α.ØP-set.off-PFV.ND-2|3SGS$_A$
           'Where are you setting off to?'

6.5 Imperative sentences

Imperatives in Nama make use of a large proportion of the verb morphology described in chapter 4, leading to a plethora of forms. (There are no dedicated imperative morphemes.) First, there are two types of imperatives: immediate and future. Verbs in both of these types differ in form according to the number of addressees, aspect (perfective vs imperfective), the transitivity of the verb, and (for transitive verbs) the number of patients or beneficiaries affected by the action. Each of these can also be marked for whether the commanded action is towards or near the speaker, or away or far from the speaker, and each can have benefactive forms for actions being done for someone. There are also third person imperatives.

6.5.1 Immediate imperatives

Immediate imperatives involve someone telling an addressee or addressees to do something then and there. In Nama, with only a few exceptions, it is only transitive and A-aligned intransitive verbs that have these imperative forms. As in other languages, the 2nd person A or S$_A$ argument referring to the addressee(s) (e.g. 'you' in English) is not normally expressed overtly in imperative sentences. But in Nama it is still indexed on the verb with one of the usual 2/3 person A-/S$_A$-indexing suffixes. In the case of imperatives, this suffix indicates 2nd person and number: -Ø singular, -*i* plural (i.e. 3+) (with some exceptions, see below) or -*nd* dual (2). Here are some examples:

(692)   a.   *wèrár*   *terang.*   (NB5:6)
               wèrár   t-werè-Ø-ang-Ø
               wallaby   β.3SGP-hold-PFV.ND-INC-2|3SGA
               '(you SG) hold the wallaby.'

6.5 Imperative sentences ▬▬ **235**

    b. *wèrár   terangi.*   (NB5:6)
       wèrár   t-werè-Ø-ang-i
       wallaby β.3sGP-hold-PFV.ND-INC-2|3PLA
       '(you PL) hold the wallaby.'

    c. *wèrár   terend.*   (NB5:6)
       wèrár   t-werè-e-nd
       wallaby β.2sGP-hold-PFV.DU-2|3DUA
       '(you DU) hold the wallaby.'

The imperatives in these examples are perfective, using the recent inceptive tense (section 4.8.1). As they emphasise the start of an action; better translations in the examples would be something like 'start holding the wallaby'. The distinction between punctual and durative found in the perfective inceptive tenses with singular A or S_A arguments is also found in imperatives for some verbs, where the absence of the inceptive suffix -*ang* indicates a more punctual action. For example, compare the following to (692a) above:

(693) *wèrár   terè.*   (NB5:6)
      wèrár   t-werè-Ø-Ø
      wallaby β.2sGP-hold-PFV.ND-2|3sGA
      '(you SG) grab the wallaby.'

Here is another pair of examples comparing punctual and durative:

(694) a. *ghi   taufè.*   (NB5:7)
       ghi   t-waufè-Ø-Ø
       bark,torch β.3sGP-blow-PFV.ND-2|3PLA
       '(you SG) blow on the bark torch (e.g. once quickly to extinguish the flame).'

    b. *ghi   taufang.*   (NB5:7)
       ghi   t-waufè-Ø-ang-Ø
       bark,torch β.3sGP-blow-PFV.ND-INC-2|3PLA
       'you (SG) blow on the bark torch (e.g. continuously to increase the flame).'

This distinction does not occur for plural perfective imperatives (i.e. those with three or more addressees), for which the inceptive marker -*ang* always follows the verb stem, followed by the perfective plural A/S_A-indexing suffix -*i*, as in (692b). It also does not occur for dual perfective imperatives (those with 2 addressees), for which the dual perfective marker -*e* rather than the inceptive marker follows the verb stem, followed by the dual perfective A/S_A-indexing suffix -*nd* as in (692c).

236 —— 6 Simple sentences

Immediate imperatives can also be imperfective. While the perfective imperatives emphasise the start or the punctuality of the action, imperfective imperatives emphasise the continuity of the action. These examples compare perfective (695a) and imperfective (695b) imperatives:

(695) a. *sáláme   tumbé.*   (NB5:9)
          sáláme   t-wumbé-Ø-Ø
          cloth   β.3SGP-tie-PFV.ND-2|3SGA.IRR
          'start tying the cloth.'
       b. *sáláme   tumbeta.*   (NB5:9)
          sáláme   t-wumbè-ta-Ø
          cloth   β.3SGP-tie-IPFV.ND-2|3SGA.IRR
          'tie up the cloth.'

Example (695b) shows that for singular imperfective imperatives, the verb is in the imperfective recent tense with irrealis aspect (section 4.7.1). However, for plural and dual imperfective imperatives, the verb is in the inceptive imperfective aspect (section 4.9), with the suffix *-tang* followed by the plural A/S$_A$-indexing suffix *-i* for plural imperatives (again, with some exceptions), and the dual perfective marker *-e* followed by the dual A/S$_A$-indexing suffix *-nd* for dual imperatives. This is shown in (696b) and (697b), compared to the perfective in (696a) and (697a).

(696) a. *sáláme   tumbangi.*   (NB5:9)
          sáláme   t-wumbè-Ø-ang-i
          cloth   β.3SGP-tie-PFV.ND-INC-2|3PLA
          '(you PL) start tying the cloth.'
       b. *sáláme   tumbetangi.*   (NB5:9)
          sáláme   t-wumbè-tang-i
          cloth   β.3SGP-tie-INC.IPFV-2|3PLA
          '(you PL) tie up the cloth.'

(697) a. *sáláme   tumbend.*   (NB5:9)
          sáláme   t-wumbè-e-nd
          cloth   β.3SGP-tie-PFV.DU-2|3DUA
          '(you DU) start tying the cloth.'
       b. *sáláme   tumbetangend.*   (NB3:4)
          sáláme   t-wumbè-tang-e-nd
          cloth   β.3SGP-tie-INC.IPFV-PFV.DU-2|3DUA
          '(you DU) tie up the cloth.'

As seen in the above examples, prefixes for immediate imperatives, both perfecive and imperfective, come from the β set. All the preceding examples have transitive verbs. For A-aligned intransitive verbs, the prefix is ØP *k-*, as shown here with the antipassive verb *atar/* 'dig', perfective in (698a) and imperfective in (698b).

(698) a. *katar.*   (NB5:6)
        k-atar-Ø-Ø
        β.ØP-dig-PFV.ND-2|3SGS$_A$
        '(you SG) start digging.'
    b. *katanangi.*   (NB5:6)
        k-atar-tang-i
        β.ØP-dig-INC.IPFV-2|3PLS$_A$
        '(you PL) keep on digging.'

An example of an imperfective imperative from the naturalistic data:

(699) *ei*   *mèinyotyo*   *ghakèr-ghakèr*   *kuflitangi.*   (YD11(3))
      ei   mèinyotyo   ghakèr~ghakèr   k-ufèli-tang-i
      hey   all   male~PL   β.ØP-get.in-INC.IPFV-2|3PLS$_A$
      'hey, all you men, jump in.'

For transitive verbs, the prefixes, as usual, indicate the person and number (singular vs nonsingular) of the P argument. All the preceding examples of transitive verbs had singular P arguments. Following are examples of imperatives, comparing those with singular P arguments to those with nonsingular P arguments. All have 2nd person singular addressees. Example (700) shows perfective imperatives and (701) imperfective imperatives.

(700) a. *wèrár*   *tènfè.*   (NB6:48)
        wèrár   t-nèfè-Ø-Ø
        wallaby β.3SGP-cut-PFV.ND-2|3SGA
        '(you SG) start cutting up the wallaby.'
    b. *wèrár*   *tánfè.*   (NB6:48)
        wèrár   tá-nèfè-Ø-Ø
        wallaby β.2|3NSGP-cut-PFV.ND-2|3SGA
        '(you SG) start cutting up the wallabies (3+).'

238 —— 6 Simple sentences

(701) a. *wèrár   tènfeta.*   (NB6:48)
      wèrár   t-nèfè-ta-Ø
      wallaby   β.3sGP-cut-IPFV.ND-2|3sGA.IRR
      '(you SG) cut up the wallaby.'

    b. *wèrár   tánfeta.*   (NB6:48)
      wèrár   tá-nèfè-ta-Ø
      wallaby   β.2|3NSGP-cut-IPFV.ND-2|3sGA.IRR
      '(you SG) cut up the wallabies (3+).'

Note that the nondual perfective/imperfective suffixes give the reading of a plural
(3+) P argument rather than dual. (See below for how dual P arguments as opposed
to plural are indicated.) Similarly, the following examples show plural imperatives
with singular P arguments (in a.) and nonsingular P arguments (in b.):

(702) a. *yèkw   tètarngi.*   (NB7:20)
      yèkw   t-tar-Ø-ang-i
      hole   β.3sGP-dig-PFV.ND-INC-2|3PLA
      '(you PL) start digging a hole.'

    b. *yèkw   tatarngi.*   (NB7:20)
      yèkw   ta-tar-Ø-ang-i
      hole   β.2|3NSGP-dig-PFV.ND-INC-2|3PLA
      '(you PL) start digging holes (3+).'

(703) a. *yèkw   tètanangi.*   (NB7:20)
      yèkw   t-tar-tang-i
      hole   β.3sGP-dig-INC.IPFV-2|3PLA
      '(you PL) dig the hole.'

    b. *yèkw   tatanangi.*   (NB7:20)
      yèkw   ta-tar-tang-i
      hole   β.2|3NSGP-dig-INC.IPFV-2|3PLA
      '(you PL) dig the holes (3+).'

Here are some examples from naturalistic recordings:

(704) *ághè   tawanmo.*   (DC25)
      ághè   ta-wanmo-Ø-Ø
      dog   β.2|3NSGP-call-PFV.ND-2|3sGA
      '(you SG) call the dogs.'

6.5 Imperative sentences — **239**

(705) *mè   tèrèmna.*   (B&B:172)
mè   t-rèman-ta-Ø
CONT β.3sGP-mix-IPFV.ND-2|3sGA.IRR
'(you sG) keep on mixing it.'

(706) *wèikor   tasamngongi.*   (DC11)
wèikor   t-wasamngo-Ø-ang-i
again   β.3sGP-extend.action-PFV.ND-INC-2|3PLA
'(you PL) continue doing it again.'

The prefix can also index non-third person P arguments, as in this example:

(707) *kwèyárangi.*   (B&B:249)
kw-yárè-Ø-ang-i
β.1sGP-hear-PFV.ND-INC-2|3PLA
'(you PL) listen to me.'

Now we will turn to imperatives where either the addressee or the P argument or
both are dual. Those with dual addressees (i.e. dual imperatives) that have a singular P argument are relatively straightforward in both perfective and imperfective
imperatives:

(708) a. *sáláme   tumbend.*   (NB5:9)
sáláme   t-wumbè-e-nd
cloth   β.3sGP-tie-PFV.DU-2|3DUA
'(you DU) start tying the cloth.'
b. *sáláme   tumbetangend.*   (NB7:22)
sáláme   t-wumbè-tang-e-nd
cloth   β.3sGP-tie-INC.IPFV-PFV.DU-2|3DUA
'(you DU) tie the cloth.'

For dual imperfective imperatives, the expected form is with the inceptive imperfective suffix -*tang*, as in (697b) and the following:

(709) *wèrár   tifnangend.*   (NB10:22)
wèrár   t-wifár-tang-e-nd
wallaby β.3sGP-chase-INC.IPFV-PFV.DU-2|3DUA
'(you DU) chase the wallaby.'

240 —— 6 Simple sentences

However, a different suffix *-wang* (2DU.IMP) is also often used with other verbs:[44]

(710) *áuyè* *tèfenwangend.*   (NB3:33)
   áuyè t-fenè-wang-e-nd
   cassowary β.3SGP-feed-2DU.IMP-PFV.DU-2|3DUA
   '(you DU) feed the cassowary.'

(711) *yèkw* *tètarwangend.*   (NB7:20)
   yèkw t-tar-wang-e-nd
   hole β.3SGP-dig-2DU.IMP-PFV.DU-2|3DUA
   '(you DU) dig the hole.'

Examples from naturalistic recordings follow with *-tang* (712) and with *-wang* (713):

(712) *tèfandangend.*   (DC02)
   t-fanda-tang-e-nd
   β.3SGP-look.at-INC.IPFV-PFV.DU-2|3DUA
   '(you DU) look at him.'

(713) *si* *taitowangend.*   (DC01)
   si t-waito-wang-e-nd
   talk β.3SGP-tell-2DU.IMP-PFV.DU-2|3DUA
   '(you DU) tell something.'

For singular imperatives with a dual P argument as opposed to plural one (3+), this is distinguished in the perfective by the presence of the dual perfective marker *-e* and in the imperfective by the Ø dual imperfective marker plus the dual argument (DA) marker *-wè*, as (714b):

(714) a. *wèrár* *tánfe.*   (NB6:48)
     wèrár tá-nèfè-e-Ø
     wallaby β.2|3NSGP-cut-PFV.DU-2|3SGA
     '(you SG) start cutting the wallabies (2).'
   b. *wèrár* *tánfèwè.*   (NB6:48)
     wèrár tá-nèfè-Ø-wè-Ø
     wallaby β.2|3NSGP-cut-IPFV.DU-DA-2|3SGA
     '(you SG) cut up the wallabies (2).'

---

44 It is possible that this suffix originates from the combination of the dual argument marker *-wè* and the inceptive suffix *-ang.*

6.5 Imperative sentences ── **241**

For dual and plural imperatives with a dual P argument, things are more complicated (as described for non-imperatives in sections 4.8.1 and 4.9). Dual imperatives
generally have the same the form for both a dual P argument and a plural P argument. In addition, this form is also used for plural imperatives with a dual P argument – using the dual A/S_A-indexing suffix -*nd* instead of the expected plural suffix
-*i*. Examples (715a) and (716a) illustrate this with perfective imperatives, and (715b)
and (716b) with imperfective ones:

(715) a. *sáláme   tawumbend.*   (NB5:9)
     sáláme   ta-wumbè-e-nd
     cloth   β.2|3NSGP-tie-PFV.DU-2|3DUA
     '(you DU) start tying the cloths (2 or 3+)
     or '(you PL) start tying the cloths (2).'
    b. *sáláme   tawumbetangend.*   (NB5:9)
     sáláme   ta-wumbè-tang-e-nd
     cloth   β.2|3NSGP-tie-INC.IPFV-PFV.DU-2|3DUA
     '(you DU) tie up the cloths (2 or 3+).'
     or '(you PL) tie up the cloths (2).'

(716) a. *yèkw   tatarend.*   (NB7:20)
     yèkw   ta-tar-e-nd
     hole   β.2|3NSGP-dig-PFV.DU-2|3DUA
     '(you DU) start digging holes (2 or 3+).'
     or '(you PL) start digging holes (2).'
    b. *yèkw   tatarwangend.*   (NB7:20)
     yèkw   ta-tar-wang-e-nd
     hole   β.2|3NSGP-dig-2DU.IMP-PFV.DU-2|3DUA
     '(you DU) dig holes (2 or 3+).'
     or '(you PL) dig holes (2).'

Table 6.1 shows the immediate imperative forms (without any additional affixes)
for the transitive verb *tar*/ 'dig' (all from NB6:20):

6.5.2 Future imperatives

As the name indicates, future imperatives tell someone to do something not imme-
diately but some time in the future, usually the following day or later. Future imperatives all use the current imperfective forms of the verb. This means that they have
the α set of prefixes and the following A or S_A indexing suffixes: Ø for 2nd person

242 — 6 Simple sentences

Table 6.1: Immediate imperative forms for *tar*/ 'dig'.

	SG (1) P argument	**PL (3+) P argument**	**DU (2) P argument**
Perfective			
SG (1 addressee)	*tètar*	*tatar*	*tatare*
PL (3+ addressees)	*tètarngi*	*tatarngi*	*tatarend*
DU (2 addressees)	*tètarend*	*tatarend*	*tatarend*
Imperfective			
SG (1 addressee)	*tètana*	*tatana*	*tatarwange*
PL (3+ addressees)	*tètanangi*	*tatanangi*	*tatarwangend*
DU (2 addressees)	*tètarwangend*	*tatarwangend*	*tatarwangend*

singular irrealis: *-t* for 2nd person dual and *-t-i* for 2nd person plural (3+). Here are examples comparing immediate and future imperatives with the A-aligned intransitive verb *awarawer*/ 'be careful':

(717) a. *kawarawena.*   (NB5:43)
   k-awarawer-ta-Ø
   β.ØP-be.careful-IPFV.ND-2|3SGS$_A$.IRR
   '(you SG) be careful (now).'

  b. *nawarawena.*   (NB5:43)
   n-awarawer-ta-Ø
   α.ØP-be.careful-IPFV.ND-2|3SGS$_A$.IRR
   '(you SG) be careful (later).'

And examples with the transitive verb *wúmbár*/ 'wash' (all from NB6:35–38):

(718) a. *sosfen* *túmbna.*
   sosfen   t-wúmbár-ta-Ø
   cooking.pot β.3SGP-wash-IPFV.ND-2|3SGA.IRR
   '(you SG) wash the cooking pot (now).'

  b. *sosfen* *yúmbna.*
   sosfen   y-wúmbár-ta-Ø
   cooking.pot α.3SGP-wash-IPFV.ND-2|3SGA.IRR
   '(you SG) wash the cooking pot (later).'

6.5 Imperative sentences — **243**

(719) a. *sosfen*   *túmbárend.*
       sosfen   t-wúmbár-e-nd
       cooking.pot   β.3sGP-wash-PFV.DU-2DUA
       '(you DU) start washing the cooking pot (now).'
   b. *sosfen*   *yúmbárèt.*
       sosfen   y-wúmbár-Ø-t
       cooking.pot   α.3sGP-wash-IPFV.DU-2|3NSGA
       '(you DU) wash the cooking pot (later).'

(720) a. *sosfen*   *táwúmbnangi.*
       sosfen   tá-wúmbár-tang-i
       cooking.pot   β.2|3NSGP-wash-INC.IPFV-2|3PLA
       '(you PL) wash the cooking pots (now).'
   b. *sosfen*   *ewúmbnati.*
       sosfen   e-wúmbár-ta-t-i
       cooking.pot   α.2|3NSGP-wash-IPFV.ND-2|3NSGA-2PL
       '(you PL) wash the cooking pots (later).'

Table 6.2 compares the simple immediate imperative forms (imperfective) and future imperative forms for the transitive verb \fenè/ 'feed' (from NB3:33):

Table 6.2: Immediate and future imperative forms for \fenè/ 'feed'.

	SG (1) P argument	PL (3+) P argument	DU (2) P argument
Immediate (imperfective)			
SG (1 addressee)	*tèfeneta*	*táfeneta*	*tátfenwange*
PL (3+ addressees)	*tèfenetangi*	*táfenetangi*	*táfenwangend*
DU (2 addressees)	*tèfenwangend*	*táfenwangend*	*táfenwangend*
Future			
SG (1 addressee)	*yèfeneta*	*efeneta*	*efenwè*
PL (3+ addressees)	*yèfenetati*	*efenetati*	*efenèt*
DU (2 addressees)	*yèfenèt*	*efenèt*	*efenèt*

Following are some examples of future imperatives from the naturalistic data:

(721) *afane*   *si*   *náyáretati.*   (PT:81)
       afè-ne   si   n-áyárè-ta-t-i
       father-GEN   story   α.ØP-listen-IPFV.ND-2|3NSGS$_A$-2PL
       '(you PL) listen to father's story.'

244 — 6 Simple sentences

(722)   *tane*   *mèrès*   *yarawena,*
        tane   mèrès   y-warawer-ta-Ø
        1SG.GEN girl   α.3SGP-take.care.of-IPFV.ND-2|3SG$_A$.IRR
        mer   *mènjagh*   *yèmnja.*   (B&B:234)
        mer   mènja-gh   y-mènnja-ta-Ø
        good   advise--NOM α.3SGP-guide-IPFV.ND-2|3SGA.IRR
        '(you SG) take care of my girl, guide her with good advice.'

(723)   *Ben*   *kor*   *yanjengèt.*   (KK11)
        Ben   kor   y-wanjeng-Ø-t
        Ben   EMPH   α.3SGP-be.in.charge-IPFV.DU-2|3NSGA
        '(you DU) take care of Ben.'

(724)   *efanda*   *ekmeta.*   (PT:195)
        e-fanda-ta-Ø   e-kèmè-ta-Ø
        α.2|3NSGP-look.at-IPFV.ND-2|3SGA.IRR   α.2|3NSGP-lay.down-IPFV.ND-2|3SGA.IRR
        '(you SG) look at them, lay them down.'

6.5.3 Negative imperatives

Negative imperatives are formed by preposing the negative marker *yau* to the verb. In the data, they occur only with future imperatives:

(725)   *yau*   *mámághèrae*   *nafarota.*   (NB7:90)
        yau   mámághèrae   n-afaro-ta-Ø
        NEG   untidily   α.ØP-write-IPFV.ND-2|3SGS$_A$.IRR
        '(you SG) don't write untidily.'

(726)   *yau*   *ye*   *nendati.*   (NB2:17)
        yau   ye   n-eng-ta-t-i.
        NEG   crying   α.ØP-cry-IPFV.ND-2|3NSGS$_A$-2PL
        '(you PL) don't cry.'

(727)   *fèmofem,* *yau* *menmen* *kaka* *ewasèrmbot.*   (DC16)
        fèmofem, yau menmen kaka e-wasèrmbo-Ø-t
        2NSG.ERG NEG bird   near   α.2|3NSGP-spoil-IPFV.DU-2|3NSGA
        'you two, don't spoil the birds nearby.'

(728) *wagif   yaufi   yau   táf   eneta.*   (DC07)
wagif   yaufi   yau   táf   e-ne-ta-Ø
fish   bad   NEG   OBL2   α.2|3NSGP-eat-IPFV.ND-2|3SGA.IRR
'(you SG) don't eat bad fish.'

6.5.4 Imperatives with deictic prefixes

Imperatives can also have a deictic prefix: the venitive prefix *n-* (VEN) 'towards or near the speaker' or the andative prefix *ng-* (AND) 'away or far from the speaker'(see section 4.4). The use of this prefix is crucial in the frequent imperative forms for 'come' and 'go' which use the verb *afèl*/ 'move a short distance':

(729) a.   *kènafèl.*   (NB5:20)\
k-n-afèl-Ø-Ø
β.ØP-VEN-move-PFV.ND-2|3SGS$_A$
'(you SG) come (here).'
b.   *kèngafèl.*   (NB5:20)\
k-ng-afèl-Ø-Ø
β.ØP-AND-move-PFV.ND-2|3SGSA
'(you SG) go (away).'

As mentioned in section 4.4, with A-aligned intransitive verbs, the ØP prefix *n-* or *k-* may be deleted preceding the andative prefix *ng-*, for example:

(730) *ngaflangi.*   (NB5:20)
ng-afèl-Ø-ang-i
AND-move-PFV.ND-INC-2|3PLS$_A$
'(you PL) go (away).'

Further examples with the venitive prefix *n-*:

(731) *kènoraya*   *ndauot.*   (DC06)
k-n-oray-ta-Ø   ndau-ot
β.ØP-VEN-say-IPFV.ND-2|3SGS$_A$.IRR   garden-PURP
'(you SG) talk about going to the garden.'

246 —— 6 Simple sentences

(732) *nde   kènretangend.*   (DC02)
        nde   k-n-ère-tang-e-nd
        this.way β.ØP-VEN-look-INC.IPFV-PFV.DU-2|3DUS$_A$
        '(you DU) look this way.'

(733) *ás   wau ámb yau yènmer.*   (DC29)
        ás   wau ámb yau y-n-wèrè-Ø-Ø
        coconut ripe any   NEG α.3SGP-VEN-hold-PFV.ND-2|3SG$_A$
        '(you SG) don't grab any ripe coconut.'

Further examples with the andative prefix:

(734) *kor   nèngawarawena.*   (DC07)
        kor   n-ng-awarawer-ta-Ø
        careful α.ØP-AND-be.careful-IPFV.ND-2|3SG$_A$.IRR
        '(you SG) be careful.'

(735) *tangfandange.*   (PT:201)
        ta-ng-fanda-tang-e-Ø
        β.NSGP-AND-look.at-INC.IPFV-PFV.DU-2|3SGA
        '(you SG) look at them (2)'.

(736) *mbaghayan kor   táf   ngófam.*   (NB3:12).
        mbaghè-an kor   táf   ng-ófamè-Ø-Ø
        bridge-LOC2 careful OBL2 VEN-cross-PFV.ND-2|3SG$_A$
        '(you SG) cross carefully on the bridge.

6.5.5 Special imperative forms

Nama has some special imperative forms. The most commonly heard are *awe* 'come', *so* 'stop, wait' and *somè* 'wait, stand by'. They can be used with singular, plural or dual addressees.

In contrast, the special imperative form of the copula 'be' has different forms form for different numbers: *anèm* SG, *aem* PL and *aere* DL:

(737) a. *mer ambum anèm.*   (NB2:92)
            mer   ambum anèm
            good child be.IMP.SG
            '(you SG) be a good child.'

6.5 Imperative sentences — **247**

b. *mer   ambum   aem.*   (NB2:92)
    mer   ambum   aem
    good child   be.IMP.PL
    '(you PL) be good children.'

c. *mer   ambum   aere.*   (NB2:92)
    mer   ambum   aere
    good child   be.IMP.DU
    '(you DU) be good children.'

6.5.6 Benefactive imperatives

Nama imperatives also make use of the applicative prefix *wa-* (section 4.5.2) to
make the command benefactive – i.e. to do something for someone – for example:

(738) *kwatfo.*   (NB4:68)
       kw-wa-tèfo-Ø-Ø
       β.1SGP-APP-break.off.a.piece-PFV.ND-2|3SGA
       '(you SG) break off a piece for me.'

(739) *yèkw   tatanangi.*   (NB7:20)
       yèkw   t-wa-tar-tang-i
       hole   β.3SGP-APP-dig-INC.IPFV-2|3PLA
       '(you PL) dig a hole for him.'

(740) *ás   tawafarorend.*   (NB3:35)
       ás   ta-wa-faror-e-nd
       coconut β.2|3NSGP-APP-split-INC.IPFV-2DUA
       '(you DU) split a coconut for them.'

(741) *tènmáwúmbár.*   (NB6:36)
       tèn-wá-wúmbár-Ø-Ø
       β.1NSGP-APP-wash-PFV.ND-2|3SGA
       '(you SG) wash it for us.'

Like with other imperatives, a distinction can sometimes be made between perfective and imperfective – for example:

248 —— 6 Simple sentences

(742) a. *ás*   *kwafarorngi.*   (NB3:35)
      ás   kw-wa-faror-Ø-ang-i
      coconut β.1SGP-APP-split-PFV.ND-INC-3PLA
      '(you PL) split a coconut for me.'
    b. *ás*   *kwafaronangi.*   (NB3:35)
      ás   kw-wa-faror-tang-i
      coconut β.1SGP-APP-split-INC.IPFV-3PLA
      '(you PL) split coconuts for me.'

As described in section 4.5.2, the P-indexing prefix which precedes applicative prefix indexes the person and number of the beneficiary or recipient (the indirect object in traditional grammatical terms), not of the P argument (the direct object) that is normally indexed in a transitive verb. Sometimes the semantics of the verb combined with its aspect give an indication of the number of the direct object. For example, the verb *faror*/ 'split' is semantically punctual, so the perfective imperative in (742a) implies a single action on one coconut. In contrast, the imperfective imperative (with the inceptive imperfective suffix *-tang*) in (742b) implies that the splitting may be continuous, and therefore involving more than one coconut. However, the following examples with *tar*/ 'dig' could refer to a singular or plural direct object:

(743) a. *yèkw* *kwatar.*   (NB7:20)
      yèkw kw-wa-tar-Ø-Ø
      hole   β.1SGP-APP-dig-PFV.ND-2|3SGA
      '(you SG) start digging one hole or many holes for me.'
    b. *yèkw* *kwatana.*   (NB7:20)
      yèkw kw-wa-tar-ta-Ø
      hole   β.1SGP-APP-dig-IPFV.ND-2|3SGA.IRR
      '(you SG) dig one hole or many holes for me.'

For singular benefactive imperatives, though, a dual direct object, as opposed to singular or plural, is indicated by dual perfectivity marking:

(744) a. *yèkw* *kwatare.*   (NB7:20)
      yèkw kw-wa-tar-e-Ø
      hole   β.1SGP-APP-dig-PFV.DU-2|3SGA
      '(you SG) start digging two holes for me.'
    b. *yèkw* *kwatarwè.*   (NB7:20)
      yèkw kw-wa-tar-Ø-wè-Ø
      hole   β.1SGP-APP-dig-IPFV.DU-DA-2|3SGA
      '(you SG) dig two holes for me.'

On the other hand, dual perfectivity marking is also used with dual beneficiaries. So the following example is ambiguous as to whether the indirect or direct object is dual:

(745) *ás*   *tawafarore.*   (NB3:36)
       ás   ta-wa-faror-e-Ø
       coconut β.2|3NSGP-APP-split-PFV.DU-2|3SG$_A$
       '(you SG) split two coconuts for them (3+).'
       or '(you SG) split a coconut or coconuts for them (2).'

Similarly, the following example could be used for both dual and plural imperatives and could refer to dual or plural beneficiaries and, if it is a dual imperative, to a direct object of any number:

(746) *ás*   *tawafarorend*   (NB3:36)
       ás   ta-wa-faror-e-nd
       coconut β.2|3NSGP-APP-split-PFV.DU-2DUA
       '(you DU) split any number of coconuts for them (2 or 3+).'
       or '(you PL) split 2 coconuts for them (2 or 3+).'
       or '(you PL) split any number of coconuts for them (2).'

Benefactive imperatives can be both immediate, as in the examples above, and future, as in the following:

(747) *yèkw*   *watanati.*   (NB7:21)
       yèkw   w-wa-tar-ta-t-i
       hole   α.1SGP-APP-dig-IPFV.ND-2|3NSGS$_A$-2PL
       '(you PL) dig a hole (or holes) for me (later).'

(748) *yáf*   *yèngwarsa.*   (NB5:96)
       yáf   y-ng-wa-rèsa-Ø-Ø
       basket   α.3SGP-AND-APP-carry-PFV.ND-2|3SGA
       '(you SG) carry the basket (or baskets) (away) for him (later).'

Deictic prefixes can occur with benefactive imperatives, as in (748) and the following:

(749) a. *kwènmarsa.*   (NB5:96–7)
           kw-n-wa-rèsa-ta-Ø
           β.1SGP-VEN-APP-carry-IPFV.ND-2|3SGA.IRR
           '(you SG) carry it here for me.'

250 —— 6 Simple sentences

b. *tángwarsa.* (NB5:96–7)
   tá-ng-wa-rèsa-ta-Ø
   β.2|3NSGP-AND-APP-carry-IPFV.ND-2|3SG_A.IRR
   '(you SG) carry it away for them.'

And finally, there are autobenefactive imperatives (apparently, only for SG imperatives) using the autobenefactive prefix *o-* (section 4.5.3):

(750) *yèkw   kotar.* (NB7:21)
      yèkw   k-o-tar-Ø-Ø
      hole   β.ØP-AUTO-dig-PFV.ND-2|3SGS_A
      '(you SG) start digging a hole (or holes) for yourself.'

(751) *káye   táf   nowúmbna.* (NB6:38)
      káye   táf   n-o-wúmbár-ta-Ø
      tomorrow   OBL2   α.ØP-AUTO-dig-IPFV.ND-2|3SGS_A.IRR
      '(you SG) wash it (or them) for yourself tomorrow.'

6.5.7 Third person imperatives

Third person imperatives tell the addressee(s) to tell someone else (the third person or persons) to do something. In Nama these are expressed by making the noun or pronoun referring to the third person(s) the A or S_A argument so that the A- or S_A-indexing suffix on the verb refers to it rather than to the addressee(s) as in other imperatives. Thus the addressees can be of any number. Because 2nd person and 3rd person forms are almost always identical, the actual form of the command is the same as that of a statement, and only the context makes it imperative.

For a singular third person, the form of the verb differs from that of the usual imperative, using the inceptive imperfective – for example:

(752) a. *kásáfota.* (NB5:84)
         k-ásáfo-ta-Ø
         β.ØP-work-IPFV.ND-2|3SGS_A.IRR
         '(you SG) work.'
      b. *fá   kásáfotangè.* (NB5:84)
         fá   k-ásáfo-tang-è
         3.ABS   β.ØP-work-INC.IPFV-2|3SGS_A
         'tell him/her to work.'

(753) a. *tèrsa.* (NB6:40)
t-rèsa-Ø-Ø
β.3SGP-carry-PFV.ND-2|3SGA
'(you SG) carry it.'

b. *yèmo tèrsangè.* (NB6:40)
yèmo t-rèsa-tang-è
3SG.ERG β.3SGP-carry-INC.IPFV-2|3SGA
'tell her/him to carry it.'

For dual or plural third persons, the form of the verb is the same as that for regular imperatives with dual or plural addressees:

(754) a. *kásáfotangi.* (NB5:84)
k-ásáfo-tang-i
β.ØP-work-INC.IPFV-2|3PLS$_A$
'(you PL) work.'

b. *fá kásáfotangi.* (NB5:84)
fá k-ásáfo-tang-i
3.ABS β.ØP-work-INC.IPFV-2|3PLS$_A$
'tell them (3+) to work.'

(755) a. *tèrswangend.* (NB6:40)
t-rèsa-wang-e-nd
β.3SGP-carry-2DU.IMP-PFV.DU-2|3DUA
'(you DU) carry it.'

b. *yèmofem tèrswangend.* (NB6:40)
yèmofem t-rèsa-wang-e-nd
3NSG.ERG β.3SGP-carry-2DU.IMP-PFV.DU-2|3DUA
'tell them (2) to carry it.'

6.6 Minor sentences

Minor sentences (or pro-sentences) are those made up of a single word or a short phrase, without a verb. They have a complete meaning on their own, but depend on the context of the utterance for this meaning. They can be of two types: The first includes single words or phrases that could be expanded to full sentences, such as answers to questions – e.g.:

252 —— 6 Simple sentences

(756) dialogue (DC14)
    speaker 1: *Efe   yètkwèn   yèm?*
                efe   yètkwèn   y-m
                who.ABS   name   a.3SGS$_P$-COP.ND
                'What's (his) name?"
    speaker 2: *Ndárfi.*
                Ndárfi.

Here, the answer could be considered part of possible longer answer, e.g. *His name is Ndárfi.* Another example is *Nuam,* said by children wanting water. This is short for *Nuam wèramete.* 'I'm thirsty.' (lit. 'Water is affecting me.'), as shown in example (630) above.

Other minor sentences include greetings and farewells, expressions of politeness, interjections and discourse markers.

Greetings and farewells in Nama, as in English, are made up of the word meaning 'good' (*mer*) and the times of the day: *mer akw* 'good morning', *mer efogh* 'good day', *mer sèite* 'good afternoon/evening', *mer sèmbár* 'good night'. The expression *yawo* is used for 'farewell' or 'good-bye. (This word is also found throughout the Torres Strait, and is thought to have come from the Kala Lagaw Ya language of the Western Torees Strait islands.)

Words and expressions used for politeness include *kánjo* and *eso* for 'thanks. (Like *yawo*, *eso* appears to come from the Kala Lagaw Ya language and is widespread across the southern part of PNG and the Torres Strait.) Also, *mambanè* can mean 'sorry' or 'please'.

Interjections, as in other languages, form a word class made up of words that are used to express feelings, reactions or emotions, but are generally not grammatically related to other parts of a sentence. Some examples are:

aa (lengthened /a/)	'ah!' (expression of realisation)
ayau	'look out!'
ái	'yikes!' (strong expression of surprise or shock)
árè	'ouch!'
ei	'hey!' (emphasis or reacting to something unexpected)
fèu	'this/that is the one!'
kalfal	'bless you!' (said after someone sneezes)
karèm	'(said in anger) you do it your own way!'
oo (lengthened /o/)	'oh' (mild expression of recognition, surprise, disappointment or uncertainty, also used in joking)

6.6 Minor sentences —— **253**

rèrri	'stuck!' (said when something gets stuck or hooked on something)
so	'wait, stop'[45]

The terms for groups of six called out in yam counting – *ñambi, yèndè*, etc (section 1.1.5) – can also be considered interjections.

An example comes from a recording of the Picture Task (section 1.1.3) – where consultants have to describe what is happening in a set of pictures (drawings) and then put the pictures in order to make a coherent story. One picture shows a man striking a woman. One of the speakers suggests that another picture, showing the man drinking alcohol, should come before. The second speaker makes the following three utterances approximately 3 seconds apart:

(757) a. *wèri nu.*   (PT:281–283)
       'alcohol.'

     b. **Aa.**
       'ah!'

     c. *O   efalo.*
       yes   true
       'Yes, true.'

Two other examples show mild surprise:

(758) a. *yène syèm. . .*   (PT:375–76)
       yène s=y-m
       DEM   PROX2=α.3SGS_P-COP.ND
       'this is it. . .'

     b. [5 seconds later] **Oo, yau!**
                                'Oh, no!'

(759) dialogue   (DC14)
       speaker 1: *Endene   ambum?*
                  ende-ne   ambum
                  who.DAT-GEN child
                  'Whose child?'

---

45 Note that this form is homophonous with the future marker *so*. The word for 'wait' and the future marker are also homophonous (/kwa/) in neighbouring Komnzo (Döhler 2018:122).

254 —— 6 Simple sentences

speaker 2: *Pitane.*
            Pitè-ne
            Peter-GEN
            'Peter's.'
speaker 1: ***Oo* *Pitoene!***
            oo Pitè-o-e-ne
            oh Peter-SG-DAT-GEN
            'Oh, Peter's!'

The next example shows reaction to something unexpected:

(760) ***Ei,* *yabun* *mèngo!***   (DC21)
       hey big   house
       'Hey, what a big house!'

Many of these interjections can also be part of complete exclamative sentences, as demonstrated in section 6.7 below.

Discourse markers are words or short phrases that relate to the structure and flow of conversation. Although, some lexical items in Nama occur only as discourse markers and belong to the discourse particle word class, lexical items from other word classes can also function as discourse markers. In this analysis, there are four kinds of discourse markers based on their functions: providing an evaluative response, framing an interaction, drawing the attention of a participant and managing the flow of the discourse. The discourse markers that have one of the first three of these functions can occur as minor sentences. (Those that manage the flow of discourse do not occur on their own and are described in section 6.8 below.)

Evaluative responses indicate agreement, disagreement or degree of knowledge, following on from a preceding utterance. The negative marker *yau* is used for 'no' indicating refusal or disagreement. There is no single word in Nama meaning 'yes'. The expression *mm* (lengthened /m/) indicates agreement with a proposition, *ma* indicates approval and *o* indicates acknowledgement, acceptance or agreement to do something (similar to 'okay'). In answer to a question, the expression *ai* means 'I don't know'.[46] Some examples:

---

[46] There is no agreement in linguistics about what word class evaluative responses such as 'yes' and 'no' belong to, with classifications ranging from adverbs to interjections.

6.6 Minor sentences — **255**

(761) dialogue   (DC21)
    speaker 1: *Ferafam*   *yau* *nèmáútáne?*
                 fe-rafè-am   yau nèn-wáútár-ta-e
                 2SG-father-ERG NEG α.2SGP-help-IPFV.ND-2|3SGA
                 'Your father's not helping you?'
    speaker 2: **Yau.**
                 'No.'

(762) **Mm,** *Tinane.*   (DC24)
    mm Tinè-ne
    yes Tina-GEN
    'Yes, (it's) Tina's'.

(763) dialogue   (PT:109–110)
    speaker 1: *Fèm*   *mat*   *ánde* *nde*   *yèm*   *yènè* *nu?.*
                 fèm   mat   ánde nde   y-m   yènè nu
                 2.ABS knowing where like.this α.3SGSP-COP.ND DEM water
                 'Do you know where that water is like this ?'
    speaker 2: **ai.**
                 'I don't know.'

As just mentioned, words from various word classes in Nama can function as discourse markers to frame interaction. For example, *fèyotaro* 'enough' is used to frame the end of an interaction, like 'okay' in English, as in this example:

(764) dialogue   (DC15)
    speaker 1: *Mer* *njaran*   *wèi* *yenmorang*
                 mer   njar-an   wèi yen-mor-ang
                 good shade-LOC2 now α.1NSGSP-stay-INC
                 ákiafene
                 áki-af-e-ne
                 grandfather-PL-DAT-GEN
                 mènè *yam*   *awafarègh*   *sifayè.*
                 mènè yam   awafar-gh   sifayè
                 thing tradition be.covered-NOM place
                 'We're sitting now in good shade, forefathers' (what's it called) traditional shady place.'
    speaker 2: **Fèyotaro.**
                 'OK.'

256 —— 6 Simple sentences

Sometimes this is combined with other words, as in the following:

(765) *Eso* **fèyotaro** *mer.* (DC34)
    thanks **OK** good
    'Thanks, OK, good.'

But some lexical items with this function belong only to the word class of discourse particle. One example is *bèra,* which is used to acknowledge or confirm a statement made by another, as in the following:

(766) dialogue (DC12)
    speaker 1: *Kès* *yifo.*
               kès y-wifo-Ø-Ø
               there α.3SGP-finish-PFV.ND-2|3SGA
               'There, you've finished it.'
    speaker 2: **Bèra.**
               'Yep.'

Three examples of discourse particles used to draw the attention of a participant are *iri* 'see that', *kèsa* 'just here' and *kèsou* 'just over there'. These last two appear to be derived from the immediacy adverb *kès,* described in section 5.3.6. Example (590) illustrates the use of *kèsou.* In the following, two participants are taking part in the Picture Task. Speaker 1 is trying to identify what's going on in a picture, while Speaker 2 is interrupting and appears to be disengaging from the task. Speaker 1 attempts to draw her back in with the use of these two discourse particles in separate utterances and then after 2 seconds continues with the task.

(767) dialogue (PT:641–46)
    speaker 1: *Yèna* *syèmronamèng.*
               yèna s=y-mèror-ta-m-ng
               here PROX2=α.3SGP-quarrel-IPFV.ND-REM.DLT-2|3SGA
               'Here he quarrelled with her.'
    speaker 2: *Tèk* *kès* *yènè* *yènèrsawè.*
               tèk kès yènè yèn-rèsa-Ø-wè-Ø
               long.time there DEM α.1NSGP-carry-IPFV.DU-DA-2|3SGA
               'This is taking us (2) a long time.'
    speaker 1: **Iri.**
               'See that.'

speaker 2:	*Mato*	*wèkeye.*	
	mato	wèkeye	
	INV	quick	
	'Try to be quick.'		
speaker 1:	**Kèsa**		
	'Just here.'		
speaker 1:	*Yènè*	*mato*	*yèna*
	yènè	mato	yèna
	DEM	possibly	here

syèmronamènd. . .
s=y-mèror-ta-m-nd
PROX2=α.3SGP-quarrel-IPFV.ND-REM.DLT-2|3NSGA
'Maybe they quarrelled with her here. . .'

6.7 Exclamative sentences

Exclamative sentences express strong emotion, surprise or excitement, or forcefully emphasise a statement. They are sometimes used for joking. Some exclamative sentences are distinguished only by rising intonation at the beginning or middle of the utterance for example:

(768)   *Yènd*   *mè*   *yènfande!*   (DC20)
        yènd   mè   yèn-fanda-ta-e
        1.ABS   CONT   α.1NSGP-look.at-IPFV.ND-2|3SGA
        'It's still looking at us!'

(769)   *Fèm*   *yènè*   *fnèm*   *fái*   *mámá-ambagro!*   (DC19)
        fèm   yènè   f=n-m   fái   mámá-ambag-ro
        2.ABS   DEM   PROX1=α.2SGS$_P$-COP.ND   EMPH   exceeding-humbug-RSTR
        'You're just a real bighead!'

However, most begin or end with an interjection. Here are some examples:

(770)   **Aa**   *wèri*   *tèfnár*   *em.*   (PT:67)
        aa   wèri   tèfnár   e-m
        ah   drunkenness   practitioner   α.2|3NSGS$_P$-COP.ND
        'Ah, they're drunkards!'

(771) **Oo** *sèno tè enwanangi!* (PT:97)
oo sèno tè e-n-wanè-Ø-ang-i
oh POSB PRF α.2|3NSGP-VEN-take-PFV.ND-INC-2|3PLA
'Oh, maybe they already brought them!'

(772) **oo** *fèm mat ánde yènè yèm!* (PT:74)
oo fèm mat ánde yènè y-m
oh 2.ABS knowing where DEM α.3SGS_P-COP.ND
'(joking) oh, you know where that is!'

(773) **Ái,** *mèrès yèsmete.* (PT:140)
ái, mèrès y-sèmè-ta-e
yikes girl α.3SGP-hit-IPFV.ND-2|3SGA
'Yikes, he hit the girl!'

(774) *Kimb úrran náwám* **rèrri!** (NB7:44)
kimb úrèr-an n-áwám-Ø-Ø **rèrri**
pig trap-INC2 α.ØP-be caught-PFV.ND-2|3SGS_A **stuck**
'The pig is caught in the trap now!'

6.8 Discourse markers in simple sentences

Discourse markers that always appear in full sentences to manage the flow of discourse are described here.

Again, words of other classes can function as this type of discourse marker – for example, *mènè* 'thing' being used like 'whatchamacallit' in example (764) above and (776) below. Other words and phrases frequently used as discourse markers are *fái* and *fètè*, both meaning 'really, actually' and both used for emphasis (see examples 555–559 and 576–578 in chapter 5). Additional examples are *wèi* 'now', *fèyo* 'then' and *kès yèm(o)* 'that's it' (lit. 'it's just there'):

(775) *Yènè akwan* **wèi** *kuwanongèm.* (YD11(3))
yènè akw-an **wèi** k-uwano-Ø-ang-m
DEM morning-LOC2 **now** β.ØP-set.off-PFV.ND-INC-1NSGS_A
'In the morning now we set off.'

6.8 Discourse markers in simple sentences — **259**

(776) *Yènè* **mènè** *nèwè* *wèi* *nokawetan.* (MK13(1))
yènè **mènè** nèwè wèi n-okawè-ta-èn
DEM **thing** catfish now α.ØP-pull-IPFV.ND-1SG$_A$
'Now I was pulling this whatchamacallit, catfish.'

(777) *sìf* *so* *nákèrnat* **kès** **yèm.** (DC19)
(=595) sìf so n-ákèrár-ta-t **kès** **y-m**
hair FUT α.ØP-burn-IPFV.ND-2|3NSG$_A$ **just.there** **α.3SG$_P$-COP.ND**
'the hair will burn just so!'

(778) **fèyo** *yanai* *fètè*
fèyo y-wanè-Ø-ay-Ø fètè
then α.3SGP-take-PFV.ND-REM.PUNC-2|3SGA really
yèngwanai
y-ng-wanè-Ø-ay-Ø
α.3SGP-take-PFV.ND-REM.PUNC-2|3SGA
fái **kès** **yèm.** (ND13)
fái **kès** **y-m**
EMPH **there** **α.3SG$_P$-COP.ND**
'then she took her, really took her far away.'

Conjunctive adverbs also function as discourse markers, referring to an earlier sentence or bringing together two independent sentences. These include *yènamamèn* 'that's why, for this reason' and *(yènè) mènayèn* 'that's why, for this/that purpose':

(779) *Krúfèr-krúfèram* *wèi* *fwènramete;*
krúfèr~krúfèr-am wèi f=w-n-ramè-ta-e
extreme.cold-ERG now PROX2=α.1SGP-VEN-affect-IPFV.ND-2|3SGA
yènamamèn *nafalan.* (DC15)
yènamamèn n-a-fal-ta-èn
for.this.reason α.ØP-REFL-warm-IPFV.ND-1SG$_A$
'The extreme cold has affected me; for this reason, I'm warming myself (near the fire).'

260 —— 6 Simple sentences

(780) *mámáyamafè*   *tèm;*   **yènamamèn;** *kès*
mámáyam-afè   t-m   **yènamamèn**   kès
extreme.behaviour-COM β.3SGSP-COP.ND **that's.why** just.there
yèsmetan.   (MD99(2))
y-sèmè-ta-èn
α.3SGP-kill-IPFV.ND-1SGA
'he was very badly behaved; that's why I killed him.'

(781) *yènè*   *efoghkaf*   *fem;*   *yènè*   **mènayan**
yènè   efo-gh-kaf   f=e-m   yènè   **mènayan**
DEM   finish-NOM-ATTR2   PROX1=α.2|3NSGSP-COP.ND   DEM   **for.this.purpose**
kor   *nènangote.*   (PT:484)
kor   n-n-ango-ta-e
again   α.ØP-VEN-return-IPFV.ND-2|3SGSₐ
'Those are finished; for this purpose he's returning again.'

(782) **mènayan**   *yènè*   *fyúároi.*   (FD13)
mènayan   yènè   f=y-wúáro-Ø-ay-Ø
that's.why   DEM   PROX=α.3SGP-set.fire.to-PFV.ND-REM.PUNC-2|3SGA
'that's why he set fire to it.'

However, some discourse particles function only as discourse markers as part of a
longer utterance. These include, *e, wèt, si* and *ee(ee).* The discourse particle *e* 'hey'
is used to get another speaker's attention;

(783) *e*   *Nánjár*   *mer*   *rokár*   *fnènmásmete.*   (DC19)
e   Nánjár   mer   rokár   f=n-n-wá-sèmè-ta-e
hey   Nanjar   good   thing   PROX1=α.2SGP-VEN-APP-kill-IPFV.ND-2|3SGA
'hey, Nanjar, it's good he killed the thing for you (in hunting).'

(784) *e*   *álet*   *nufngo*   *ánde*   *ndenae*
e   ále-t   n-ufngo-Ø-Ø   ánde   ndenae
hey   hunting-ALL   α.ØP-start-PFV.ND-2|3SGSₐ   where   this.way
fan   *tambèn.*   (DC25)
fan   tambèn
clear place   side
'hey, start going hunting this way, where the clear place is.'

The particle *wét* functions similarly to 'so' in English:

(785) | *táf* | *enèm* | | ***wèt*** | *wagh* | *tosè* | |
táf	e-nèm		**wèt**	wagh	tosè	
that.time	α.2|3NSGS_P-come.ND		**so**	singsing	small	
wèngwawermndati					*ta.*	(MK13(4))
w-ng-wa-w-ermáng-ta-t-i					ta	
α.1SGP-AND-APP-TR-dance-IPFV.ND-2|3NSGA-2PL					2SG.DAT	

'so you all come then and dance a little singsing for me.'

(786) | ***Wèt*** | *fèm* | *árkamamèn* | *fifi* | *nèm.* | (TE13) |
| **wèt** | fèm | árkè-mè-mèn | fifi | n-m | |
| **so** | 2.ABS | which-PERL-ORIG | truly | α.2SGS_P-COP.ND | |

'So that's where you're really from.'

The discourse particle *si*, which occurs at the end of an utterance, indicates a question:[47]

(787) | *yèna* | *so* | *tèlawan* | | ***si?*** | (PT:237) |
| yèna | so | t-lawè-Ø-èn | | **si** | |
| here | FUT | β.2\|3NSGP-put.inside-PFV.ND-1SGA | | **QUES** | |

'I'll put them here?'

(788) | *E* | *efe* | *yènè* | *emor* | | ***si?*** | (PT:66) |
| e | efe | yènè | e-mor | | **si** | |
| hey | who.ABS | DEM | α.2\|3NSGS_P-stay | | **QUES** | |

'Hey, who are they sitting there?'

The particle *ee* (prolonged /e/) indicates the continuation of an action in story telling. The length of the pronunciation of the marker is often proportional to the length of time the action took place – i.e. the longer the *ee* is stretched out the longer the prolonged action it describes.

(789) | *fèyo* | *wèyotam* | | ***eee*** | *kèmèghfaf.* | (MK13(1)) |
| fèyo | w-yo-ta-m-Ø | | **eee** | kèmègh-faf | |
| the | α.1SGP-travel-IPFV.ND-REM.DLT-2\|3SGA | | **PROL** | sleeping-ASS | |

'then I travelled all the way to the sleeping place.'

---

47 An alternative analysis is that this is the word *si* 'talk, story' functioning as a discourse marker.

(790) kènewonau   ***eeee***   *mbandot.*   (MD13(1))
k-newor-ta-w-Ø   **eeee**   mband-ot
β.ØP-descend-IPFV.ND-REM.DUR-3SGSA   **PROL**   ground-ALL
'she was coming down down to the ground.'

(791) tènngèrwèn   ***eeee***   *Kwèmbainmèngo.*   (TE11(1))
tèn-ngèr-wèn   **eeee**   Kwèmbainmèngo
β.1NSGS_P-go.DU-REM.DUR.PA   **PROL**   Kwèmbainmèngo
'we (2) went all the way to Kwèmbainmèngo.'

6.9 Morphological discord as a stylistic device

We have seen that in Nama, the S argument of a P-aligned intransitive verb is indexed by a prefix indicating person and number – normally singular (SG) vs nonsingular (NSG). For nonsingular referents, dual number is distinguished from plural by a suffix *-ar* and, as all P-aligned intransitives are perfective, the dual perfective marker *-e*.

But in simple sentences with apparent morphological discord, a singular prefix can co-occur with the dual suffix and dual perfective marker. This is interpreted as a stylistic devise that is used to signal that a state is out of the ordinary, usually in size or number. It may be interpreted as an extraordinarily large plural – e.g.:

(792) Áuyè   *yakáyare.*   (PC/MD:1.5.19)
áuyè   y-wakáy(o)-**ar-e**
cassowary   α.3SGS_P-be.standing-**DU.PA-PFV.DU**
'Very many cassowaries are standing.'

(793) Fès   *yèmángare.*   (NB10:24)
fès   y-mángo-**ar-e**
firewood   α.3SGS_P-be.piled.up-**DU.PA-PFV.DU**
'The firewood is piled up in very many heaps.'

Or it could be a small number or size:

(794) Mèngon   *wèkmare.*   (NB10:42)
mèngo-n   w-kèmè-**ar-e**
house-LOC1   α.1SGS_P-be.lying.down-**DU.PA-PFV.DU**
'I'm the only one sleeping in the house.'

(795) *Ambum   mèngon   **yèkmare.*** (NB10:42)
      ambum   mèngo-n   y-kèmè-**ar-e**
      child   house-LOC1   α.**3SG$_P$**-be.lying.down-**DU.PA-PFV.DU**
      'Only the child is sleeping in the house.'
      or 'The small child is sleeping in the house.'

Discord with the *-ang* inceptive suffix, which normally does not occur with duals, indicates that the state lasted for a long time:

(796) *kámnjongèn   **wèmorangre.*** (MK13(1))
      k-ámnjo-Ø-ang-èn   w-mor-**ang-ar-e**
      β.ØP-sit-PFV.ND-INC-1SG$_A$ α.**1SG$_P$**-stay-**INC-DU.PA-PFV.DU**
      'I sat and stayed for a very long time.'

(797) *Sèsafne   mè   **yèsauèrngre.*** (NB2:31)
      sèsafne   mè   y-sauèr-**ang-ar-e**
      door   CONT   α.**3SG$_P$**-be.open-**INC-DU.PA-PFV.DU**
      'The door has been open for a long time.'

A similar phenomenon occurs with the copula, with singular S$_P$-indexing prefixes being added to the dual copula \r\ plus the dual perfective marker *-e*. In these cases, the interpretation is diminutive – unusually small in number or size.

(798) *Yáf   kèrte   **yère.*** (NB10:42)
      yáf   kèrte   y-r-e
      basket   heavy   α.**3SG$_P$-COP.DU-PFV.DU**
      'The basket is heavy (for its small size).'

(799) *Yèndro   mèngon   **wère.*** (NB10:42)
      yènd-ro   mèngo-n   w-r-e
      1.ABS-EXCL   house-LOC1   α.**1SG$_P$-COP.DU-PFV.DU**
      'I alone am in the house.'

(800) *Wem   ndau   nèmbrar   fifi   **yère.*** (PC/MD:1.5.19)
      wem   ndau   nèmbrar   fifi   y-r-e
      yam   garden   small   really   α.**3SG$_P$-COP.DU-PFV.DU**
      'The yam garden is exceptionally small.'

264 —— 6 Simple sentences

(See the discussion in section 8.2.4.)

In at least one instance, morphological discord is also found with verb types other than P-aligned intransitives. With the agentless transitive verb \wáwáliman/ 'go in a large group', a singular P-indexing prefix co-occurs with dual argument (DA) marker -wè, giving the interpretation that very many are going or went in the group:

(801) *Wáwálimanwè*   *Moetèt.*   (PC/MD:1.5.19)
     w-wáwáliman-Ø-**wè**   Moet-t
     α.1sGP-go.in.large.group-IPFV.DU-**DA**   Morehead-ALL
     'Very many of us are going to Morehead.'

(802) *Táwálimanwè.*   (NB10:33)
     t-wáwáliman-Ø-**wè**
     β.3sGP-go.in.large.group-IPFV.DU-**DA**
     'They all (very many) went yesterday (or the day before).'

Another way of expressing the existence of an unusually large number is to use the singular form of the verb \ngèm/ 'go' instead of the copula when it is clear from the context that the subject is plural:

(803) *Ár*   *yabun* *yèngèm.*   (PC/MD:1.5.19)
     ár   yabun **y**-ngèm
     person big   **α.3sGS$_P$**-go.ND
     'There were very many people.'

(804) *Wanjen*   *wagif* *yèngèm.*   (PC/MD:1.5.19)
     wanje-n   wagif **y**-ngèm
     river-LOC1 fish **α.3sGS$_P$**-go.ND
     'There are very many fish in the river.'

It could also be that words with the *r=* clitic indicating 'extraordinary' and ana-lysed as mirative (described in section 5.4.2) are actually displaying morphological discord: The dual copula \r/ is combined with the nondual copula \m/.

7 Compound and complex sentences

This chapter describes compound and complex sentences in Nama, which are both made up of more than one clause, and by definition (see chapter 6), have more than one verb phrase. Compound sentences are comprised of more than one independent clause, while complex sentences have an independent clause and one or more dependent (or subordinate) clauses. The first section covers compound sentences, and the following sections deal with the various types of complex sentences classified by type of subordinate clause: adverbial clause, relative clause, complement clause and focus marking clause.

7.1 Compound sentences

As already mentioned, compound sentences consist of two or more independent clauses. They can be joined by one of the coordinating conjunctions introduced in section 6.1.2: *a* 'and' (conjunctive) and *o* 'or' (alternative):

(805) *fèyo   nuot   kaflangè   a   nu*
      fèyo   nu-ot   k-afèli-Ø-ang-è   a   nu
      then   water-ALL β.ØP-move-PFV.ND-INC-2|3SGS$_A$ **and** water
      tènmukangè.   (MD13(1))
      t-n-wukè-Ø-ang-è
      β.3SGP-VEN-fetch-PFV.ND-INC-2|3SGA
      'then she went to the river and fetched water.'

(806) *yerayènd   a   limanègh*
      y-werè-Ø-ay-nd   a   limán-gh
      α.3SGP-hold-PFV.ND-REM.PUNC-2|3NSGA **and** pull-NOM
      yafngoyènd.   (YD11(2))
      y-wafngo-Ø-ay-nd
      α.3SGP-start-PFV.ND-REM.PUNC-2|3NSGA
      'they caught it and started pulling it.'

https://doi.org/10.1515/9783111077017-007

266 —— 7 Compound and complex sentences

(807) *yèna so yènkmare*   **o** *so*
       yèna so yèn-kèmè-ar-e   **o** so
       here FUT α.1NSGS$_P$-be.lying.down-DU.PA-PFV.DU **or** FUT
       nangowèm?   (TE11(1))
       n-ango-Ø-wè-m
       α.ØP-return-IPFV-DU-DA-1NSGS$_A$
       'will we (2) sleep here or will we go back?'

(808) *wagh so yufanat*   **o** *si*
       wagh so y-wufar-ta-t   **o** si
       singsing FUT α.3SGP-sing.out-IPFV.ND-2|3NSGA   **or** story
       so norayat.   (B&B:167)
       so n-oray-ta-t
       FUT α.ØP-say-IPFV.ND-2|3NSGS$_A$
       'they will sing out a singsing or tell stories.'

Coordination can also be unmarked, as in comparative constructions:

(809) *Tane mèngo yabun yèm;*   *fene syèm*
       Tane mèngo yabun y-m;   fene s=y-m
       1SG.GEN house big α.3SGS$_P$-COP.ND 2SG.GEN PROX2=α.3SGS$_P$-COP.ND
       nèmbne yèm   (NB1:8)
       nèmbne y-m
       small α.3SGS$_P$-COP.ND
       'My house is bigger than yours.' (lit. 'My house is big; yours, it's small.')

Note, however, that comparative constructions do not occur spontaneously in narratives or conversations.[48]

Unmarked coordination can also occur with clause chaining, as in the following:

---

[48] The same is true for superlative constructions, which are expressed with a simple sentence:
Yáne mèngo yabun fifi yèm   *mèngotuan.*   (NB1:8)
yáne mèngo yabun fifi y-m   mèngotu-an.
3SG.GEN house big very α.3SGS$_P$-COP.ND village-LOC2
'His house is the biggest in the village.' (lit. 'His house is very big in the village.')

(810) *yánjo   ambum-afayo   amaf-afayo   wèi   kètè   tek-ofnar-e*
yánjo   ambum-afè-yo   amaf-afè-yo   wèi   kètè   tek-ofnar-e
3SG.own   child-COM-EXCL   wife-COM-EXCL   now   there   long.time-PRIV-INST
nafrendamènd
n-afreng-ta-m-nd
α.ØP-prepare-IPFV.ND-REM.DLT-2|3NSGS_A
nuwanoyènd. . .   (MD99(1))
n-uwano-Ø-ay-nd
α.ØP-set.off-PFV.ND-REM.PUNC-2|3NSGS_A
'His own children and wife quickly got ready and set off. . .'

(811) *susi   tèsúfangèn   kwès   wèi*
susi   t-súfè-Ø-ang-èn   kwès   wèi
fishing.line   β.3SGP-pull.out-PFV.ND-INC-1SGA   bait   now
turawangèn   yèngwitan.   (MK13(1))
t-wurawè-Ø-ang-èn   y-ng-wi-ta-èn
β.3SGP-join-PFV.ND-INC-1SGA   α.3SGP-AND-throw-IPFV.ND-1SGA
'I pulled out the fishing line, put bait on (the hook) and threw it out.'

7.2 Complex sentences with adverbial clauses

Adverbial clauses in Nama consist of at least an inflected verb and an adverbial subordinating conjunction. There are six types of adverbial clauses: time, place, concessive, purpose, conditional and reason.

7.2.1 Adverbial clauses of time

Adverbial subordinate clauses of time are expressed with five temporal subordinating conjunctions. The first is *njam* 'when' (which also functions as an interrogative adverb):

(812) **njam**   *kènangotawèng*   *so*
njam   k-n-ango-ta-w-ng   so
when   β.ØP-VEN-return-IPFV.ND-REM.DUR-2|3SGS_A   FUT
tènmaufetau   *yènè   rokár.*   (DC34)
t-n-waufè-ta-w-Ø   yènè   rokár
β.3SGP-VEN-blow-IPFV.ND-REM.DUR-2|3SGA   DEM   thing
'when he was returning he would blow this thing [a flute].'

268 —— 7 Compound and complex sentences

(813) *Darut*   **njam**   *tèngèn*   *enjine*
       Daru-t   **njam**   t-ngèn   enjine
       Daru-ALL   **when**   β.3sGSₚ-go.ND sickness
       yanè.   (NB10:4)
       y-wanè-Ø-Ø
       α.3sGP-take-PFV.ND-2|3sgA
       'When he was going to Daru, he got sick.'

(814) **Njam** *fès*   *yèfarone,*   *mbilè*
       njam fès   y-faror-ta-e   mbilè
       when firewood   α.3sGP-split-IPFV.ND-2|3sgA axe
       náráfár.   (NB10:4)
       n-áráfár-Ø-Ø
       α.ØP-break-PFV.ND-2|3sGSₐ
       'When (he was) splitting the firewood, the axe broke.'

Note that the preceding example could also be expressed with a simple sentence
using the nominalised form of the verb and the locative case marker *-an*:

(815) *Fès*   *farorghan,*   *mbilè* *náráfár.*   (NB10:4)
       fès   faror-gh-**an** mbilè   n-áráfár-Ø-Ø
       firewood   split-NOM-LOC2   axe   α.ØP-break-PFV.ND-2|3sgSₐ
       'When splitting the firewood, the axe broke.'

Related to *njam* is *njamanyo* 'just when, as soon as'. This could be a morphologically
complex form (see Table 3.15).

(816) **Njamanyo** *fá*   *nurtoyènd*
       njamanyo fá   n-urto-Ø-ay-nd
       as.soon.as 3.ABS   α.ØP-came.out-PFV.ND-REM.PUNC-2|3NSGSₐ
       ngarèndta,   *árèm*   *táf*   *yayamngoyènd*
       ngarènd-ta   áre-m   táf   y-wayamngo-Ø-ay-nd
       boat-ABL   people-ERG   then   α.3sGP-recognise-PFV.ND-REM.PUNC-2|3NSGA
       Yesu.   (Mark 6:54)
       Yesu
       Jesus
       'As soon as they got out of the boat, people recognised Jesus.'

(817) *fèyo yèkw yètanamèng*   *a kès*
fèyo yèkw y-tar-ta-m-ng   a kès
then hole α.3sGP-dig-IPFV.ND-REM.DLT-2|3sGA and just.there
yèkw **njamanyo** *nefoi*   *ambum*
yèkw **njamanyo** n-efo-Ø-ay-Ø   ambum
hole **as.soon.as** α.ØP-finish-PFV.ND-REM.PUNC-2|3sGS$_A$ child
yerayèng   *yèkwèn*
y-werè-Ø-ay-ng   yèkw-n
α.3sGP-hold-PFV.ND-REM.PUNC-2|3sGA hole-LOC1
tènyawangè.   (MD99(2))
t-nyawè-Ø-ang-è
β3sGP-put.inside-PFV.ND-INC-2|3sgA
'then he dug a hole and as soon as he finished, he grabbed his son and put him in the hole.'

The third temporal coordinating conjunction is *sèfè*, which also functions as an adverb meaning 'at this time' (see examples 344 and 832). It is used to express concurrent actions – i.e. with the meaning 'while' – for example:

(818) **sèfè** *náwifnat*   *yènd*
sèfè n-áwifár-ta-t   yènd
while α.ØP-fight-IPFV.ND-2|3NSGS$_A$ 1.ABS
nuwanon.   (NB7:97)
n-uwano-Ø-èn
α.ØP-set.off-PFV.ND-1sGS$_A$
'while they were fighting, I set off.'

(819) *fèmofem fá syèngèfrangongi*   **sèfè**
fèmofem fá s=y-ng-frango-Ø-ang-i   **sèfè**
2NSG.ERG 3.ABS PROX2=α.3sGP-AND-leave-PFV.ND-INC-2|3PLA **while**
nèngwaitote.   (ND13)
n-ng-èwaito-ta-e
α.ØP-AND-follow-IPFV.ND-2|3sGS$_A$
'you people left her while she was following.'

270 — 7 Compound and complex sentences

(820) Yosi   *fèm*   *safèt*   *njam* *nèngtorèng*

       Yosi   fèm   saf-t   njam n-ng-ètor-Ø-ng

       Yoshie 2.ABS clear.place-ALL when α.ØP-AND-go.out-PFV.ND-2|3SGS$_A$

       yènd *tebol-an* **sèfè** *wèmor.*   (MD11(1))

       yènd tebol-an **sèfè** w-mor

       1.ABS table-LOC2 **while** α.1SGS$_P$-stay

       'Yoshie, you went outside while I was at the table.'

The fourth temporal subordinating conjunction is *kètanotyo* 'until'. This also could be analysed as morphologically complex (see Table 3.15).

(821) *yènè* *sèmbár* *snáslamènd*

       yènè sèmbár s=n-á-sèla-Ø-m-nd

       DEM night   PROX2=α.ØP-REFL-keep.awake-IPFV.DU-REM.DLT-2|3NSGS$_A$

       kètanotyo *efáureai.*   (MD13(1))

       kètanotyo e-fáuèr-e-ay-Ø

       until   α.2|3NSGP-dawn.on-PRF.DU-2|3SGA

       'that night they kept each other awake until dawn (lit. 'it dawned on them').'

(822) *yènmorangèrmèn*   **kètanotyo** *sèite*

       yèn-mor-ang-èrmèn   **kètanotyo** sèite

       α.1NSGS$_P$-stay-INC-REM.DLT.PA **until**   evening

       náyátoi.   (MK13(4))

       n-áyáto-Ø-ay-Ø

       α.ØP-end.up-PFV.ND-REM.PUNC-2|3SGS$_A$

       'we stayed until it became evening.'

The discourse marker *eee*, indicating prolonged action (section 6.8), also functions as a temporal subordinating conjunction, with a meaning like 'for a long time until'.

(823) *kènangotawèn*   *ee*   *kufarèn*

       k-n-ango-ta-w-èn   *ee*   k-ufar-Ø-èn

       β.ØP-return-IPFV.ND-REM.DUR-1SGS$_A$ **PROL** β.ØP-arrive-PFV.ND-1SGS$_A$

       mèngotuot.   (MK13(2))

       mèngotu-ot.

       village-ALL

       'I was going back (for a long time) until I reached the village.'

7.2 Complex sentences with adverbial clauses — **271**

(824) *kètan*   *wèi*   *kásáfowèm*   ***eeee***   *seite*
      kètan   wèi   k-ásáfo-Ø-w-m   **eeee**   seite
      there   now   β.ØP-work-IPFV.DU-REM.DUR-1NSGS$_A$ **PROL**   evening
      náyátoi.   (TE11(1)))
      n-áyáto-Ø-ay-Ø
      α.ØP-end.up-PFV.ND-REM.PUNC-2|3SGS$_A$
      'we (2) were working there now continuing until it eventually became evening.'

(825) *Yènyote*   *yènyote*   ***eeee***
      yèn-yo-ta-e   yèn-yo-ta-e   ***eeee***
      α.1NSGP-travel-IPFV.ND-2|3SGA   α.1NSGP-travel-IPFV.ND-2|3SGA   **PROL**
      yabun   *bout*   *tinjongèm.*   (YD11(3))
      yabun   bout   t-winjo-Ø-ang-m
      big   boat   β.3SGP-see-PFV.ND-INC-1NSGA
      'We travel and travel until we saw a big boat.'

The discourse marker *eee* (section 6.8) can occur along with *kètanotyo* 'until'-

(826) *kètè*   *tanyawangèrwèn*   ***kètanotyo***   *eee*   *Mawai*
      kètè   ta-nyawè-ang-èrwèn   **kètanotyo**   eee   Mawai
      there   β.2|3NSGS$_P$-be.inside-INC-REM.DUR.PA   **until**   **PROL**   Mawai
      olman   *yèfène*   *kot*   *efogh*
      olman   yèfène   kot   efogh
      old.man   3NSG.GEN   court   day
      nufaryèng.   (MD99(1))
      n-ufar-Ø-ay-ng
      α.ØP-arrive-PFV.ND-REM.PUNC-2|3SGS$_A$
      'they were inside there (in prison) a long time until Mawai and the old man's
      court date arrived.'

The final temporal subordinating conjunction is *kètanmayo* 'since', again possibly
morphologically complex (Table 3.15):

(827) ***Kètanmayo***   *fá*   *snawerai,*
      kètanmayo   fá   s=n-awerè-Ø-ay-Ø
      since   3.ABS   PROX2=α.ØP-get.married-PFV.ND-REM.PUNC-2|3SGS$_A$
      yèm;   *wèri nu*   *yau*   *enete.*   (NB6:81)
      y-mo   wèri nu   yau   e-ne-ta-e
      α.3SGS$_P$-COP.ND   alcohol   NEG   α.2|3NSGP-drink-IPFV.ND-2|3sgA
      'Since he got married, he doesn't drink alcohol.'

272 —— 7 Compound and complex sentences

Note that the temporal nominals *soramè* 'later, afterwards' and *fronde* 'earlier, first' function as conjunctive adverbs but not as subordinating conjunctions corresponding to 'after' and 'before', as in these examples:

(828) *Amam* *basiaue* *tèneghawèng;*   *yènè* *basiau;*
  amè-m basiau-e t-negh-ta-w-ng   yènè basiau
  mother-ERG soup-INST β.3SGP-cook-IPFV.ND-REM.DUR-2|3SGA DEM soup
  soramè *amam* *njamke*
  soramè amè-m njamke
  later mother-ERG food
  ewaflitam   *merekinan.*   (MD93)
  e-wafèli-ta-m-Ø   merekin-an
  α.2|3NSGP-put.in-IPFV.ND-REM.DLT-2|3SGA plate-LOC2
  'Mother cooked soup; afterwards, she served this soup in a plate.'

(829) *Tárnau*   **fronde;** *njam* *rènaghe*
  tá-rèna-ta-w-Ø   **fronde** njam rèna-gh-e
  β.2|3NSGP-scrape-IPFV.ND-REM.DUR-2|3SGA **first** when scrape-NOM-INST
  ewifoi,
  e-wifo-Ø-ay-Ø
  α.2|3NSGP:finish-PFV.ND-REM.PUNC-2|3SGA
  táwúmbnau.   (TE11(2))
  tá-wúmbár-ta-w-Ø
  β.2|3NSGP-wash-IPFV.ND-REM.DUR-2|3SGA
  'She scraped them first; when she finished scraping, she washed them.'

7.2.2 Adverbial clauses of place

Adverbial clauses of place are introduced by the interrogative adverb *ánde* 'where' and its alternative *ándan*, which in this context function as subordinating conjunctions:

(830) *yèngmormèn*   ***ánde*** *fá*
  y-ngèmo-èrmèn   **ánde** fá
  α.3SGSₚ-go.ND-REM.DLT.PA **where** 3.ABS
  káwúmánwèt.   (MD99(1))
  k-áwúmán-Ø-w-t
  β.ØP-wrestle-IPFV.DU-REM.DUR-2|3NSGA
  'he went where they (2) were wrestling.'

7.2 Complex sentences with adverbial clauses — **273**

(831) *kènangowèm*   *kètan* *yáf* ***ánde***
k-n-ango-Ø-w-m   kètan yáf **ánde**
β.ØP-return--IPFV.DU-DUR-1NSGS$_A$ there basket **where**
tákmangèrwèng.   (TE11(2))
tá-kèmè-ang-èrwèng
β.2|3NSGSP-be.lying-INC-REM.DUR.PA
'we (2) returned there where the baskets were lying.'

(832) *sèfè*   *yèroghat*   ***ándan***
sèfè   y-rogha-ta-t   **ándan**
at.this.time α.3SGP-drag-IPFV.ND-2|3NSGA **where**
yármbnangi.   (PT:278)
y-wármbèn-Ø-ang-i
α.3SGP-put.inside-PFV.ND-INC-2|3PLA
'now they're dragging him to where they put him inside.'

Ánde can also function as a relative adverb (see the end of section 7.3 below).

7.2.3 Concessive adverbial clauses

Concessive (or contrastive) adverbial clauses use the subordinating conjunction
wènde 'but':

(833) *Nu*   *káye*   *afoghare*   *tèmo,*   ***wènde*** *sènonjo*
nu   káye   afo-gh-are   t-mo   **wènde** sènonjo
water yesterday be.full-NOM-ATTR1 β.3SGS$_P$-COP.ND **but** today
sotro   *yèm;*   (NB2:18)
sot-ro   y-m
empty-RSTR α.3SGS$_P$-COP.ND
'Yesterday it was full of water but today it's empty.'

274 — 7 Compound and complex sentences

(834) *fá*   *yènmormèn*   *nèkwafè*   *besi*
        fá   y-nèmo-èrmèn   nèkw-afè   besi
        3.ABS α.3SGSₚ-come.ND-REM.DLT.PA   anger-COM spear
        enwanai   *yènd* *yau* **wènde** *nèkw*
        e-n-wanè-Ø-ay-Ø   yènd yau **wènde** nèkw
        α.2|3NSGP-VEN-take-PFV.ND-REM.PUNC-2|3SGA 1.ABS NEG **but**   anger
        kwèmorwèn.   (FD13)
        kw-mo-èrwèn
        β.1SGSₚ-COP.ND-REM.DUR.PA
        'He came angrily carrying spears, but I wasn't angry.'

Note that the contrastive coordinating conjunction is often placed within the concessive clause rather than initially, as in (834).

Contrastive *wénde* can also be followed by *nde* 'like this/that', which also functions as a subordinating conjunction (see section 7.4.1 below). It gives a stronger contrast with the meaning 'but instead':

(835) *kètè* *tèmorwèn*   *áuwèghafè*   *yánjoyo*
        kètè t-mo-èrwèn   áuwè-gh-afè   yánjo-yo
        there β.3SGSₚ-COP.ND-REM.DUR.PA be.happy-NOM-COM 3SG.own-EXCL
        mèrèn *kènjún* **wènde** **nde** *sènonjo* *diburayan* *kès*
        mèrèn kènjún **wènde** **nde** sènonjo diburè-yan kès
        family inside **but**   **that** today   prison-LOC2 just.there
        yèm.   (PT-764)
        y-m
        α.3SGSₚ-COP.ND
        'He was happy there with his family, but instead now he's in prison.'

(836) *Yènd* *fene*   *si*   *ásáfoghèt*   *fètè* *fnufngon*
        yènd fene   si   ásáfo-gh-t   fètè f=n-ufngo-Ø-èn
        1.ABS 2SG.GEN talk work-NOM-ALL VAL PROX1=α.ØP-start-PFV.ND-1SGₐ
        wènde **nde** *kwèngèm*   *kembnet.*   (NB7:56)
        wènde **nde** kw-ngèm   kembne-t
        but   **that** β.1SGSₚ-go.ND game-ALL
        'I should have started the work you told me to do; but instead I went to the game.'

A subordinate clause with the concessive conjunction *wénde* also can occur in combination with the attemptive modal *yèta* in the independent clause to emphasise the contrast:

7.2 Complex sentences with adverbial clauses — **275**

(837) | _Fá_ | **_yèta_** | _yabun_ | _ár_ | _yèm;_ | | **_wènde_** | **_nde_** | _yúwek_
| --- | --- | --- | --- | --- | --- | --- | --- | ---
| fá | **yèta** | yabun | ár | y-m | | **wènde** | **nde** | yúwek
| 3.ABS | **ATT** | big | man | α.3SGS_P-COP.ND | | **but** | **that** | simple

_ár_ _yèm._ (NB5:94)
ár y-m
man α.3SGS_P-COP.ND
'Although he's a big man, he's a simple man.'

7.2.4 Adverbial clauses of purpose

The subordinating conjunction _mènat_ 'in order to, so that' introduces adverbial clauses of purpose. Again, this could also be a morphologically complex form (see Table 3.15).

(838) | _rokár-rokár_ | _ekmang_ | | **_mènat_** | _ár_
| --- | --- | --- | --- | ---
| rokár~PL | e-kèmè-ang | | **mènat** | ár
| things | α.2\|3NSGS_P-be.lying.down-INC | | **so.that** | people

sái	_endmè_	_enèm_		_so_
sái	end-mè	e-nèm		so
INT	road-PERL	α.2|3NSGS_P-come.ND		FUT

yèngwarsat. (PT-90)
y-ng-wa-rèsa-ta-t.
α.3SGP-AND-APP-buy-IPFV.ND-2\|3NSGA
'things are lying there so that people coming along the road will buy them for him.'

(839) | _yèkw_ | _yètanamèng_ | | _ámb_ | _efoghèn_ | **_mènat_** | _yènè_
| --- | --- | --- | --- | --- | --- | ---
| yèkw | y-tar-ta-m-ng | | ámb | efogh-n | **mènat** | yènè
| hole | α.3SGP-dig-IPFV.ND-REM.DLT-2\|3SGA | | some | day-LOC1 | **so.that** | DEM

yèkwèn	_sia_	_tènyawangai_		_yáne_
yèkwèn	sia	t-nyawè-Ø-ang-ay-Ø		yáne
hole:LOC1	INT	β.3SGP-put.inside-PFV.ND-INC-REM.PUNC-2|3SGA		3SG.GEN

ambum. (MD11(2))
ambum.
child
'he dug a hole one day so that he could put his son in that hole.'

Mènat is also sometimes used to introduce a mental predicate with the verb \emán\ 'think':

276 —— 7 Compound and complex sentences

(840)   *njam* *mani* *yafrotan*   *so* *nèngemnan*
       njam mani y-wafro-ta-èn   so n-ng-emán-ta-èn
       when money α.3sGP-make-IPFV.ND-1sGA FUT α.ØP-AND-think-IPFV.ND-1sGS$_A$
       mènát *mèngo* *so* *yèitan.*   (FD11)
       mènát mèngo so y-y-ta-èn.
       so.that house FUT α.3sGP-build-IPFV.ND-1sGA
       'When I make money, I will think about building a house.'

(841)   *kemne*   ***mènát*** *so*
       k-emán-ta-e   **mènát** so
       β.ØP-think-IPFV.ND-2|3sGS$_A$ **so.that** FUT
       taramangè.   (PT:49)
       t-wa-ramè-Ø-ang-è
       β.3sGP-APP-give-PFV.ND-INC-2|3sGA
       'he thought about giving it to him.'

Instead of using subordination, purpose clauses are often expressed with the nominalised form of the verb, followed by the allative suffix -*t*:

(842)   *Yèmo* *wiwi*   *yefnote*   ***mènat***
       Yèmo wiwi   y-wefno-ta-e   **mènat**
       3sG.ERG mango.tree α.3sGP-bend.down-IPFV.ND-3sGA **in.order.to**
       wiwi *wau* *frárghèt.*   (NB5-82)
       wiwi wau fèrár-gh-t.
       mango ripe pick-NOM-ALL
       'He's pulling down (the branch) of the mango tree to pick the ripe mangoes.'

(843)   *ghèwi* *kètanmè*   *ninyè* *ausam*
       ghèwi kètanmè   ninyè ausè-m
       piss   from.there witch old.woman-ERG
       fyènèfroryèng
       f=y-n-fror-Ø-ay-ng
       PROX1=α.3sG-VEN-spray-PFV.ND-REM.PUNC-2|3sGA
       mènat   *yèta* *fái*   *fès*   *wumbèrtoghèt.*   (ND13)
       mènat   yèta fái   fès   wumbèrto-gh-t.
       in.order.to ATT EMPH fire extinguish-NOM-ALL
       'The old witch sprayed piss from there in order to put out the fire.'

7.2 Complex sentences with adverbial clauses — **277**

Adverbial subordinate clauses of purpose can also be expressed not with a subordinating conjunction but rather with the modal *siya/sái* which indicates a goal or intention (section 5.3.2). Here it has the meaning 'in order to':

(844)   *Sáf afraye wuyogh yènmafngoi*
        Sáf   afraye wuyo-gh y-n-wafngo-Ø-ay-Ø
        face slowly lift-NOM α.3SGP-VEN-start-PFV.ND-REM.PUNC-2|3SGA
        fandaghèt   sái tinjongai.   (MD13(2))
        fanda-gh-t   **sái** t-winjo-Ø-ang-ay-Ø
        look.at-NOM-ALL **INT** β.3SGP-see-PFV.ND-INC-REM.PUNC-2|3SGA
        'He started to lift his face slowly in order to see her.'

7.2.5 Conditional adverbial clauses

Perhaps because the intentional modal *sái/siya* marks unrealised propositions or not yet achieved goals, it is also used with a hypothetical meaning in conditional adverbial clauses. Some examples:

(845)   *Fèm **siya** nèngèm,   efe   so tinjo?*   (NB7:79)
        fèm   **siya** n-ngèm   efe   so t-winjo-Ø-Ø
        2.ASB **INT** α.2SG$_P$-go.ND who.ABS FUT β.3SGP-see-PFV.ND-2|3SGA
        'If you go, who will you see?'

(846)   *Fá   **sái** narota   ághae,*
        fá   **sái** n-aro-ta-Ø   ághe-e
        3.ABS **INT** ØP-go.hunting-IPFV.ND-2|3SGA.IRR dog-INST
        wèrár   so   yèsmete.   (NB7:11)
        wèrár   so   y-sèmè-ta-e
        wallaby FUT α.3SGP-kill-IPFV.ND-2|3SGA
        'If he goes hunting with dogs, he'll kill a wallaby.'

(847)   *Fèmo **sia** tinjo   fainár,   táf*
        fèmo   **sia** t-winjo-Ø-Ø   fainár   táf
        2SG.ERG **INT** β.3SGP-see-PFV.ND-2|3SGA pineapple OBL1
        yèneta.   (NB5:77)
        y-ne-ta-Ø
        α.3SGP-eat-IPFV.ND-2|3SGA.IRR
        'If you see the pineapple, you must eat it.'

278 —— 7 Compound and complex sentences

The subordinating conjunction of time *njam* 'when' can also be used in conditional clauses as in the following. (Note the use of irrealis aspect in (849), as in (846) above.)

(848) *Ámb   ndenè   korayawèt*   *"yau   njam*
      ámb   ndenè   k-oray-ta-w-t   yau   njam
      some   like.this   β.ØP-say-IPFV.ND-REM.DUR-2|3NSGS$_A$ NEG   if
      kurtongi   *sófam*   *so*
      k-urto-Ø-ang-i   sóf-am   so
      β.ØP-get.out-PFV.ND-INC-2|3PLS$_A$ wave-ERG   FUT
      ewalitote".   (YD11(3))
      e-walito-ta-e
      α.2|3NSGP-sink-IPFV.ND-2|3SGA
      'Some (people) said "if you don't get out [of the water], a wave will sink you".'

(849) *Njam   Fransis   nambna*   *yènd   so   fronde*
      njam   Fransis   n-ambar-ta-Ø   yènd   so   fronde
      if   Francis   α.ØP-play-IPFV.ND-2|3SGS$_A$.IRR   1.ABS   FUT   first
      yènnèm.   (NB5:95)
      yèn-nèm
      α.1NSGS$_P$-come.ND
      'If Francis plays, we will win.'

Similarly, counterfactual clauses use *njam*, but with the remote tense:

(850) *Njam   Fransis   kambnau*   *yènd   so   fronde*
      njam   Fransis   k-ambar-ta-w-Ø   yènd   so   fronde
      if   Francis   β.ØP-play-IPFV.ND-REM.DUR-2|3SGS$_A$ 1.ABS   FUT   first
      tènnmorwèn   (NB5:95)
      tèn-nèmo-èrwèn
      β.1NSGS$_P$-come-REM.DUR.PA
      'If Francis had played, we would have won.'

7.2 Complex sentences with adverbial clauses — **279**

(851)   *Yau* **njam** *ausam*   *terangai*
(=487) yau **njam** ausè-m   t-werè-Ø-ang-ay-Ø
      NEG **if**   old.woman-ERG β.3SGP-hold-PFV.ND-INC-REM.PUNC-2|3SGA
      sèkw so   káráfnau   *yènd so*
      sèkw so   k-áráfár-ta-w-Ø   yènd so
      boat FUT β.ØP-break-IPFV.ND-REM.DUR-2|3SGS_A   1.ABS FUT
      kolitotawèm.   (YD11(3))
      k-olito-ta-w-m
      β.ØP-sink-IPFV.ND-REM.DUR-1NSGS_A
      'If the old woman had not held the canoe, it would have broken and we
      would have sunk.'

7.2.6 Adverbial clauses of reason

Adverbial clauses of reason are introduced by the subordinating conjunction
mènamèn 'because', once again possibly morphologically complex (Table 3.15).

(852)   *Dafi   nèkwafè   kamndangè*   **mènamèn**
      Dafi   nèkw-afè   k-amndè-Ø-ang-è   **mènamèn**
      Duffy anger-COM β.ØP-become-PFV.ND-INC-2|3SGS_A **because**
      yu   nayu tènmorwèn.   (YD11(3))
      yu   nayu tèn-mo-èrwèn
      place far   β.1NSGS_P-COP.ND-REM.DUR.PA
      'Duffy got angry because the place was so far away.'

(853)   *so   nangowèm*   **mènamèn** *ambum tèrtèr*
      so   n-ango-Ø-wè-m   **mènamèn** ambum tèrtèr
      FUT α.ØP-return-IPFV-DU-DA-1NSGSA **because** child   small.one
      kètè em   *mèngo-n.*   (TE11(1))
      kètè e-m   mèngo-n
      there α.2|3NSGS_P-COP.ND house-LOC1
      'we'll be returning because the small children are there at home.'

Adverbial clauses of reason are also used in apprehensive constructions.

(854)   *Fèmo*   *ngangè*   *táúmbna*   **mènamèn**
       fèmo   ngangè   tá-wúmbár-ta-Ø   **mènamèn**
       2SG.ERG   hand   β.2|3NSGP-wash-IPFV.ND-2|3SGA.IRR   **because**
       enjine   *mètè*   *tanè.*   (NB5:94)
       enjine   mètè   t-wanè-Ø-Ø
       sickness   POT1   β.2SGP-take-PFV.ND-2|3SGA
       'Wash your hands; otherwise you might get sick.'

Note that a reason can also be expressed by a simple sentence with a nominalised verb and the originative case marker:

(855)   *yènè*   *ghakèr*   *sèmègh**mèn***   *yerangi –*   *amaf*
       yènè   ghakèr   sèmè-gh-**mèn**   y-werè-Ø-ang-i   amaf
       DEM   young.man   hit-NOM-ORIG   α.3SGP-hold-PFV.ND-INC-2|3PLA   woman
       *sèmègh**mèn**.*   (PT:369,371)
       sèmè-gh-**mèn**
       hit-NOM-ORIG
       'they (the police) are holding the young man because of the beating – because of hitting the woman.'

7.3 Complex sentences with relative clauses

Relative clauses are embedded (subordinate) clauses that modify a nominal. Most Nama relative clauses use the Relative Pronoun strategy for relativisation (see Siegel 2019). This strategy is common in European languages but typologically unusual elsewhere (see section 8.1). According to Comrie (1998, 2006), the Relative Pronoun strategy has three defining characteristics: (1) the relativised nominal (the head) stays outside the relative clause; (2) the head is indicated by a pronoun that shows its semantic or syntactic role in the relative clause (by case marking or an adposition); and (3) this pronoun appears at the beginning of the relative clause. It is also pointed out (Comrie 2006: 137) that that relative pronoun does not have to be formally distinct from other types of pronouns, and "it is more frequent than not" that it is identical to an interrogative pronoun (e.g. English *who, which*) or a demonstrative pronoun (e.g. German *der*).

In Nama, the relativised nominal usually stays outside the relative clause and is indicated by a relative pronoun that shows its semantic or syntactic role in the

7.3 Complex sentences with relative clauses —— **281**

relative clause. This occurs most often at the beginning of the relative clause. There are five relative pronouns. Three indicate animate (usually human) nominals, two of these indicating syntactic roles (*emo* 'who ERGATIVE' and *efe* 'who(m) ABSOLUTIVE') and one a semantic role (*ende* 'to/for whom DATIVE'). Two other relative pronouns – *árkè* 'which' and *nèmè* 'what' – refer to either a non-human or attributive nominal. These are all identical in form to the interrogative pronouns (section 6.4.2).

Relative clauses occur relatively infrequently in Nama conversations and narratives, and when they do occur, they are often in very complex constructions. Therefore, the majority of examples used here to illustrate relative clauses come from elicitation. Further examples from naturalistic conversations and narratives are given below. (The relative clause is shown in square brackets.)

Here are examples in which the relative pronouns indicate human referents:

(856) *Yènè ár   [emo   wiwi   yanai]*   *tè*
      yènè ár   **emo**   wiwi   y-wanè-Ø-ay-Ø   tè
      DEM   man   **who.ERG** mango   a.3SGP-take-PFV.ND-REM.PUNC-2|3SGA   PRF
      nènufar.   (NB5:93)
      n-n-ufar-Ø-Ø
      a.ØP-VEN-arrive-PFV.ND-2|3SGS$_A$
      'The man who stole the mango has already arrived.'

(857) *Yèndo   yinjon   ár   [efe   túmèn*
      yèndo   y-winjo-Ø-èn   ár   **efe**   túmèn
      1SG.ERG   a.3SGP-see-PFV.ND-1SGA   man   **who.ABS** always
      náwátárne].   (NB5:94)
      n-áwátárán-ta-e
      a.ØP-complain-IPFV.ND-2|3SGA
      'I saw the man who is always complaining.'

(858) *Yèndo   yènè ár   yinjon   [efe   fèmo*
      yèndo   yènè ár   y-winjo-Ø-èn   **efe**   fèmo
      1SG.ERG   DEM   man   a.3SGP-see-PFV.ND1SGA   **who.ABS** 2SG.ERG
      yèsèm].   (RC-Qs)
      y-sèmè-Ø-Ø
      a.3SGP-hit-PFV.ND-2|3SGA
      'I saw the man who you hit.'

282 —— 7 Compound and complex sentences

(859) *Yèndo yènè amaf yinjon* **[*ende* *fèmo***
yèndo yènè amaf y-winjo-Ø-èn **ende** fèmo
1SG.ERG DEM woman ɑ.3SGP-see-PFV.ND-1SGA **who.DAT** 2SG.ERG
yame yaram.] (RC-Qs)
yame y-wa-ramè-Ø-Ø
mat ɑ.3SGP-APP-give-PFV.ND-2|3SGA
'I saw the woman who you gave the mat to.'

In these examples, the relative pronouns indicate inanimate referents:

(860) *Yèndo náifè yinjon* **[*árkè* *Mawai-e***
yèndo náifè y-winjo-Ø-èn **árkè** Mawai-e
1SG.ERG knife ɑ.3SGP-see-PFV.ND-1SGA **which** Mawai-DAT
yaramai]. (NB5:93)
y-wa-ramè-Ø-ay-Ø
ɑ.3SGP-APP-give-PFV.ND-REM.PUNC-2|3SGA
'I saw the knife which you gave to Mawai.'

(861) *Defidene mat* **[*nèmè* *náyárayèn]*** *yau*
Defid-e-ne mat **nèmè** n-áyárè-Ø-ay-èn yau
David-DAT-GEN report **what** ɑ.ØP-hear-PFV.ND-REM.PUNC-1SG$_A$ NEG
kètan nátmai. (TE13)
kètan n-átèm-Ø-ay-Ø
there ɑ.ØP-join-PFV.ND-REM.PUNC-2|3SG$_A$
'David's report that I heard doesn't fit there (i.e. doesn't seem right).'

All of the relative pronouns except *efe* 'who ABSOLUTIVE' can also take other case
suffixes – for example, the genitive *-ne*:

(862) *Yènd yinjon* *yènè ár* **[*ende-ne* *amaf***
Yènd y-winjo-Ø-èn yènè ár **ende-ne** amaf
1SG.ERG ɑ.3SGP-see-PFV.ND-1SGA DEM man **who.DAT-GEN** wife
násáfote Darun]. (NB5:93)
n-ásáfo-ta-e Daru-èn
ɑ.ØP-work-IPFV.ND-2|3SG$_A$ Daru-LOC1
'I saw the man whose wife works in Daru.'

The following examples demonstrate relative pronouns with the purposive, perlative and comitative suffixes:

7.3 Complex sentences with relative clauses — **283**

(863) *Yènè ár* [*endeot áuyè yèmánjne*]
yènè ár ende-ot áuyè y-mánjár-ta-e
DEM man **who.DAT-PURP** cassowary α.3SG-feed-IPFV.ND-2|3SGA
Darun yèmor. (RC-Qs)
Daru-n y-mor
Daru-LOC1 α.3SGS$_P$-stay
'The man who she's feeding the cassowary for lives in Daru.'

(864) *Yènè tètkafwe* [*árkamè tènnmorwèn*] *nu*
yènè tètkafwe **árkè-mè** tèn-nèmo-èrwèn nu
DEM creek **which-PERL** β.1NSGS$_P$-come-REM.DUR.PA water
kèfram tè yafaryèng. (RC-Qs)
kèfèr-am tè y-wafar-Ø-ay-ng
flood PRF α.3SG-cover-PFV.ND-REM.PUNC-2|3SGA
'The creek which we were coming along was flooded.'

(865) *Yènè ár yèm;* [*emofè Tonifè*
yènè ár y-m **emo-fè** Toni-fè
DEM man α.3SGSP-COP.ND **who-COM** Tony-COM
enre]. (RC-Qs)
e-nèr-e
α.2|3NSGS$_P$-come.DU-PFV.DU
'That's the man who Tony came with.'

Note that the comitative suffix goes with the ergative relative pronoun *emo* even though its grammatical function in the relative clause is absolutive. The same phenomenon occurs with the interrogative pronoun:

(866) *Emofè enèm.*
emo-fè e-nèm
who.ERG-COM α.2|3NSGS$_P$-come.ND
'Who did you come with?'

Similarly, the ergative pronominal nonsingular ending *fem* occurs with the ergative relative pronoun where one would expect the absolutive:

284 —— 7 Compound and complex sentences

(867)   *Tane*   *futárfet*   *wèi*
            tane   futár-f-e-t   wèi
            1SG.GEN   friends-PL-DAT-ALL   then
            táwánfetawèt   **[*emofem***
            tá-wá-nèfè-ta-w-t   **emofem**
            α.2|3NSGP-APP-cut-IPDV.ND-REM.DUR-2|3NSGA   **who.ERG.NSG**
            wagh   *kerèmndawèt].*   (MK13(3))
            wagh   k-erèmáng-ta-w-t
            singsing   β.ØP-dance-IPDV.ND-REM.DUR-2|3NSGS_A
            'They were cutting (the pig) for my friends who were dancing in the singsing.'

Other examples from recorded narratives and conversations follow:

(868)   *Yènè*   *tot*   *ambumom*   *mbingongo*   *yènè*
            yènè   tot   ambum-o-m   mbingongo   yènè
            DEM   young   youth-SG-ERG   jews.harp   DEM
            fyútáne   **[*emo***   *mbingongo-mèn*
            f=y-wútár-ta-e   **emo**   mbingongo-mèn
            PROX1=α.3SG-play-IPFV.ND-2|3SGA   **who.ERG**   jews.harp-ORIG
            si   *fyaitote].*   (DC33)
            si   f=y-waito-ta-e
            story   PROX1=α.3SG-tell-IPFV.ND-2|3SGA
            'This young man who just told a story about the jews-harp just played the jews-harp.'

(869)   . . .*yènè*   *árro*   *so*   *eronjan*
            yènè   ár-ro   so   e-ronja-ta-èn
            DEM   people-RSTR   FUT   α.2|3NSGP-look.for-IPFV.ND-1SGA
            [*efe*   *fifi*   *efaloye*   *násáfotat].*   (FD11(1))
            efe   fifi   efaloye   n-ásáfo-ta-t
            who.ABS   very   honestly   α.ØP-work-IPFV.ND-2|3NSGSA
            '. . .I will look for only people who work very honestly.'

(870)   . . .*yènè*   *árèm*   *yau*   *njam*   *tènetau*   *ámb*
            yènè   ár-m   yau   njam   t-ne-ta-w-Ø   ámb
            DEM   man-ERG   NEG   when   β.3SG-eat-IPDV.ND-REM.DUR-2|3SGA   some
            [*ende*   *yèmo*   *tèkmangai].*   (ND11(1))
            ende   yèmo   t-kèmè-Ø-ang-ay-Ø
            who.DAT   3SG.ERG   β.3SG-lay.down-PFV.ND-INC-REM.PUNC-2|3SGA
            '. . .when the man for whom he laid it down [the food] wasn't eating it.'

7.3 Complex sentences with relative clauses —— **285**

(871) *fèyo   mèrès   tosè   tútorngè*
fèyo   mèrès   tosè   t-wútor-Ø-ang-è
then   girl   young.one   β.3SG-take.out-PFV.ND-INC-2|3SGA
ewirone   mèinyotyo   [nèmè
e-wiror-ta-e   mèinyotyo   **nèmè**
α.2|3NSGP-take.off-IPFV.ND-2|3SGA   all   **what**
*tèfayotawèt   faru-faru].   (ND13)
t-fayo-ta-w-t   farufaru
β.3SG-put.on.top-IPFV.ND-REM.DUR-2|3NSGA   piled.up
'then the young girl took off everything that they piled up on top of her.'

(872) *Yèta   end   tèronjawèm   [árkamè   so*
yèta   end   t-ronja-ta-w-m   **árkè-mè   so**
ATT   road   β.3SG-look.for-IPFV.ND-REM.DUR-1NSGA   **which-PERL   FUT**
*kèlawangèm].   (YD11(3))
k-èlawè-Ø-ang-m
β.ØP-go.inside-PFV.ND-INC-1NSGA
'We were trying to look for a way along which to go inside.'

(See further discussion of the Relative Pronoun strategy in section 8.1.)
    While the Relative Pronoun construction clearly exists in Nama, some varia-
tion in word order can occur – i.e. the relative pronoun can be preceded by another
element of the relative clause – for example, compare the following two examples
to (858) and (860) above:

(873) *Yèndo   yènè   ár   yinjon   [fèmo   efe*
yèndo   yènè   ár   y-winjo-Ø- èn   fèmo   **efe**
1SG.ERG   DEM   man   α.3SGP-see-PFV.ND-1SGA   2SG.ERG   **who.ABS**
*yèsèm].   (RC-Qs)
y-sèmè-Ø-Ø
α.3SGP-hit-PFV.ND-2|3SGA
'I saw the man who you hit.'

(874) *Yèndo   náifè   yinjon   [Mawai-e   árkè*
yèndo   náifè   y-winjo-Ø-èn   Mawai-e   **árkè**
1SG.ERG   knife   α.3SGP-see-PFV.ND-1SGA   Mawai-DAT   **which**
*yaramai].   (NB5:93)
y-wa-ramè-Ø-ay-Ø
α.3SGP-APP-give-PFV.ND-REM.PUNC-2|3SGA
'I saw the knife which you gave to Mawai.'

286 —— 7 Compound and complex sentences

Nama also uses a non-reduction strategy for relativisation – i.e. internally headed relative clauses. This appears to be limited to use with the relative pronoun *árke* 'which' indicating an attributive nominal:

(875)   [*Árke*   *yáfèn*   *wem*   *yèfèl*]   *yabun*
         árke   yáf-n   wem   y-fèli-Ø-Ø   yabun
         which   basket-LOC1   yam   α.3sGP-put.inside-PFV.ND-2|3sGA   big
         yèm.   (NB8:69)
         y-m
         α.3sGS$_P$-COP.ND
         'The basket you put the yam in is big.'

(876)   *Yèmo*   *tètarne*   [*árkè*   *náifae*   *wèn*
         Yèmo   t-taran-ta-e   **árkè**   náifè-e   wèn
         3sG.ERG   β.3sGP-lose-IPFV.ND-2|3sGA   which   knife-INST   wood
         tafane].   (RC-Qs)
         ta-far-ta-e
         β.2|3NSGP-carve-IPFV.ND-2|3sGA
         'He lost the knife which he carved wood with.'

In addition to relative pronouns, the interrogative adverb *ánde* 'where' can function as a relative adverb to modify a nominal indicating a location – for example:

(877)   *Kètè*   *Rouku*   *mèngotuan*   [*ánde*   *tane*   *bafafe*
         kètè   Rouku   mèngotu-an   **ánde**   tane   bafè-af-e
         there   Rouku   village-LOC1   **where**   1sG.GEN   uncle-PL-DAT
         emor]. . .   (YD11(2))
         e-mor
         α.2|3NSGS$_P$-stay
         'There in Roku village where my uncles live. . .'

(878)   *mbèro-faf*   *nufareayèm*   [*ánde*   *ndènd*
         mbèro-faf   n-ufar-e-ay-m   **ánde**   ndènd
         muddy.place-ASS   α.ØP-arrive-PFV.DU-REM.PUNC-1NSGS$_A$ **where**   worm
         so   *tatarwèm*   *kwakoyot.*   (TE11(2))
         so   ta-tar-Ø-wè-m   kwako-yot
         FUT   β.2|3NSG-dig-IPFV.DU-DA-1NSGA   bait-PURP
         'we (2) arrived at the muddy place where we would dig worms for bait.'

There are no examples of *njam* 'when' or *nèmamèn* 'why' being used as relative adverbs.

7.4 Complex sentences with complement clauses

Complement clauses (also referred to as clausal arguments) in Nama are of three types: those introduced by the complementiser *nde* 'that', those without a complementiser, and those that begin with an interrogative/relative pronoun or adverb. (The complement clauses are shown in square brackets.)

7.4.1 Complement clauses introduced by *nde* 'that'

Complement clauses introduced by *nde* 'that' are commonly found with mental predicates – i.e. following verbs referring to thinking or perception. Some examples are:

(879) *nawinjeayènd*   [***nde***   *fènatayo*
n-a-winjo-e-ay-nd   **nde**   fènatayo
α.ØP-REFL-see-PFV.DU-REM.PUNC-2|3NSGS_A   **that**   different
tárwèn].   (MD13(1))
tá-r-wèn
β.2|3NSGS_P-COP.DU-REM.DUR.PA
'they (2) saw in each other that they were different.'

(880) *fá*   *mafina*   *tèmorwèn*   [***nde***   *mèrès*   *tukèn*
fá   mafina   t-mo-èrwèn   **nde**   mèrès   tuk-n
3.ABS   not.knowing   β.3SGS_P-COP.ND-REM.DUR.PA   **that**   girl   top-LOC1
yèm].   (MD13(1))
y-m
α.3SGS_P-COP.ND
'he didn't know that the girl was at the top (of the tree).'

(881)   *fèyo yènè mèrès tosè*   *syèm*   *kèma*
          fèyo yènè mèrès tosè   s=y-m   kèma
          then DEM girl   young.one PROX2=α.3SGSₚ-COP.ND idea
          námyoi   **[*nde***   *tane*   *alè*
          n-ámyo-Ø-ay-Ø   **nde**   tane   alè
          α.ØP-accept-PFV.ND-REM.PUNC-2|3SGSₐ   **that** 1SG.GEN grandmother
          fètè   *yèm*].   (ND13)
          fètè   y-m
          really α.3SGSP-COP.ND
          'then this young girl, she accepted that she (the witch) was really her grand-mother.'

The complementiser *nde* is also used to introduce indirect speech, following verbs meaning 'tell', 'say', etc.:

(882)   *Árèm*   *yènamè*   *swèmndetat*   **[*nde*** *yu*
          ár-m   yènamè   s=w-mèndè-ta-t   **nde** yu
          people-ERG from.here PROX2-α.1SGP-tell-IPFV.ND-2|3NSGA **that** place
          farorègh *syèm*].   (KS11)
          faror-gh   s=y-m
          split-NOM PROX2=α.3SGSₚ-COP.ND
          'People from here are telling me that the place is splitting.'

(883)   *Kominiti*   *tè*   *norayai*   **[*nde*** *yènamè*
          kominiti   tè   n-oray-Ø-ay-Ø   **nde** yènamè
          community PRF α.ØP-say-PFV.ND-REM.PUNC-2|3SGA **that** along.here
          watamegh *mèngo* *so*   *yèm*.]   (FD13)
          watame-gh mèngo so   y-m
          teach-NOM house FUT α.3SGSₚ-COP.ND
          'The community already said that the school building will be along here.'

(884)   *yènamamèn* *norayan*   **[*nde*** *yènd* *yúrè* *fifi*
          yènamamèn n-oray-ta-èn   **nde** yènd yúrè fifi
          that's.why α.ØP-say-IPFV.ND-1SGSₐ   **that** 1.ABS owner real
          yènèm].   (KS11)
          yèn-m
          α.1NSGSₚ-COP.ND
          'that's why I'm saying that we are the true owners.'

As shown in section 7.2.2, the complementiser *nde* also follows the subordinating conjunction *wènde* 'but'. This might be interpreted as a clausal complement of an adverbial phrase with *wènde* as a contrastive adverb. A parallel construction would be one with the adverb *ndenanit* 'similar to/like this':

(885) njamandyo tinjongè   **ndenánit** [*nde* tikfam
       njamandyo t-winjo-Ø-ang-è   **ndenánit** nde   tikèf-am
       as.soon.as β.3SG-see-PFV.ND-INC-2|3SGA **like.this** that   heart-ERG
       kár yèfrango.]   (MD13(1))
       kár y-frango-Ø-Ø
       DUB α.3SGP-leave-PFV.ND-2|3SGA
       'as soon as he saw her, it was like his heart might leave him.'

More frequently, a clausal complement with *nde* follows the adverb *ndenè* 'like this/that' with both mental predicates and quotatives:

(886) tinjongè   **ndenè**   [*nde* yabun kiwál   kár
       t-winjo-Ø-ang-è   **ndenè**   nde   yabun kiwál   kár
       β.3SG-see-PFV.ND-INC-2|3SGA **like.this** that big   west.wind   DUB
       nènuwano.]   (ND13)
       n-n-uwano-Ø-Ø
       α.ØP-VEN-set.off-PFV.ND-2|3SGS$_A$
       'she saw that a big westerly wind might have started.'

(887) yárafè   mat   namndai   **ndenè**
       yá-rafè   mat   n-amndè-Ø-ay-Ø   **ndenè**
       3SG-father knowing   α.ØP-become-PFV.ND-REM.PUNC-2|3SGS$_A$ **like.this**
       [*nde* yèmo   yam   yátmè   tè   tafrote.]   (MD11(2))
       nde yèmo   yam   yá-tmè   tè   t-wafro-ta-e
       that 3SG.ERG   action 3SG-mother PRF β.3SG-do-IPFV.ND-2|3SGA
       'his father came to know that he had done this action to his mother.'

290 —— 7 Compound and complex sentences

(888)   *amaf* *yèmndayèn*   **ndenè**   [*káye*
        amaf y-mèndè-Ø-ay-èn   **ndenè**   káye
        wife α.3sGP-tell-PFV.ND-REM.PUNC-1sGA **like this**   tomorrow
        swèm   *yènd* **nde** *so*   *wèyota*
        s=w-m   yènd **nde** so   w-yo-ta-Ø
        PROX2=α.1sGS$_P$-COP.ND 1.ABS **that** FUT α.1sGP-travel-IPFV.ND-2|3sGA.IRR
        ála-t.]   (MK13(2))
        ála-t
        hunting-ALL
        'I told my wife that tomorrow I'll go hunting.'

7.4.2 Complement clauses without a complementiser

Some direct quotatives occur with the adverb *ndenè* 'like this/that' but without the complementiser *nde* as in the following:

(889) *yátmam*   **ndenè** *emndete*   [*"afane*
        yá-tmè-am   **ndenè** e-mèndè-ta-e   afè-ne
        3sG-mother-ERG **like.this** α.2|3NSGP-tell-IPFV.ND-2|3sGA father-GEN
        si   *náyáretati*   *táf"*].   (PT:81)
        si   n-áyárè-ta-t-i   táf
        story α.ØP-listen-IPFV.ND-2|3NSGS$_A$-2PL OBL2
        'His mother is telling them "you should listen to father's story".'

(890) *yèmo* *wèi* **ndenè** *yènèmndeai*
        yèmo wèi **ndenè** yèn-mèndè-e-ay-Ø
        3sG.ERG now **like.this** α.1NSGP-tell-PFV.DU-REM.PUNC-2|3sGA
        [*"nayèt* *eyowè?"*].   (TE11(1))
        nayèt   e-yo-wè-Ø
        where   α.2|3NSGP-travel-DA-2|3sGA
        'now he said to us (2) "where are you (2) going?".'

Other complement clauses, including quotations (direct and indrect), occur without an adverb or any other marking – for example:

(891) *yáfindamèn*   [*mbi*
       y-wáfing-ta-m-èn   mbi
       α.3SGP-wait-IPFV.ND-REM.DLT-1SGA   sago
       nèitam   *kwartruan*].   (NP11)
       n-èy-ta-m-Ø   kwartru-an
       α.ØP-settle-IPFV.ND-REM.DLT-2|3SGS_A container-LOC1
       'I waited for the sago to settle in the container.'

(892) [*áuyè*   *yèfenetan*]   *amam*
        áuyè   y-fenè-ta-èn   amè-am
        cassowary   α.3SGP-feed-IPFV.ND-1SGA   mother-ERG
       wèngèmndè.   (DC20)
       w-ng-mèndè-Ø-Ø
       α.1SGP-AND-tell-PFV.ND-2|3SGA
       'mother told me to feed the cassowary.'

(893) *ADC-m*   *tènèmndangè*   [*ár*   *so*
        ADC-m   tèn-mèndè-Ø-ang-è   ár   so
        ACD-ERG   β.1NSGP-tell-PFV.ND-INC-2|3SGA   people   FUT
       enèm   *kètandmè*   *Mosbi*].   (MD99(1))
       e-nèm   kètandmè   Mosbi.
       α.2|3NSGS_P-come.ND   from.there   Moresby
       'The ADC [Assistant District Commissioner] told us that people will come
       from Port Moresby.'

7.4.3 Complement clauses with an interrogative/relative pronoun or adverb

The final type of complement clause is introduced by a word which functions elsewhere as an interrogative or relative pronoun, or an adverb (sections 6.4 and 7.3):

(894) *mwighayan*   *kemnangè*   [**emo**
(=78)   mwighè-yan   k-emán-Ø-ang-è   **emo**
        mind-LOC2   β.ØP-think-PFV.ND-INC-2|3SGS_A **who.ERG**
       wèláite].   (MD13(1))
       w-lái-ta-e
       α.1SG-shoot.at-IPFV.ND-2|3SGA
       'in his mind he thought who is shooting at me.'

292 — 7 Compound and complex sentences

(895)   *fèyo* *kamotangend*   [*fèm*   ***efe***
(=339) *fèyo* k-a-mo-tang-e-nd   fèm   **efe**
        then β.ØP-REFL-ask-INC.IPFV-PFV.DU-2|3DUA 2.ABS **who.ABS**
        nèm].   (MD13(1))
        n-m
        α.2SGS_P-COP.ND
        'then they started asking each other who are you.'

(896)   *yènè* *mato* *tèfanda*   [*efe*
        yènè mato t-fanda-ta-Ø   **efe**
        DEM INV β.3SGP-look.at-IPFV.ND-2|3SGA.IRR **who.ABS**
        yènè *yèm*].   (PT:605)
        yènè y-m
        DEM α.3SGSP-COP.ND
        'try and look at who that is.'

(897)   *korayangè*   [*nèmè* *tèm*
        k-oray-Ø-ang-è   **nèmè** t-m
        β.ØP-say-PFV.ND-INC-2|3SGS_A **what** β.3SGS_P-COP.ND
        kemnangè].   (ND13)
        k-emán-Ø-ang-è
        β.ØP-think-PFV.ND-INC-2|3SGS_A
        'she said what it was she thought.'

The interrogative/relative pronoun can also be marked with nominal morphology:

(898)   *fá*   *ámrorègh* *wèi* *kufngend*   *tèfen-tèfen*
        fá   ámror-gh wèi k-ufngo-e-nd   tèfentèfen
        3.ABS argue-NOM now β.ØP-start-PFV.DU-2|3DUS$_A$ blaming
        kawafngend   [*endene* *tèfen*
        k-a-wafngo-e-nd   ende-ne fault
        β.ØP-REFL-start-PFV.DU-2|3DUS$_A$ **who.DAT-GEN** fault
        yèm].   (MD99(1))
        y-m
        α.3SGS_P-COP.ND
        'The two of them started arguing now, and started blaming [the one] whose fault it is.'

7.4 Complex sentences with complement clauses ⸺ **293**

(899) Yèndo   yèronjan   [**nèmayan**   wem
       yèndo   y-ronja-ta-èn   **nèmè-yan**   wem
       1SG.ERG α.3SGP-look.for-IPFV.ND-1SGA **what-LOC2** yam
       tèfèl].   (NB8:69)
       t-fèli-Ø-Ø
       β.3SGP-put.inside-PFV.ND-2|3SGA
       'I'm looking for what you put the yam in.'

These examples use interrogative adverbs (section 6.4.2):

(900) yèmofem   fifi   so   yúrtotat   so
       yèmofem   fifi   so   y-wúrto-ta-t   so
       3NSG.ERG   body   FUT   α.3SGP-take.out-IPFV.ND-2|3NSGA FUT
       yèngfandat   [**nèmamèn**   kèr
       y-ng-fanda-ta-t   **nèmamèn**   kèr
       α.3SGP-AND-look.at-IPFV.ND-2|3NSGA **why**   dead
       tèmong].   (MD99(1))
       t-mo-ang
       β.3SGSP-COP.ND-INC
       'they will take out the body to look at why he died.'

(901) yèndon   ámb   si   so   nafrotan   a
       yèndon   ámb   si   so   n-afro-ta-èn   a
       1SG.ERG   some   story FUT   α.ØP-report-IPFV.ND-1SGSA and
       norayan   mènamèn   áki-áki   kèraf   [**ndernae**
       n-oray-ta-n   mènamèn   ákiáki   kèr-af **ndernae**
       α.ØP-talk-IPFV.ND-1SGSA about   ancestors dead-PL **how**
       támorangèrwèng].   (ND11)
       tá-mor-ang-èrwèng
       β.2|3NSGSP-stay-INC-REM.DUR.PA
       'I will tell a story and talk about how the ancestors lived.'

(902) Jonèm   nufènmete   [**nderè**   nayu   wèrár
       Jon-m   n-ufènmán-ta-e   **nderè**   nayu   wèrár
       John-ERG α.ØP-estimate-IPFV.ND-2|3SGSA **how**   far   wallaby
       yèmorang].   (NB8:13)
       y-mor-ang
       α.3SGSP-stay-INC
       'John is estimating how distant the wallaby is.'

294 —— 7 Compound and complex sentences

Interrogative pronouns functioning as indefinite pronouns also commonly occur
in complement clauses:

(903) [*emo*⁴⁹   *ámbitè*   *njam*
       emo   ámbitè   njam
       who.ERG married.without.exchange when
       kawerangè]   *kètè* *mè* *so* *yènè* *ár*
       k-awerè-Ø-ang-è   kètè   mè   so   yènè   ár
       β.ØP-get.married-PFV.ND-INC-2|3SGS_A   there   CONT FUT   DEM   people
       yámnendat.   *yámblalat.*   (ND11)
       y-wámneng-ta-t   yá-mblalè-t
       α.3SGP-ask.for-IPFV.ND-2|3NSGA   3SG-exchange.girl-ALL
       'whoever is without an exchange partner when they get married, they will
       keep asking him for his exchange girl.'

(904) *Yènd* *fronde* *wèm;*   [*efe*
       yènd   fronde   w-m   **efe**
       1.ABS   first   α.1SGS_P-COP.ND **who.ABS**
       nènorayat   *yau*] *tane* *soramè*
       n-n-oray-ta-t   yau   tane   soramè
       α.ØP-VEN-say-IPFV.ND-2|3NSGS_A   NEG   1SG.GEN later
       enèm.   (KS11)
       e-nèm
       α.2|3NSGS_P-come.ND
       'I am the first; whoever says not came after me.'

(905) [*yáne* *yaufi* **nèmè** *tèmorwèn*]
       yáne   yaufi   **nèmè** t-mo-èrwèn
       3SG.GEN bad   **what** β.3SGS_P-COP.ND-REM.DUR.PA
       tèngwawitawèn.   (FD113)
       t-ng-wawi-ta-w-èn
       β.3SGP-AND-forgive-IPFV.ND-REM.DUR-1SGA
       'I was forgiving whatever his bad behaviour was.'

---

49 It is not clear why the ergative form of the interrogative pronoun occurs in this context.

(906) [**nèmè** njamke tánetau]   yènè
      nèmè njamke tá-ne-ta-w-Ø   yènè
      what food β.2|3NSGP-eat-IPFV.ND-REM.DUR-2|3SGA   DEM
      tánwughánau   *wagifam yènè*
      tá-n-wughár-ta-w-Ø   wagif-am yènè
      β.2|3NSGP-VEN-submerge-IPFV.ND-REM.DUR-2|3SGA fish-ERG DEM
      ftánetawèt.   (MD13(1))
      f=tá-ne-ta-w-t
      PROX2=β.2|3NSGP-eat-IPFV.ND-REM.DUR-2|3NSGA
      'whatever food she was eating she dropped in the water and the fish just ate it.'

7.5 Complex sentences with focus marking clauses

A subordinate clause with the nondual copula \m/ (or its variants \mo/ or \mon/) or the dual copula \r/ is used to emphasise or focus on an argument of a sentence. Such constructions are the functional equivalent of cleft sentences or left dislocation in English. The overt argument in the focus marking clause is extracted from the main clause, but unlike in English, it is not referred to with a pronoun or relative pronoun in the main clause; rather it is indexed by an A/S_A-indexing suffix or P-indexing prefix.

In some instances, the copula in the focus marking clause occurs following the immediacy adverb *kès* 'just here/there' (section 5.3.6):

(907) [*yènè rabè wèn* **kès** **em**]   *yèndo so*
      yènè rabè wèn **kès** **e-m**   yèndo so
      DEM rubber tree **just.there** **α.2|3NSGSP-COP.ND** 1SG.ERG FUT
      enfetan   *naifae.*   (YD11(3))
      e-nèfè-ta-èn   naifè-e
      α.2|3NSGSP-cut-IPFV.ND-1SGA   knife-INST
      'those rubber trees, I will cut them down with a bushknife.'

(908) [*fá* **kès** **yèmo**]   *yènèm.*   (ND13)
      fá **kès** **y-mo**   y-nèm
      3.ABS **just.there** **α.3SGSP-COP.ND**   α.3SGSP-come.ND
      'she's the one who came.'

296 —— 7 Compound and complex sentences

(909) [*susi*   *ewumbèwan*   **kès**
      susi   e-wumbè-Ø-wè-èn   **kès**
      fishing.line   α.2|3NSGP-tie.up-IPFV.DU-DA-1SGA   **just.there**
      ere]   *tawifen.*   (MK13(1))
      e-r-e   ta-wifo-e-èn
      α.2|3NSGSₚ-COP.DU-PFV-DU   β.2|3NSGP-finish-PFV.DU-1SGA
      'the (2) fishing lines I was tying up, I finished them.'

But far more frequently, focus constructions use the copula with the locational proximal clitic *s=*, which most likely originated from the immediacy adverb *kès*. The majority of the examples in the data are with the 3rd person singular form of the copula in the current tense:

(910) [*tane*   *ausè*   ***syèm***]   *mè*   *kès*
      tane   ausè   **s=y-m**   mè   kès
      1SG.GEN   old.woman   **PROX2=α.3SGSₚ-COP.ND**   CONT   just.there
      yakái.   (B&B:78)
      y-wakáy(o)
      α.3SGSₚ-be.standing
      'my grandmother, she's still standing.'

(911) [*yènè*   *áuyè*   ***syèm***]   *nandimbweot*   *so*
      yènè   áuyè   **s=y-m**   nandimbwe-ot   so
      DEM   cassowary   **PROX2=α.3SGSₚ-COP.ND**   money-PURP   FUT
      yèrsan   Moet   tesenot
      y-rèsa-ta-èn   Moet   tesen-ot
      α.3SGP-carry-IPFV.ND-1SGA   Morehead   station-PURP
      maketot.   (MD99(1))
      maket-ot
      market-PURP
      'This cassowary, I'll take it to Morehead station to the market to sell it.'

(912) *Tèmndangè*   [*tane*   *mèngo*   *syèm*]
      t-mèndè-Ø-ang-è   tane   mèngo   **s=y-m**
      β.3SGP-tell-PFV.ND-INC-2|3SGA   1SG.GEN   house   **PROX2=α.3SGSₚ-COP.ND**
      nayu   *yèm.*   (MD13(1))
      nayu   y-m
      far   α.3SGSₚ-COP.ND
      'He told her: my house, it's far away.'

7.5 Complex sentences with focus marking clauses — **297**

(913) [*yènè*   *yawar*   **syèmo*]**   *kètèf*
      yènè   yawar   **s=y-mo**   kètè=f
      DEM   black.palm   **PROX2=α.3SGSₚ-COP.ND**   there=PROX1
      yèfotamèng   *Kamblèm.*   (SK11(1))
      y-fo-ta-m-ng   Kambèl-m
      α.3SGP-chop.down-IPFV.ND-REM.DLT-2|3SGA   Kambèl-ERG
      'this black palm, Kambal just chopped it down there.'

The same focus construction is also used with other tenses and other persons:

(914) [*Yènè*   *mèrès*   **stèmorwèn*]**   *yabun*   *wèn*
      yènè   mèrès   **s=t-mo-èrwèn**   yabun   wèn
      DEM   girl   **PROX2=β.3SGSₚ-COP.ND-REM.DUR.PA**   big   tree
      tukèn   *tèfayangèrwèn.*   (MD13(1))
      tuk-n   t-fayo-ang-èrwèn
      top-LOC1   β.3SGSₚ-be.on.top-INC-REM.DUR.PA
      'This girl, she was on top of a big tree.'

(915) [*yènd*   **swèm*]**   *afè*   *yau*   *wèm;*   [*yènd*
      yènd   **s=w-m**   afè   yau   w-m   yènd
      1.ABS   **PROX2=α.1SGSₚ-COP.ND**   father   NEG   α.1SGSₚ-COP.ND   1.ABS
      swèm*]   *mè*   *watamegh*   *tèfnárro*
      s=w-m   mè   watame-gh   tèfnár-ro
      PROX2=α.1SGSₚ-COP.ND   CONT   teach-NOM   practitioner-RSTR
      wèm.   (FD13)
      w-m
      α.1SGSₚ-COP.ND
      'Me, I'm not a father; me, I'm still only a teacher.'

(916) *yèndo*   *tèmndan*   *"ok*   *yèna*   [*fèm*
      yèndo   t-mèndè-Ø-èn   ok   yèna   fèm
      1SG.ERG   β.3SGP-tell-PFV.ND-1SGA   ok   here   2.ABS
      snèm]**   *kuwano".*   (FD13)
      s=n-m   k-uwano-Ø-Ø
      PROX2=α.2SGSₚ-COP.ND   β.ØP-set.off-PFV.ND-2|3SGSₐ
      'I told him "ok, you, you set off.'

298 —— 7 Compound and complex sentences

(917) [*wèrár* **sem**]   *mènde* *yau*
       wèrár   **s=e-m**   mènde   yau
       wallaby **PROX2=α.2|3NSGS$_P$-COP.ND** liking NEG
       nánfete.   (SK11(2))
       n-ánfè-ta-e
       α.ØP-taste-IPFV.ND-2|3SGSA
       'the wallabies, they don't taste good.'

However, it may be that the 3rd person singular form – *kès yèm(o)* or *syèm(o)/
syèmon* – is becoming grammaticalised as a focus marker, as it is used with other
persons and numbers. Here are three examples:

(918) [*yènè...* *mwighè* *ndimbal* *ndimbal* **kès**   **yèm**]   *yau*
       yènè   mwighè   ndimbal   ndimbal   **kès**   **y-m**   NEG
       DEM   thought   big   big   **just.there** **α.3SGS$_P$-COP.ND**
       so *tawawalindangè.*   (B&B:142)
       SO   ta-wawaling-tang-è
       FUT   β.2|3NSGP-forget-INC.IPFV-2|3SGA
       'These big big memories, you won't forget them.'

(919) [*fèm* **syèm**]   *ninyafè* *fèfè* *ere*
       fèm   **s=y-m**   ninyè-afè fèfè   e-r-e
       2.ABS **PROX2=α.3SGS$_P$-COP.ND**   witch-COM INCH1   α.2|3NSGS$_P$-COP.DU-PFV.DU
       unghèt.   (ND13)
       unè-gh-t
       net.fish-NOM-ALL
       'you, you were with the witch netting fish.'

(920) [wagif **syèmon**]
       wagif   **s=y-mon**
       fish   **PROX2=α.3SGS$_P$-COP.ND**
       tawèlongai.   (ND13)
       ta-wèlo-Ø-ang-ay-Ø
       β.2|3NSGP-swallow-PFV.ND-INC-2|3SGA
       'the fish, she swallowed them.'

8 Typological implications

The examination of the Nama language, along with other previously undocumented languages of southern New Guinea, has revealed many features that underline the importance of language documentation to language typology. For example, it was thought that no existing language has a senary (or base 6) counting system (Plank 2009). But, as described in section 1.1.3, Nama and other languages of the Yam family have such a system for ceremonial counting of yams. This short chapter looks at other features of Nama that have implications for linguistic typology in that they are unusual or even unique, or that they contradict accepted generalisations about linguistic properties and their distribution.

8.1 Relative clauses

Section 7.3 demonstrated that Nama uses the Relative Pronoun strategy in relative clauses (Comrie 1998, 2006). This has significant implications regarding the geographic distribution of this strategy, as described in Siegel (2019). Comrie (1998: 61) stated that:

> relative clauses formed using the relative pronoun strategy are quite exceptional outside Europe, except as a recent result of the influence of European languages. . . The relative pronoun strategy thus seems to be a remarkable areal typological feature of European languages, especially the standard written languages.

This statement was quoted by Haspelmath (2001: 1495), who considered this strategy to be a salient feature of Standard Average European. Comrie (2006: 136) later reiterated that "the Relative Pronoun strategy is, by and large, if not exclusively, restricted in its areal coverage to languages spoken in Europe, plus languages that have been in areal contact with languages of Europe." Comrie asserted that this is an areal rather than genetic feature as Indo-European languages outside Europe (e.g. Persian) do not have it while some non-Indo-European languages in Europe (e.g. Hungarian) do have it.

This view is repeated in the World Atlas of Language Structures (WALS): "Note that this strategy stands out as being typically European since it is not found in Indo-European languages spoken outside Europe, and is exceptional more generally outside Europe" (Comrie & Kuteva 2013a, 2013b). Maps 122A and 123A in WALS show that the Relative Pronoun strategy predominates in Europe in relativisation of both subjects and obliques, and only one language outside the European region

https://doi.org/10.1515/9783111077017-008

300 — 8 Typological implications

uses this strategy: Acoma, a Pueblo language of the Keres family, spoken in New Mexico (USA).[50]

Clearly, views about the geographic distribution of the Relative Pronoun strategy may need to be reconsidered on the basis of the data from Nama and other Yam family languages.

8.2 Verbs

Nama verbs have several significant features with regard to the following: tense, aspect (specifically perfectivity), grammatical number distinction and morphological discord as a stylistic device.

8.2.1 Tense

Chapter 4 (section 4.7.1) describes how in Nama the imperfective current tense covers habitual, progressive and iterative actions or events that are occurring at the time of speaking. This would normally be called present tense. However, in Nama, this tense also covers actions or events that occurred earlier in the day. This also occurs in the closely related languages, Nen and Nambo, where it is labelled "non-prehodiernal tense" (Evans 2015b; Kashima 2020). Other languages have a tense for the current day, termed HODIERNAL. These include the Bantu languages Haya, Luganda and Mwera (Comrie 1985: 90,93; Bybee, Perkins & Pagliuca 1994: 101), Ancash Quechua (Comrie 1985: 94), and Ngkolmpu, another Yam language of southern New Guinea (Carroll 2017: 184). In these languages, however, the tense is past hodiernal, referring to actions or events that occurred before the time of speaking but earlier in the same day. As far as I am aware, no other language outside the Nambu subgroup of the Yam family has merged the present and past hodiernal tenses into one category, as we find with the current tense in Nama.

---

50 However, personal communication from Bernard Comrie and Tania Kuteva (1 May 2018) states that after more detailed analysis of their source, they now believe that Acoma does not manifest the Relative Pronoun strategy, but rather the Gap strategy. A correction has been posted to WALS <<http://blog.wals.info/datapoint-122a-wals_code_aco/comment-page-1/#comment-174512>.

8.2.2 Perfectivity

As shown in section 4.6.1, Nama, like other languages, makes an important distinction between perfective and imperfective aspect. As in other languages, the category of imperfective pays attention to the internal structure of an event, encompassing progressive, habitual and iterative aspects. Perfective focusses on the event as an unanalysed whole, most often on a punctual event or on the boundaries of a nonpunctual event. According to most accounts of perfective aspect, the boundary of the nonpunctual event that is focussed on is its end point, located in the past (e.g. Dahl 1985: 78). In WALS, Dahl and Velupillai (2013) characterise the perfective distinction as "used exclusively or almost exclusively for single completed events in the past". They go on to say: "To be interpreted as a perfective... a form should be the default way of referring to a completed event in the language in question." As we have seen in Nama, however, perfective aspect marks punctual events and the commencement of non-punctual events, encompassing inceptive and inchoative aspects. This focus on inception rather than completion in perfective marking contradicts the generally accepted view.

Another notable feature in Nama is also related to the perfective-imperfective distinction. This is the combination of perfective and imperfective inflections to produce inceptive imperfective aspect (section 4.9). The combination of perfective and imperfective occurs in other languages, such as Georgian, Bulgarian and Russian, but with a stem and an affix. For example, in Bulgarian, an imperfective suffix is added to a perfective stem to show the whole as ongoing or habitual (Miller 2006: 150). And in Russian, a "perfectivizing" prefix is added to an imperfective stem to give the nuance of "shortness of duration" (Dahl 1985: 77). Also, in Nama's neighbouring language, Komnzo, iterative aspect is indicated by prefixes that normally mark imperfective aspects occurring with verb stems that are normally perfective (Döhler 2018: 183, 241). In Nama, however, the imperfective suffix *-ta* and the perfective inceptive suffix *-ang* appear to be combined and occur together on one stem, along with the perfective suffix *-e* when the verb has an argument with two referents.

The most remarkable characteristic of perfective-imperfective marking is the use of dual number to distinguish the two major aspectual categories. Although not a common category, dual grammatical number is found in many modern languages (Corbett 2000). In Slovene and Modern Standard Arabic, for example, dual number is marked on nouns, adjectives and pronouns, and referenced by verbal inflections or special forms of the verb. In Oceanic Austronesian languages, such as Hawaiian and Fijian, it is marked only in personal pronouns. As noted in section 4.6.1, Nama is unusual in that dual number is generally not marked on nouns and not distinguished from plural (3 or more) in pronouns or the verbal prefixes that index P arguments. These usually distinguish only between singular and nonsingular, which includes dual and plural. Only in some perfective tenses and aspects (and in the inceptive imperfective) are there separate dual and plural suffixes indexing A or S_A arguments. However, more striking is the fact that semantic space for grammatical number is divided into two categories: dual (2), and nondual, incorporating both singular and plural (3 or more). This is discussed in the following subsection.

8.2.3 Dual versus nondual

The dual-nondual distinction in Nama is also relevant to other areas of verb marking besides perfectivity. P-aligned intransitives are all perfective, but those that have an S argument with a dual referent have special marking. As shown in section 4.11.1, in addition to the dual perfective suffix *-e*, there is another suffix *-ar* that indicates dual. This suffix, however, does not occur with the copula. Instead, there are two different forms of the copula; \r/ DUAL and \m(o)/ NONDUAL (section 4.11.2). This is also true for the related forms for 'come' and 'go': \nèr/ 'come.DU', \nèm(o)/ 'come.ND and \ngèr/ 'go.DU', \ngèm(o)/ 'go.ND.

To my knowledge, dual versus nondual has not been reported in the literature as a grammatical distinction in core areas of grammar, as it is in Nama and in closely related Nen (e.g. Evans & Levinson 2009) and Nambo (Kashima 2020).[51] In grammatical number systems, four types are generally recognised: (1) no number distinction; (2) singular/plural distinction; (3) singular/dual/plural distinction and (4) singular/dual/trial (or paucal)/plural distinction. Two of these have a dual category, but it does not contrast with only a single other category – nondual, which conflates singular and plural. Only these four types are considered in framing typological universals for grammatical categories based on number – for example, the implicational hierarchy based on Greenberg's Universal 34 (Greenberg 1963):

Singular < Dual < Trial < Plural

Croft (2003, 2004) uses the semantic map model to show the division of the cardinal number conceptual space for the different types grammatical distinctions based on number (see Figure 8.1):

---

51 The only other language I am aware of that has this distinction is Kiowa, a Native American language (Wonderly, Gibson & Kirk 1954), but only for a small group of nouns (class III). These nouns are implicitly dual, but when they are suffixed with the the morpheme marking "inverse", they become nondual – i.e. either singular or plural (3+). (Thanks go to an anonymous reviewer for alerting me to this article.)

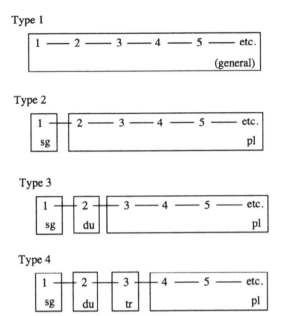

Figure 8.1: Four types of grammatical distinctions based on number (Croft 2004).

To this, we can add a map of the dual/nondual distinction found in Nama and related languages (Figure 8.2).

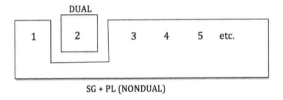

Figure 8.2: Dual/nondual grammatical distinction.

Figure 8.1 shows that in the four previously attested types of grammatical distinctions based on number partitions between categories, when they occur, are between consecutive positive integers. This property has been taken as granted and is intrinsic to the implicational hierarchy referred to above. It has also been intrinsic to universals proposed to exclude "bizarre types" (Croft 2003: 51) of grammatical marking systems. For example, Greenberg (1969) noted that out of an infinite number of logically possible systems of grammatical number categories, only a few have been empirically attested in language. (Greenberg considered only types 1–3.) He proposed generalisations (in the form of two axioms) that would account for these systems but prohibit those that are not attested. However, although not explicitly stated, these axioms depend on the partition between categories being between consecutive positive integers.

It is clear from Figure 8.2, that the partition between the categories of dual and nondual is not between consecutive positive integers, and that nonconsecutive cardinal numbers (1 and 3+) are included in the same category. Consequently, this grammatical feature of Nama appears to warrant a re-evaluation of universals that have been proposed concerning grammatical categories based on number.

8.2.4 Morphological discord

Section 6.9 showed how in Nama morphological discord with verbal inflections is used as a stylistic device. A similar phenomenon occurs in English, where morphological discord in the verb phrase is used for joking or stylistic effect. Examples are "Let's went" (popularised by Poncho in the 1950s TV series *The Cisco Kid*), and a heading on an internet discussion list: "3D TV, I has one . . . I likes it".[52] (This is perhaps inspired by the speech of Gollum/Sméagol in *The Hobbit* and *Lord of the Rings* – e.g.: "I wants it" and "We hates it forever!" .)

The morphological discord in Nama is found mostly with stative (P-aligned intransitive) verbs to signal that something is out of the ordinary – i.e. remarkably large or small in size or number, or lasting an unusually long time. The discord generally involves singular P-indexing prefix co-occurring with the dual suffix and dual perfective marker.

Other languages closely related to Nama also juxtapose discordant inflections: Komnzo (Döhler 2018), Nen (Evans 2019b) and Nambo (Kashima 2020). In these languages, however, this juxtaposition is analysed as the grammatical means of marking a numerical category referred to as "large plural" or "greater plural". Unlike in Nama, the discord does not appear to be salient, and the construction does not have interpretations other than large or greater plural.

Evans (2019b: 112) shows that greater plural in Nen is indicated on some stative verbs (which he calls "positionals") by the combination of singular and dual inflections. This is basically the same as what occurs with P-aligned verbs in Nama. In Nen, however, marking of greater plural occurs with a wider range of verb types, and in most cases this is done similarly by combining other verb inflections that previously did not co-occur because of semantic constraints. For example, with

---

52 Post by mattthomas, 1 October 2012, on the RationalSkepticism.org website. http://www.rational skepticism.org/the-arts/3d-tv-i-has-one-i-likes-it-t34886-20.html, retrieved on 28 October, 2018.

"middle verbs", which are equivalent to A-aligned intransitives in Nama, greater plural is indicated as follows: A nonsingular undergoer (U) prefix (P-indexing prefix in Nama) that normally occurs with transitive verbs is used instead of the invariant *n-* prefix (marked M in Nen, ØP in Nama) that is normally used for middle verbs. This co-occurs with the singular "actor suffix" (marked A in Nen, and S_A in Nama), as illustrated in the following examples:

(921)   Nen (Evans 2019b: 114)
    a. *amni   naebndat.*
       amni   n-aebn-da-t
       bird   M:α-fly-IPFV.ND-2|3NSGA
       'some birds are flying.'
    b. *amni   yawaebnde.*
       amni   yaw-aebn-d-e
       bird   2|3nsgU:α-fly-IPFV.ND-2|3sGA
       'many birds are flying.'

According to Evans (2019b: 101), the category of greater plural is a recent innovation in Nen that increased "semantic expressivity" by the addition of a fourth category of number indicated by verbal inflections. This was the result of what he calls "opportunistic recombination of existing forms". For example, with regard to positionals in Nen, since a singular prefix and a dual suffix did not previously co-occur because of semantic clash, this combination was available to mark a new grammatical category – i.e. greater number. Evans also states (p.101) that in such constructions, the original meanings or functions of the individual combined morphemes are suppressed.

To find out if such morphological discord occurs in other languages and if it can be conventionalised with a particular meaning or function, I posted a query on the Lingtyp discussion list for the Association of Linguistic Typology[53] (27 November 2018). Randy LaPolla replied (28 November 2018) with examples from the Rawang language of Kachin State, Myanmar (Tibeto-Burman). In this language there are markers for four degrees of past tense: a couple of hours ago; more than a couple of hours ago but today; within the last few days (recent past); and remote past), plus there is nonpast. He noted that discord was used in some cases for evidential purposes – specifically that nonpast declarative marking was used for a past event to show direct experience.

---

[53] https://listserv.linguistlist.org/pipermail/lingtyp/2018-November/. The relevant posts are in the thread "Functional discord?".

306 —— 8 Typological implications

The following example from a conversation was provided, in which the speaker was talking about someone saying something they heard someone else say. In reporting what the other person said, the remote past marker *yv̀ng* co-occurs with the nonpast declarative marker *-ē* to show that the person who the speaker heard say the words actually heard the original person – i.e. that it wasn't hearsay:

(922) Rawang (Randy LaPolla, Lingtyp list, 28.11.2018)

...vyà	yv̀ng-ē	wā
say	REMOTE.PAST-NONPAST	say

In another example (elicited), the recent past marker *dø̀r* co-occurs with the nonpast declarative marker:

(923) Rawang (Randy LaPolla, Lingtyp list, 28.11.2018)

dī	dø̀r-ē
go/walk	RECENT.PAST-NONPAST

'(he) was walking (I saw him).'

David Gil also replied (29 November 2018) with an example from Hebrew. He explained that Hebrew verbs have seven morphological paradigms, called *binyanim*, that express categories of voice, aksionsart and others. Some of these form two pairs productively related to each other as active and passive. The passive is characterised by the presence of a *-u-a-* vocalic pattern. For example, the root *k-t-b* 'write' forms the basis for the following (the cited forms are all past 3rd person singular masculine):

(924)   *kitev* 'address' (active) ~ *kutav* 'be addressed' (passive)

Another *binyan* expresses various medial categories such as reflexive, or in the case of *k-t-b*, reciprocal:

(925)   *hitkatev* 'write to each other'

The root *p-t-r* can have the meaning 'fire' (as in 'to dismiss from a job'). Exhibiting the same *binyan* as in (924) we have:

(926)   *piter* 'fire' (active) ~ *putar* 'be fired' (passive)

Exhibiting the same *binyan* as in (925), this time with a medial/reflexive meaning, the form would be:

8.2 Verbs — **307**

(927)   *hitpater* 'resign'

Gil demonstrated that, for this specific root, and for one particular context, Hebrew recently innovated an 8th *binyan*, derived by blending the features of two different pre-existing *binyanim* that normally do not co-occur: the medial as in (927) and the passive in (926):

(928)   *hitputar*

According to Gil, this seems to be in "a playful, almost jocular manner" and is "applied to those not uncommon situations in which politicians are forced to submit their resignation: officially, the politicians are resigning of their own free will, but in practice they are being fired". Gil concluded: "As far as I know, this innovated *binyan* applies just to the single root p-t-r, though I wouldn't be surprised if it extends, or perhaps already has extended, to other roots."

In a later post to the list (27 November 2018), Moshe Taube provided an additional example:

(929)   *hitnudav* 'he was made to volunteer'.

And subsequently, Gil also provided another example (p.c., 6 September 2021):

(930)   *hitguyas* 'he was made to enlist (i.e. he was drafted)'

A similar phenomenon has been reported for Mandarin Chinese by Kunze (2011).[54] The passive marker 被 *bei* is used on the internet in "grammatically impossible situations". For example, it occurs with the intransitive verb 自杀 *zisha* 'to commit suicide'. According to Kunze (2011:12), "the phrase *bei zisha* 被自杀 (*be* suicid*ized*) sounds logically and linguistically absurd". But it was used online to discuss the officially announced suicide of a man who was put in prison after reporting corruption. Many people thought he was actually forced to commit suicide or murdered in prison, and this morphological discord was used to express this view. An ungrammatical use of the passive with a similar function occurs in English with intransitive verb *disappear* – for example, in talking about the many Argentinians who "were disappeared" during the military dictatorship from 1976 to 1983.

Questions remain about the constructions in Nama and other languages that combine semantically discordant verbal marking. First, how conventionalised are

---

54 Thanks go to Randy LaPolla for bringing this to my attention.

the constructions and the meanings they convey? For example, is large or greater plural obligatorily marked by such a construction in Nen or is it a stylistic option as in Nama? Second, are the meanings in these constructions signalled by the discord (as they are in Nama, and perhaps in Hebrew)? If this is the case, the discord must be salient, which indicates that the inflections have retained their original meanings. Or are the constructions the result of the opportunistic use of combinations that were previously unused because of semantic clash (as Evans 2019b argues for Nen)? And if this is true, then have the combined morphemes lost their original individual meanings in these constructions? If this is the case, then synchronically discord is not salient and therefore not a factor in signalling meaning.

These questions may be answered by further examples of conventionalised morphological discord. It is possible that such examples exist in other languages, but have not been noticed or reported because this phenomenon has not been previously discussed in the literature. It also may be that further examples will be revealed in future language documentation.

Appendix A: Dialectal differences

Table A.1: Dialectal differences in Nama.

Mata/Daraia	Ngaraita	meaning
ále	*ála*	'hunting'
frengègh	*wúlogh*	'prepare'
fronde	*fronje*	'first'
ghèrare	*tukyuwè*	'moon'
korayè	*kuwaftè*	'say'
mèinyotyo	*bratyo*	'all'
mwinè	*mwinu*	'barren woman'
njamke	*nène*	'food'
súkáki	*nai/nainai*[55]	'sweet potato'
úrbènègh	*èlawègh*	'enter'

---

55 This is most probably a loanword from Komnzo.

https://doi.org/10.1515/9783111077017-009

Appendix B: Recordings

Note: All recordings are accessible in the PARADISEC collection https://www.para
disec.org.

Table B.1: Recordings of narratives.

	File name	Date	Place	Speaker(s)	Topic	Min	Sec
1	MD93	?-?-93	Armidale	Murry Dawi	hunting with father	2	07
2	YD11(1)	19-9-11	Morehead	Yoshie Dawi	crocodile	1	00
3	YD11(2)	19-9-11	Morehead	Yoshie Dawi	crocodile (2nd version)		47
4	FD11	24-9-11	Morehead	Francis Dawi	if I had a million kina	1	01
5	TE11(1)	23-9-11	Morehead	Tony Emoia	going to the garden	4	00
6	MD11(1)	29-9-11	Morehead	Murry Dawi	rat on Jeff's foot		41
7	NP11	23-9-11	Morehead	Naka Pawar	making sago	4	55
8	MD99(1)	?-?-99	Armidale	Murry Dawi	when Mawai killed Sefor	10	59
9	MD11(2)	19-9-11	Morehead	Murry Dawi	origin story part 1	4	11
10	KK11	24-9-11	Morehead	Kafuk Kuti	taking care of Ben	1	51
11	MD99(2)	?-?-99	Armidale	Murry Dawi	origin story part 2	5	22
12	YD11(3)	24-9-11	Morehead	Yoshie Dawi	trip to Kiunga	11	18
13	TE11(2)	29-9-11	Morehead	Tony Emoia	going fishing	3	22
14	ND11	27-9-11	Daraia	Mawai Dawi	maintaining custom marriage	6	33
15	SK11(1)	27-9-11	Daraia	Sarkai Kiana	origin story 1	6	11
16	SK11(2)	27-9-11	Daraia	Sarkai Kiana	origin story 2	6	10
17	SK11(3)	27-9-11	Daraia	Sarkai Kiana	origin story 3	1	23
18	KS11	27-9-11	Daraia	Kiria Suka	confirmation of origin story	5	06
19	ND13	13-4-13	Morehead	Mawai Dawi	witch story	17	00
20	MD13(1)	14-4-13	Morehead	Murry Dawi	reflection story	17	58
21	MD13(2)	14-4-13	Morehead	Murry Dawi	retelling of origin story	12	30
22	FD13	14-4-13	Morehead	Francis Dawi	elementary education	7	46
23	TE13	14-4-13	Morehead	Tony Emoia	complaint	2	37
24	MK13(1)	5-9-13	Ngaraita	Mokai Kafai	fishing	3	58
25	MK13(2)	5-9-13	Ngaraita	Mokai Kafai	hunting	3	06
26	MK13(3)	5-9-13	Ngaraita	Mokai Kafai	gardening	2	26
27	MK13(4)	5-9-13	Ngaraita	Mokai Kafai	pig-killing ceremony	4	28
28	BM13	5-9-13	Ngaraita	Birua Makao	*kufak* (friar bird)	1	23
29	MM13	5-9-13	Ngaraita	Matmat Mawai	taking daughter to clinic	2	50

https://doi.org/10.1515/9783111077017-010

312 —— Appendix B: Recordings

Table B.2: Daraia Clips.

File name	Speaker*	Topic	Min	Sec
DC01	Naka Pawar	washing pots		31
DC02	–	playing with toy truck		7
DC03	Siegel Dawi	climbing for coconuts		50
DC06	Francis	going to the garden		10
DC07	Mawai	departing for church meeting		55
DC09	Fiam	basket making		28
DC10	Mange	basket making		37
DC11	Balus	going to dig a well		38
DC12	Sakwi	yam gardening	2	01
DC13	Kafuk	mat making		48
DC14	Kufak	looking after a baby		24
DC15	Eli Dawi, Les Dawi	warming by a fire	1	30
DC16	?	going hunting		47
DC17	?	fetching water		13
DC18	–	playing marbles		41
DC19	Tempol, Nanjèr	cleaning an echidna		40
DC20	Ene	feeding a cassowary	2	10
DC21	Maxwell	shaping a house post		54
DC23	Mata	washing-cassava		48
DC24	Mary	cleaning a garden	1	09
DC25	Dank	going hunting with dogs		52
DC26	Ifeti, Gaita	returning from the garden		58
DC27	Bariwè, Mary	going to the garden	1	04
DC28	Warenj	basket making	1	00
DC29	Tina, Yangèm, Junior	returning from the garden	1	12
DC32	Kiria	playing a kundu drum		58
DC33	Yanki	playing a jews harp	1	32
DC34	Waframè	playing a flute		57
DC35	Mènge	demonstrating string games	1	49
DC36	Naimèr	demonstrating net fishing	1	14
DC37	Kiria, Jebèri	demonstrating initiation	1	45

*other than Murry Dawi.

Appendix C: List of grammatical morphemes

morpheme	gloss	phrase	function	section
a-	REFL	VP	reflexive/reciprocal prefix [allomorph: *á-*]	4.5.1
á-	REFL	VP	reflexive/reciprocal prefix [allomorph *a-*]	4.5.1
-af	PL	NP	plural marker [allomorph: *-f*]	3.2.1
-afè	COM	NP	comitative suffix (case marker) [allomorph: *-fè*]	3.3.5
-afnar	PRIV	NP	privative suffix (case marker) [allomorph: *-ofnar*]	3.3.6
-am	ERG	NP	ergative suffix (case marker) [allomorph: *-m*]	3.3.1
-amè	PERL	NP	perlative suffix (case marker) [allomorph: *-mè*]	3.3.9
-an	LOC2	NP	locative suffix (adessive/superessive) (case marker)	3.3.7
-ang	INC	VP	suffix indicating inceptive in perfective current, recent and remote tenses	4.8.1
-ar	DU.PA	VP	suffix indicating a dual argument for P-aligned intransitive verbs	4.11.1
-are	ATTR1	NP	attributive suffix indicating 'characterised by'	3.4.1
e-	α.2|3NSGP α.2|3NSGS_P	VP	prefix indexing 2nd or 3rd person nonsingular P argument of transitive verb or S argument of P-aligned intransitive verb; α set*	4.2.2
-e	DAT	NP	dative suffix (case marker)	3.3.2
-e	INSTR	NP	instrumental suffix (case marker)	3.3.3
-e	2|3SGA 2|3SGS_A	VP	suffix indexing 2nd or 3rd person singular A argument of transitive verb or S argument of A-aligned intransitive verb; imperfective current and recent tenses	4.2.1
-e	PFV.DU	VP	dual perfective suffix, directly following the verb stem, the dual P-aligned suffix *-ar* or the inceptive imperfectve suffix *-tang*	4.6.1
-è	2|3SGA 2|3SGS_A	VP	suffix indexing 2nd and 3rd person singular A argument of transitive verb or S argument of A-aligned intransitive verb following the perfective inceptive suffix *-ang* or inceptive imperfective suffix *-tang*	4.8.1 4.9
-èn	1SGA 1SGS_A	VP	suffix indexing 1st person singular A argument of transitive verb or S argument of A-aligned intransitive verb	4.2.1
-èrmèn(g)	REM.DLT.PA	VP	suffix indicating delimited remote tense for nondual P-aligned intransitive verbs	4.11.1

https://doi.org/10.1515/9783111077017-011

314 —— Appendix C: List of grammatical morphemes

(continued)

morpheme	gloss	phrase	function	section
-èrwèn(g)	REM.DUR.PA	VP	suffix indicating durative remote tense for nondual P-aligned intransitive verbs	4.11.1
f	PROX1	VP	temporal proximal clitic	5.4.1
-f	PL	NP	plural marker [allomorph: -af]	3.2.1
-fè	COM	NP	comitative suffix (case marker) [allomorph: -afè]	3.3.5
-faf	ASS	NP	associative suffix (case marker)	3.3.12
fái	OBL1	VP	obligative quasi-modal [alternative form: fiya]	5.3.5
fe-	2SG	NP	2nd person singular possessive prefix (relationship nouns)	3.7
fèf(è)	INCH1	VP	inchoate aspect marker	5.3.2
fèfe-	2NSG	NP	2nd person nonsingular possessive prefix (relationship nouns)	3.7
fètè	VAL	VP	validative quasi-modal	5.3.5
fèyotaro	ABIL	VP	abilitative quasi-modal	5.3.5
fiya	OBL1	VP	obligative quasi-modal [alternative form: fái]	5.3.5
-gh	NOM	VP	nominalising suffix	4.1
-i	2PL	VP	suffix indicating 2nd person plural in imperfective aspect	4.2.1
-i	2|3PLA 2|3PLS$_A$	VP	suffix indexing 2nd or 3rd person plural A argument of transitive verb or S argument of A-aligned intransitive verb in perfective current and recent tenses and in inceptive imperfective aspect	4.7.1
k-	β.ØP	VP	prefix indicating the lack of P argument for A-aligned intransitive verb; β set*	4.2.2
-kaf	ATTR2	NP	attributive suffix indicating 'recently characterised by'	3.4.1
kár	DUB	VP	modal indicating doubt (dubitative)	5.3.3
kèn-	β.2SGP β.2SGS$_P$	VP	prefix indexing 2nd person singular P argument of transitive verb or S argument of P-aligned intransitive verb; β set*	4.2.2
kw-	β.1SGP β.1SGS$_P$	VP	prefix indexing 1st person singular P argument of transitive verb or S argument of P-aligned intransitive verb; β set*	4.2.2
-m	ERG	NP	ergative suffix (case marker) [allomorph: -am]	3.3.1
-m	1NSGA 1NSGS$_A$	VP	suffix indexing 1st person nonsingular A argument of transitive verb or S argument of A-aligned intransitive verb	4.2.1
-m	REM.DLT	VP	suffix indicating delimited imperfective remote tense	4.7.2

Appendix C: List of grammatical morphemes — **315**

(continued)

morpheme	gloss	phrase	function	section
mato	INV	VP	investigative quasi-modal	5.3.5
mè	CONT	VP	continuative aspect marker	5.3.2
-mè	PERL	NP	perlative suffix (case marker) [allomorph: -amè]	3.3.9
-mèn	ORIG	NP	originative suffix (case marker)	3.3.13
-mèn	REM.DLT.PA	VP	suffix indicating delimited remote tense for dual P-aligned intransitive verbs	4.10.1
mètè	POT1	VP	modal indicating potential	5.3.3
miya	OBL.CONT	VP	modal/aspect marker indicating obligatory continuation	5.3.3
n-	α.ØP	VP	prefix indicating the lack of P argument for A-aligned intransitive verb; α set*	4.2.2
n-	VEN	VP	venitive prefix (follows P-/S_P-indexing prefix or ØP prefix)	4.3
n-/nèn-	α.2SGP α.2SGS$_P$	VP	prefix indexing 2nd person singular P argument of transitive verb or S argument of P-aligned intransitive verb; α set*	4.2.2
-n	LOC1	NP	locative suffix (incessive) (case marker)	3.3.7
-nd	2|3NSGA 2|3NSGS$_A$	VP	suffix indexing 2nd or 3rd person nonsingular A argument of transitive verb or S argument of A-aligned intransitive verb in imperfective delimited remote tense and perfective remote punctual	4.6.2
-nd	2|3DUA 2|3DUS$_A$	VP	suffix indexing 2nd or 3rd person dual A argument of transitive verb or S argument of A-aligned intransitive verb in perfective current and recent tenses, and in inceptive imperfective aspect	4.7.1
-ne	GEN	NP	genitive suffix (case marker)	3.3.4
ng-	AND	VP	andative prefix (follows P-/S_P-indexing prefix or ØP prefix, or may occur without this prefix)	4.3
-ng	2|3SGA 2|3SGS$_A$	VP	suffix indexing 2nd and 3rd person singular A argument of transitive verb or S argument of A-aligned intransitive verb directly following remote tense marking, both imperfective and perfective (-m, -w, -ay)	4.7.2
-nit	SEMB	NP	semblative suffix (case marker)	3.3.14
-o	SG	NP	singular marker (limited distribution)	3.2.2
o-	AUTO	VP	autobenefactive prefix	4.5.3
-ofnar	PRIV	NP	privative suffix (case marker) [allomorph: -afnar]	3.3.6
-ot	ALL	NP	allative suffix (case marker) [allomorph: -t]	3.3.10

316 ── Appendix C: List of grammatical morphemes

(continued)

morpheme	gloss	phrase	function	section
-ot	PURP	NP	purposive suffix (case marker) [allomorph: -yot]	3.3.11
r=	MIR	VP	mirative proclitic (limited use)	5.4.2
-ro	RSTR	NP	restrictive suffix	3.5.2
s	PROX2	VP	locational proximal clitic	5.4.1
sái	INT	VP	modal indicating intention [alternative form: siya]	5.3.3
sèno	POSB	VP	modal indicating possibility	5.3.3
siya	INT	VP	modal indicating intention [alternative form: sái]	5.3.3
so	FUT	VP	future tense marker	5.3.1
t-	β.3sGP β.3sGS$_P$	VP	prefix indexing 3rd person singular P argument of transitive verb or S argument of P-aligned intransitive verb; β set*	4.2.2
-t	ALL	NP	allative suffix (case marker) [allomorph: -ot]	3.3.10
-t	2|3NSGA 2|3NSGS$_A$	VP	suffix indexing 2nd or 3rd person singular A argument of transitive verb or S argument of A-aligned intransitive verb in imperfective aspect	4.2.1
ta-	1SG	NP	1st person singular possessive prefix (relationship nouns)	3.7
ta-	β.2|3NSGP β.2|3NSGS$_P$	VP	prefix indexing 2nd or 3rd person nonsingular P argument of transitive verb or S argument of P-aligned intransitive verb; β set* [allomorph: tá-]	4.2.2
tá-	β.2|3NSGP β.2|3NSGS$_P$	VP	prefix indexing 2nd or 3rd person nonsingular P argument of transitive verb or S argument of P-aligned intransitive verb; β set* [allomorph: ta-]	4.2.2
-ta	ABL	NP	ablative suffix (case marker)	3.3.8
-ta	IPFV.ND	VP	nondual imperfective suffix; directly following the verb stem	4.6.1
táf	OBL1	VP	obligative quasi-modal	5.3.5
-tang	INC.IPFV	VP	suffix indicating inceptive imperfective aspect	4.9
-tau	TEMP1	NP	temporal suffix (present/future)	3.4.2
-táwár	TEMP2	NP	temporal suffix (past)	3.4.2
tè	PRF	VP	perfect aspect marker	5.3.2
tèf(è)	INCH2	VP	inchoate aspect marker	5.3.2

Appendix C: List of grammatical morphemes — **317**

(continued)

morpheme	gloss	phrase	function	section
tèfe-	1NSG	NP	1st person nonsingular possessive prefix (relationship nouns)	3.7
tèn-	β.1NSGP β.1NSGS$_P$	VP	prefix indexing 1st person nonsingular P argument of transitive verb or S argument of P-aligned intransitive verb; β set*	4.2.2
tèu	PERM	VP	modal indicating permission (permissive)	5.3.3
w-	α.1SGP α.1SGS$_P$	VP	prefix indexing 1st person singular P argument of transitive verb or S argument of P-aligned intransitive verb; α set*	4.2.2
w-	TR	VP	transitiviser prefix, changing an A-aligned intransitive verb to a transitive (follows the applicative or autobenefactive prefix)	4.5.2
-w	REM.DUR	VP	suffix indicating durative imperfective remote tense	4.7.2
wa-	APP	VP	applicative prefix, changing a transitive verb to ditransitive (follows P-indexing prefix or deictic prefix) [allomorph *wá*]	4.4.2
wá-	APP	VP	applicative prefix, changing a transitive verb to ditransitive (follows P-indexing prefix or deictic prefix) [allomorph *wa*]	4.4.2
-wang	2DU.IMP	VP	2nd person dual suffix used in immediate imperatives	6.5.1
wè	POT2	VP	modal indicating potential	5.3.3
-wè	DA	VP	in dual imperfective aspect current and recent tenses, suffix directly following the verb stem (with Ø aspect marking) normally to indicate that that the P argument is dual	4.7.1
-wèn	REM.DUR.PA	VP	suffix indicating durative remote tense for dual P-aligned intransitive verbs	4.11.1
y-	α.3SGP α.3SGS$_P$	VP	prefix indexing 3rd person singular P argument of transitive verb or S argument of P-aligned intransitive verb; α set*	4.2.2
yá-	3SG	NP	3rd person singular possessive prefix (relationship nouns)	3.7
yèf	EVID	VP	evidential modal	5.3.3
yèfe-	3NSG	NP	3rd person nonsingular possessive prefix (relationship nouns)	3.7
yèn-	α.1NSGP α.1NSGS$_P$	VP	prefix indexing 1st person nonsingular P argument of transitive verb or S argument of P-aligned intransitive verb; α set*	4.2.2

318 —— Appendix C: List of grammatical morphemes

(continued)

morpheme	gloss	phrase	function	section
yènè	DEM	NP	demonstrative	5.1.3
yèta	ATT	VP	attemptive modal ('try')	5.3.3
-yo	EXCL	NP	exclusive suffix	3.5.1
-yot	PURP	NP	purposive suffix (case marker) [allomorph: -ot]	3.3.11
-Ø	IPFV.DU	VP	indicates dual imperfective aspect when directly following the verb stem and followed by a nonsingular A-/S_A-indexing suffix -m or -t (for current and recent tenses) or by an imperfective remote tense marker,-m or -w	4.6.1
-Ø	PFV.ND	VP	indicates nondual perfective aspect when directly following the verb stem and followed by a singular A-/S_A-indexing suffix or the inceptive suffix -ang	4.6.1
-Ø	2|3SGA.IRR 2|3SGS$_A$.IRR	VP	directly following the nondual imperfective suffix -ta, indicates irrealis aspect for 2nd and 3rd person singular A argument of transitive verb or S argument of A-aligned intransitive verb	4.7.1
-Ø	2|3SGA 2|3SGS$_A$	VP	at the end of a verb, indicates 2nd and 3rd person singular A argument of transitive verb or S argument of A-aligned intransitive verb; in several contexts: (1) directly following remote tense marking, both imperfective and perfective (-m, -w, -ay). (2) in perfective aspect, directly following the Ø-marked verb stem, (3) directly following the perfective inceptive suffix -ang, (4) in imperfective aspect, directly following the dual argument suffix wè	4.7.2 4.8.1 4.8.2

*P/S_P-indexing and ØP prefixes from the α set are used for current tense in all aspects and for remote delimited (imperfective), remote punctual (perfective) and remote inceptive imperfective.
P/S_P-indexing and ØP prefixes from the β set are used for recent tense in all aspects and for remote durative (imperfective) and remote inceptive (perfective). (See Table 4.13.)

Appendix D: Sample text

In this narrative [YD11(3)], Yoshie Dawi describes a journey from Morehead to Kiunga, by road and canoe, a distance of around 300 kms. (A recording of this text can be accessed at https://zenodo.org/record/7831384.)

1. *Yèndo si mènamèn so yaitotan ndernae*
yèndo si mènamèn so y-waito-ta-èn ndernae
1SG.ERG story about FUT α.3SGP-tell-IPFV.ND-1SGA how
ámb misi-misiafè nuwanoyèm. Kiungat.
ámb misimisi-afè n-uwano-Ø-ay-m Kiungè-t
some pastor-COM α.ØP-set.off-PFV.ND-REM.PUNC-1NSGS$_A$ Kiunga-ALL
'I will tell a story about how we set off for Kiunga with some pastors.'

2. *Ndernae yèrnyayèm.*
ndernae y-rèny-Ø-ay-m
how α.3SGP-start-PFV.ND-REM.PUNC-1NSGA
'How we started it.'

3. *Nuwanoyèm yènanmè trake kètanyotyo*
n-uwano-Ø-ay-m yènanmè trak-e kètanyotyo
α.ØP-set.off-PFV.ND-REM.PUNC-1NSGS$_A$ from.here vehicle-INST until
Mitarè.
Mitarè
Mitara
'We set off from here by vehicle until Mitara.'

4. *Kètè nufaryèm; kètè*
kètè n-ufar-Ø-ay-m kètè
there α.ØP-arrive-PFV.ND-REM.PUNC-1NSGS$_A$ there
yènfráreyèng.
yèn-fèrár-Ø-ay-ng
α.1NSGP-drop.off-PFV.ND-REM.PUNC-2|3SGA
'We arrived there and there we were dropped off.'

5. *Yènanmè wèi áligh nuwanoyèm.*
yènanmè wèi áligh n-uwano-Ø-ay-m
from.here now walking α.ØP-set.off-PFV.ND-REM.PUNC-1NSGS$_A$
'From here now we set off walking.'

320 — Appendix D: Sample text

6. *Mèngotu   engwaindamèm*   *Kerru;   Sekin*
   mèngotu   e-ng-waing-ta-m-m   Kerru   Seki-n
   village   ɑ.2|3NSGP-AND-pass-IPFV.ND-REM.DLT-1NSGA   Kerru   Seki-LOC1
   nákmayèm.
   n-ákèmè-Ø-ay-m
   ɑ.ØP-lie.down-PFV.ND-REM.PUNC-1NSGS$_A$
   'We passed villages like Kerru; we slept at Seki.'

7. *Ámb kwèfon   sèkwe   nèngófnamoyèm*
   ámb kwèfon[56]   sèkw-e   n-ng-ófnamo-Ø-ay-m
   next.day   canoe-INST   ɑ.ØP-AND-cross-PFV.ND-REM.PUNC-1NSGS$_A$
   witma   karfot.
   witma   karèf-ot
   side   side-PURP
   'The next day we went across by canoe to the other side (of the river).'

8. *Tènngèmorwèn*   *wèi   ámb   tètkafwe*
   tèn-ngèmo-èrwèn   wèi   ámb   tètkafwe
   β.1NSGS$_P$-go.ND-REM.DUR.PA   now   another   creek
   nèngófnamoyèm.
   n-ng-ófnamo-Ø-ay-m
   ɑ.ØP-AND-cross-PFV.ND-REM.PUNC-1NSGS$_A$
   'We went now and crossed another creek.'

9. *Kètè   wèi   ámb   sèkwe*
   kètè   wèi   ámb   sèkw-e
   there   now   another   canoe-INST
   nèngófnamoyèm.
   n-ng-ófnamo-Ø-ay-m
   ɑ.ØP-AND-cross-PFV.ND-REM.PUNC-1NSGS$_A$
   'There now we crossed with another canoe.'

---

56 *Ámb kwèfon* is a set expression meaning 'the next day'. The word *kwèfon* does not occur on its own.

Appendix D: Sample text **321**

10. *Áligh yènyotam,*
    áligh   yèn-yo-ta-m-Ø[57]
    walking   α.1NSGP-travel-IPFV.ND-REM.DLT-2|3SGA
    nufaryèm   *ámb   end   kafwet.*
    n-ufar-Ø-ay-m   ámb   end   kafwe-t
    α.ØP-arrive-PFV.ND-REM.PUNC-1NSGS$_A$ another road branch-ALL
    'We travelled by foot and arrived at a junction with another road.'

11. *Yènd   a   ámb   amaf   yètkwèn Dafi   yènd*
    yènd   a   ámb   amaf   yètkwèn Dafi   yènd
    1.ABS and   another woman   name   Duffy   1.ABS
    nuwaneayèm   *Suki   endmè.*
    n-uwano-e-ay-m   Suki   end-mè
    α.ØP-set.off-PFV.DU-REM.PUNC-1NSGS$_A$ Suki   road-PERL
    'Me and another woman named Duffy set off along the Suki road.'

12. *Yèndro.   Ár   ghakèr   enfrangeayèm.*
    yènd-ro   ár   ghakèr   e-n-frango-e-ay-m
    1.ABS-RSTR man   young.man   α.2|3NSGP-leave-PFV.DU-REM.PUNC-1NSGA
    Just us. We (2) left the men.

13. *Njam   tènngèrwèn*   *trak   endmè,   kambanam*
    njam   tèn-ngèr-wèn   trak   end-mè,   kamban-am
    when   β.1NSGS$_P$-go.DU-REM.DUR.PA   vehicle road-PERL snake-ERG
    tènngèmáunyèwèng.
    tèn-ng-wáuny-Ø-w-ng
    β.1NSGP-AND-frighten-IPFV.DU-REM.DUR-2|3SGA
    'When going on the vehicle road, a snake frightened us (2).'

14. *Ámb   kambanam   yènd   fèfè   wèrnam.*
    ámb   kamban-am   yènd   fèfè   w-rár-ta-m-Ø
    another snake-ERG   1.ABS INCH1   α.1SGP-bite-IPFV.ND-REM.DLT-2|3SGA
    'Another snake nearly bit me.'

---

57 Note that \yo/ is an agentless transitive verb, for which the P argument is semantically the subject (see section 6.1.1).

322 —— Appendix D: Sample text

15. *Yèndo   kánjo   ngangè   kafkaf   ewasáryèn.*
yèndo   kánjo   ngangè   kafkaf   e-wasár-Ø-ay-èn
1SG.ERG   luck   hand   foot   α.2|3NSGP-move-PFV.ND-REM.PUNC-1SGA
'Lucky I moved my hands and feet.'

16. *Nuwaneayèm.*
n-uwano-e-ay-m
α.ØP-set.off-PFV.DU-REM.PUNC-1NSGS_A
'We (2) set off.'

17. *Dafi   nèkwafè   kamndangè*   *mènamèn   yu*
Dafi   nèkw-afè   k-amèndè-Ø-ang-è   mènamèn   yu
Duffy   anger-COM   β.ØP-become-PFV.ND-INC-2|3SGS_A   because   place
nayu   tèmorwèn.
nayu   t-mo-èrwèn
far   β.3SGS_P-COP.ND-REM.DUR.PA
'Duffy got angry because the place was so far away.'

18. *Yáf   tawasurngè*   *kèghatèn.*
yáf   ta-wasur-Ø-ang-è   kèghat-n
basket   β.3SGP-drop-PFV.ND-INC-2|3SGA   bush-LOC1
'She dropped her baskets in the bush.'

19. *Yèndo   tákman*   *tanyo   yáf,   tuktuktae*
yèndo   tá-kèmè-Ø-èn   tanyo   yáf,   tuktuktae
1SG.ERG   β.3SGP-lay.down-PFV.ND-INC-1SGA   1SG.own   basket   face.up
tákman.
tá-kèmè-Ø-èn
β.3SGP-lay.down-PFV.ND-INC-1SGA
'I laid my own baskets down, laid them down face up.'

20. *Fá   korayangè:*   "*Yosi,   yènè   rabè   wèn   kès*
fá   k-oray-Ø-ang-è: Yosi   yènè   rabè   wèn   kès
3.ABS   β.ØP-say-PFV.ND-INC-2|3SGS_A   Yoshie   DEM   rubber   tree   there
em   *yèndo   so   enfetan*   *náifae.*"
e-m   yèndo   so   e-nèfè-ta-èn   náifè-e
α.2|3NSGS_P-COP.ND   1SG.ERG   FUT   α.2|3NSGP-cut-IPFV.ND-1SGA   knife-INST
'She said: "Yoshie, those rubber trees there, I'll cut them down with a
bushknife."'

Appendix D: Sample text — **323**

21. *Yèndo   tèmèndan:*   *"Nèmamèn, fene   yau,*
     yèndo   t-mèndè-Ø-èn   nèmamèn fene   yau
     1SG.ERG   β.3SGP-TELL-PFV.ND-1SGA why   2SG.GEN NEG
     árfene   em;   *yau enfeta.*
     ár-f-e-ne   e-m   yau e-nèfè-ta-Ø
     people-PL-DAT-GEN α.2|3NSGS_P-COP.ND NEG α.2|3NSGP-cut-IPFV.ND-2|3SGA.IRR
     'I told her: "Why, they're the people's, not yours; don't cut them.'

22. *"O   yèndo   mènamèn senfetan   yènd*
     o   yèndo   mènamèn s=e-nèfè-ta-èn   yènd
     or   1SG.ERG because   PROX2=α.2|3NSGP-cut-IPFV.ND-1SGA 1.ABS
     sakram   áligh-mèn   tè   wifo."
     sakèr-am   áligh-mèn   tè   w-wifo-Ø-Ø
     tiredness-ERG walking-ORIG PRF α.1SGP-finish-PFV.ND-2|3SGA
     "Or I'll cut them right here because I'm tired from walking."'

23. *Fèn   kufngem.*
     fèn   k-ufngo-e-m
     laughing   β.ØP-start-PFV.DU-1NSGS_A
     'We started laughing.'

24. *Kanem,   mbáfyáf kawem,*
     k-ano-e-m   mbáfyáf k-awo-e-m
     β.ØP-get.up-PFV.DU-1NSGS_A basket β.ØP-put.on-PFV.DU-1NSGS_A
     kuwanem.
     k-uwano-e-m
     β.ØP-set.off-PFV.DU-1NSGS_A
     'We got up, put on the baskets, and set off.'

25. *Yènyowè.   Mèngotu kufarem*
     yèn-yo-Ø-wè-Ø   mèngotu k-ufar-e-m
     α.1NSGP-travel-IPFV.DU-DA-2|3SGA vilage β.ØP-arrive-PFV.DU-1NSGS_A
     Enyawè.
     Enyawè.
     Enyawa
     'We travel. We arrived at a village – Enyawa.'

26. *Kètè   ámb   fútár   mèrsom   yánjo   mèngot*
Kètè   ámb   fútár   mèrès-o-m   yánjo   mèngo-t
there   some   friend   girl-SG-ERG   3SG.own   house-ALL
tènèmnde   kèmghèt.
tèn-mèndè-e-Ø   kèmègh-t
β.1NSGP-tell-PFV.DU-2|3SGA   sleeping-ALL
'There a woman friend told us to sleep at her house.'

27. *Kètè   yènkèmare.*
kètè   yèn-kèmè-ar-e
there   α.1NSGS_P-be.lying.down-DU.PA-PFV.DU
'There we sleep.'

28. *Ámb   efoghèn   kuwanem   Gigwas,   Gigwas   tesenot.*
ámb   efogh-n   k-uwano-e-m   Gigwas   Gigwas   tesen-ot
next   day-LOC1   β.ØP-set.off-PFV.DU-1NSGS_A   Gigwas   Gigwas   station-PURP
'The next day we set off for Gigwas, Gigwas station.'

29. *Kètè   Mawai   ftawinjongèm.*
kètè   Mawai   f=ta-winjo-Ø-ang-m
there   Mawai   PROX1=β.2|3NSGP-see-PFV.ND-INC-1NSGA
'There we just saw Mawai and others.'

30. *Fèyo   yènamè   kufliangèm   sembyo   sèkwe.*
fèyo   yènamè   k-ufèli-Ø-ang-m   sembyo   sèkw-e
then   from.here   β.ØP-get.in-PFV.ND-INC-1NSGS_A   two   canoe-INST
'Then from here we all got into two canoes.'

31. *Yènyote;   wanje   yabun   yèm   tèfene*
yèn-yo-ta-e   wanje   yabun   y-m   tèfene
α.1NSGP-travel-IPFV.ND-2|3SGA   river   big   α.3SGS_P-COP.ND   2NSG.GEN
Flai   wanje.
Flai   wanje
Fly   River
'We travel; the river is big, our Fly River.'

Appendix D: Sample text — **325**

32. *Yènamè*   *kuwanongèm*   *yènyote*
yènamè   k-uwano-Ø-ang-m   yèn-yo-ta-e
from.here β.ØP-set.off-PFV.ND-INC-1NSGS$_A$ α.1NSGP-travel-IPFV.ND-2|3SGA
ee   *yènyote*   *kètanotyo yu*   *yètkwèn* *Obo.*
ee   yènyote   kètanotyo yu   yètkwèn Obo
PROL α.1NSGP-travel-IPFV.ND-2|3SGA until   place   name   Obo
'From here we set off; we travel . . . we travel until a place named Obo.'

33. *Obon*   *ngákmangèm.*
Obo-n   ng-ákèmè-Ø-ang-m
Obo-LOC1 AND-lie.down-PFV.ND-INC-1NSGS$_A$
'In Obo, we went and slept.'

34. *Wèi* *yènyote*   *yènamè* *tènfangèm*
wèi yèn-yo-ta-e   yènamè tè-nèfè-Ø-ang-m
now α.1NSGP-travel-IPFV.ND-2|3SGA from.here β.3SGP-cut-PFV.ND-INC-1NSGA
yabun *sauamè.*
yabun sau-amè
big   swamp-PERL
'We travel now and from here we cut through a big swamp.'

35. *Yènè yau; yabun sau*   *stámorwèn*   *mer* *yabun.*
yènè yau yabun sau   s=tá-mo-èrwèn   mer yabun
DEM NEG big   swamp PROX2=β.2|3NSGS$_P$-COP.ND-REM.DUR.PA good big
'Not like here; they were big swamps, beautifully big.

36. *Tènyotau,*   *ngarayawèm;*
tèn-yo-ta-w-Ø   ng-araya-ta-w-m
β.1NSGP-travel-IPFV.ND-REM.DUR-2|3SGA AND-paddle-IPFV.ND-REM.DUR-1NSGS$_A$
ámb *mèngotuot* *nufaryèm.*
ámb mèngotu-ot n-ufar-Ø-ay-m
next village-ALL α.ØP-arrive-PFV.ND-REM.PUNC-1NSGS$_A$
'We were travelling, paddling away; we arrived at the next village.'

37. *Yètkwèn* *yènè* *mèngotu* *yèm*   *Wangawanga.*
yètkwèn yènè mèngotu y-m   Wangawanga
name   DEM village α.3SGS$_P$-COP.ND Wangawanga
'The name this village is Wangawanga.'

Appendix D: Sample text

38. *Kètan   kufarngèm,   kètè   yènkèm.*
    kètè-n   k-ufar-Ø-ang-m   kètè   yèn-kèmè
    there-LOC1 β.ØP-arrive-PFV.ND-INC-1NSGS$_A$ there α.1NSGS$_P$-be.lying down
    'We arrived there and sleep there.'

39. *Yènamè   kuwanongèm,   yènyote   ee*
    yènamè   k-uwano-Ø-ang-m   yèn-yo-ta-e   ee
    from.here β.ØP-set.off-PFV.ND-INC-1NSGS$_A$ α.1NSGP-travel-IPFV.ND-2|3SGA PROL
    kèfèrmangèm   nayu yu   tèmorwèn,
    k-èfèrmè-Ø-ang-m   nayu yu   t-mo-èrwèn
    β.ØP-cross-PFV.ND-INC-1NSGS$_A$ far place β.3SGS$_P$-COP.ND-REM.DUR.PA
    nayu nayu.
    nayu nayu.
    far   far
    'From here we set off and we travel until we crossed a place that was far far
    away.'

40. *Tènyotau   kètanyotyo*
    tèn-yo-ta-w-Ø   kètanyotyo
    β.1NSGP-travel-IPFV.ND-REM.DUR-2|3SGA until
    nófnamoyèm   wanje yabunan.
    n-ófnamo-Ø-ay-m   wanje yabun-an
    α.ØP-cross-PFV.ND-REM.PUNC-1NSGS$_A$ river big-LOC2
    'We were travelling until we crossed into a big river.'

41. *Orsogh   kufngongèm.*
    orso-gh   k-ufngo-Ø-ang-m
    be.lost-NOM β.ØP-started-PFV.ND-INC-1NSGS$_A$
    'We started to get lost.'

42. *Yèta end   tèronjawèm   árkamè   so*
    yèta end   t-ronja-ta-w-m   árkè-mè   so
    ATT   road β.3SGP-look.for-IPFV.ND-REM.DUR-1NSGA which-PERL FUT
    kèlawangèm,
    k-èlau-Ø-ang-m
    β.ØP-go.inside-PFV.ND-INC-1NSGS$_A$
    'We were trying to look for a way along which to go inside.'

Appendix D: Sample text — **327**

43. *nde   ngarayawèm*   *wèi*
    nde   ng-araya-ta-w-m   wèi
    so   AND-paddle-IPFV.ND-REM.DUR-1NSGS$_A$ now
    kènangotawèm   *kor,*
    k-n-ango-ta-w-m   kor
    β.ØP-VEN-return-IPFV.ND-REM.DUR-1NSGS$_A$ again
    'so we were paddling that way and coming back this way again,'

44. *ee   sembyo   ár   tawinjem*   *sèkwan*
    ee   sembyo   ár   ta-winjo-e-m   sèkw-an
    PROL   two   man   β.2|3NSGP-see-PFV.DU-1NSGA   canoe-LOC2
    tárwèn.
    tá-r-wèn.
    β.2|3NSGS$_P$-COP.DU-REM.DUR.PA
    'keeping on until we saw two men in a canoe.'

45. *Yèndfem   ngángae   tangramèm.*
    yèndfem   ngángè-e   ta-ng-ramè-Ø-m
    1NSG.ERG   hand-INST β.2|3NSGP-AND-give-IPFV.DU-1NSGA
    'We waved to them.'

46. *Fá   nènarayat*   *tèfefaf.*
    fá   n-n-araya-Ø-t   tèfe-faf
    3.ABS   α.ØP-VEN-paddle-IPFV.DU-2|3NSGS$_A$ 1NSG-ASS
    'They paddled towards us.'

47. *Tènèmndend*   *sie:*
    tèn-mèndè-e-nd   si-e
    β.1NSGP-tell-PFV.DU-2|3DUA   talk-INST
    "Edana mbamona?" [note: this quote is in Police Motu]
    edana mbamona
    what's wrong
    'They said to us: "What's wrong?"'

48. *Yènd   korayangèm:*   *"yènd   tè   nolindangèm."*
    yènd   k-oray-Ø-ang-m   yènd   tè   n-oling-tang-m
    1.ABS   β.ØP-say-PFV.ND-INC-1NSGS$_A$ 1.ABS   PRF   α.ØP-be.lost-INC.IPFV-1NSGS$_A$
    'We said: "We've just got lost."'

49. *Si   yènè   siyo   kor   korayangè:*   "*Rot   ya*

    si   yènè   si-yo   kor   k-oray-Ø-ang-è   rot   ya

    talk DEM   talk-EXCL actual β.ØP-say-PFV.ND-INC-2|3SGS$_A$ road DEM

    i   stap; hau na   yupela i   paul?" [note: this quote is in Tok Pisin]

    i   stap   hau na   yupela i   paul

    PM   stay how and   2PL   PM   in.trouble

    'This is what he said: "This is the way; how did you get in trouble?".'

50. *OK, end   tènmauyafongè.*   *Ndenè.*

    OK   end   tèn-wauyafo-Ø-ang-è   ndenè

    OK   road β.1NSGP-show-PFV.ND-INC-2|3SGA this.way

    'OK, he showed us the way. This way.'

51. *Eso   taramangèm.*

    eso   t-wa-ramè-Ø-ang-m

    thanks β.2|3SGP-APP-give-PFV.ND-INC-1NSGA

    'We gave him thanks.'

52. *Kuwanongèm   ámb   mangotuot.*

    k-uwano-Ø-ang-m   ámb   mangotu-ot

    β.ØP-set.off-PFV.ND-INC-1NSGS$_A$ another village-ALL

    'We set off to another village.'

53. *Yènamè   stangmorwèn   mer   mer   mer*

    yènamè   s=tá-ngèmo-èrwèn   mer   mer   mer

    from.here PROX2=β.2|3NSGS$_P$-go.ND-REM.DUR.PA good good good

    mer   sau   yabun.

    mer   sau   yabun

    good lake big

    'From here there were wonderful big lakes.'

54. *Yènngèmormèn   ámb   yuan*

    yèn-ngèmo-èrmèn   ámb   yu-an

    α.1NSGS$_P$-go.ND-REM.DLT.PA another place-LOC2

    tènkèmangèrwèn.

    tèn-kèmè-ang-èrwèn

    β.1NSGS$_P$-be.lying.down-INC-REM.DUR.PA

    'We went and slept in another place.'

Appendix D: Sample text — **329**

55. *Kètè   yènè   mèngotuan   mèngotu   yètkwèn   yèm   Nipan.*
    kètè   yènè   mèngotu-an   mèngotu   yètkwèn   y-m   Nipan.
    there   DEM   village-LOC2   village   name   α.3SGS_P-COP.ND   Nipan
    'In this village named Nipan.'

56. *Nipanèn   tinjongèm   efe   Moetèn*
    Nipane-n   t-winjo-Ø-ang-m   efe   Moet-n
    Nipan-LOC1   β.3SGP-see-PFV.ND-INC-1NSGA   who   Morehead-LOC1
    tèmorwèn.
    t-mo-èrwèn
    β.3SGS_P-COP.ND-REM.DUR.PA
    'In Nipan we saw someone who had been in Morehead.'

57. *Yètkwèn   ár   yèm   Tobi.*
    yètkwèn   ár   y-m   Tobi
    name   man   α.3SGS_P-COP.ND   Toby
    'The man's name is Toby.'

58. *Tobiene   mèrèn   ár   kètè   tawinjongèm.*
    Tobi-e-ne   mèrèn   ár   kètè   ta-winjo-Ø-ang-m
    Toby-DAT-GEN   family   people   there   β.2|3NSGP-see-PFV.ND-INC-1NSGA
    'We saw Toby's family there.'

59. *Yèmofem   njamke   tèfene   eneghat,   kètè*
    yèmofem   njamke   tèfene   e-negh-ta-t   kètè
    3NSG.ERG   food   1NSG.GEN   α.2|3NSGP-cook-IPFV.ND-2|3NSGA   there
    yènkèmangèrman.
    yèn-kèmè-ang-èrman
    α.1NSGS_P-be.lying.down-INC-REM.DLT.PA
    'They cooked our food, and there we slept.'

60. *Ámb kwèfon   wèi   kuwanongèm.*
    ámb kwèfon   wèi   k-uwano-Ø-ang-m
    next.day   now   β.ØP-set.off-PFV.ND-INC-1NSGS_A
    'The next day now, we set off.'

330 —— Appendix D: Sample text

61. *Yènyotam*   *yènyotam yènyotam yènyotam yènyotam.*
yèn-yo-ta-m-Ø
α.1NSGP-travel-IPFV.ND-REM.DLT-2|3SGA
'We travelled and travelled and travelled.'

62. *Sèkw tèfene*   *tèmorwèn*   *yabun kènjkam, marèt.* ....
sèkw   tèfene   t-mo-èrwèn   yabun   kènjkam,   marèt
canoe   1NSG.GEN   β.3SGS_P-COP.ND-REM.DUR.PA   big   deep   wide
'Our canoe was in a big, deep, wide [part of the river].'

63. *Yènyotam*   *nufanamèm*
yèn-yo-ta-m-Ø   n-ufar-ta-m-m
α.1NSGP-travel-IPFV.ND-REM.DLT-2|3SGA   α.ØP-arrive-IPFV.ND-REM.DLT-1NSGS_A
ámb   *wanje yèkwèn Agu wanje.*
ámb   wanje   yèkwèn   Agu   wanje
another   river   name   Agu   river
'We travelled and arrived at another river named Agu river.'

64. *Agu wanjet*   *ngènyawetawèm*   *witma wèi*
Agu   wanje-t   ng-ènyawè-ta-w-m   witma   wèi
Agu   river-ALL   AND-go.inside-IPFV.ND-REM.DUR-1NSGS_A   side   now
kètamè   *wèrèr*   *kètè*   *kènèrsau.*
kètamè   wèrèr   kètè   k-n-èrsa-ta-w-Ø
along.there   current   there   β.ØP-VEN-run-IPFV.ND-REM.DUR-2|3SGS_A
'We were going into the Agu river now and along there a current was flowing.'

65. *Tèfene*   *sèkw*   *kètè*   *wèrram*
tèfene   sèkw   kètè   wèrèr-am
1NSG.GEN   canoe   there   current-ERG
syanai   *yènamè*
s=y-wanè-Ø-ay-Ø   yènamè
PROX2-α.3SGP-take-PFV.ND-REM.PUNC-2|3SGA   from.here
yènmorman   *nde*   *sèkw*
yèn-mo-èrman   nde   sèkw
α.1NSGSP-COP.ND-REM.DLT.PA   that   canoe
namghetam   *yabun wènfaf.*
n-amghè-ta-m-Ø   yabun   wèn-faf
α.ØP-go.alongside-IPFV.ND-REM.DLT-2|3SGS_A   big   tree-ASS
'The current there took our canoe when we were there so it went alongside a
big bushy area.'

Appendix D: Sample text — **331**

66. *Ámb   ausè   tosè   yèngmormèn*   *sèkw   kor*
    ámb   ausè   tosè   y-ngèmo-èrmèn   sèkw   kor
    some old.woman small   α.3sGS$_P$-go.ND-REM.DLT.PA canoe   actually
    terangai   *wènan.*
    t-werè-Ø-ang-ay-Ø   wèn-an
    β.3sGP-hold-PFV.ND-INC-REM.PUNC-2|3sGA tree-LOC2
    'Some little old woman went and actually held the canoe in the bushy area.'

67. *Yau njam ausam*   *terangai*
    yau njam ausè-m   t-werè-Ø-ang-ay-Ø
    NEG if   old.woman-ERG   β.3sGP-hold-PFV.ND-INC-REM.PUNC-2|3sGA
    sèkw so   káráfnau   *yènd so*
    sèkw so   k-áráfár-ta-w-Ø   yènd so
    canoe FUT β.ØP-break-IPFV.ND-REM.DUR-2|3sGS$_A$ 1.ABS FUT
    kolitotawèm.
    k-olito-ta-w-m
    β.ØP-sink-IPFV.ND-REM.DUR-1NSGS$_A$
    'If the old woman had not held the canoe, it would have broken up and we
    would have sunk.'

68. *Wèikor yènngèm;*   *ámb*   *mèngotuot*
    wèikor yèn-ngèm   ámb   mèngotu-ot
    again   α.1NSGS$_P$-go.ND another village-ALL
    kufarngèm.
    k-ufar-Ø-ang-m
    β.ØP-arrive-PFV.ND-INC-1NSGS$_A$
    'We went again and arrived at another village.'

69. *Yènè yu   njam tangfandawèm*
    Yènè yu   njam ta-ng-fanda-ta-w-m
    DEM   place when β.2|3NSGP-AND-look.at-IPFV.ND-REM.DUR-1NSGA
    tangfandawèm   *mer   mèngotu.*
    ta-ng-fanda-ta-w-m   mer   mèngotu
    β.2|3NSGP-AND-look.at-IPFV.ND-REM.DUR-1NSGA good   village
    'When we looked at these places, we saw nice villages.'

70. *Tènngmorwèn;*   *yènngmormèn*   *wèn*   *kafwefaf*
    tèn-ngèmo-èrwèn   yèn-ngèmo-èrmèn   wèn   kafwe-faf
    β.1NSGS_P-go.ND-REM.DUR.PA   α.1NSGS_P-go.ND-REM.DUR.PA   tree   branch-ASS
    ngèlawetawèm.
    ng-èlawè-ta-w-m
    AND-go.inside-IPFV.ND-REM.DUR-1NSGS_A
    'We went on; we went and went inside a branch of a tree.'

71. *Wèn*   *kafwe*   *nèngèlawayèm.*   *yènd*
    Wèn   kafwe   n-ng-èlawè-Ø-ay-m   yènd
    tree   branch   α.ØP-AND-go.inside-PFV.ND-REM.PUNC-1NSGA   1.ABS
    yènamè   *nèngutnangayèn:*
    yènamè   n-ng-utár-tang-ay-èn
    from.here   α.ØP-AND-shout-INC.IPFV-REM.PUNC-1SGA
    'When we went inside the tree branch, I shouted:'

72. "*Mawai*   *kor*   *nèngáfrenda*   *wènamè;*
    Mawai   kor   n-ng-á-freng-ta-Ø   wèn-amè
    Mawai   careful   α.ØP-AND-REFL-prepare-IPFV.ND-2|3SGS_A.IRR   tree-PERL
    kofwe   *so*   *kènrauèrngè.*"
    kofwe   so   k-n-arauèr-Ø-ang-è
    neck   FUT   β.ØP-VEN-bend.down-PFV.ND-INC-2|3SGS_A
    "'Mawai, be careful and prepare yourself through the tree; bend down your
    neck.'"

73. *Mèinyotyo*   *semormèn*   *yènè*   *yabun*   *ár*
    mèinyotyo   s=e-mo-èrmèn   yènè   yabun   ár
    all   PROX2=α.2|3NSGS_P-COP.ND-REM.DLT.PA   DEM   big   man
    náwefnoyènd.
    n-áwefno-Ø-ay-nd
    α.ØP-bend.down-PFV.ND-REM.PUNC-2|3NSG
    'All those big men who were there bent down.'

74. *Mawai*   *nèngawefnoi*   *sèkwan.*
    Mawai   n-ng-awefno-Ø-ay-Ø   sèkw-an
    Mawai   α.ØP-AND-bend.down-PFV.ND-REM.PUNC-2|3SG   canoe-LOC2
    'Mawai bent down in the canoe.'

Appendix D: Sample text — **333**

75. *Wènam*   *eláitam*   *yèfene*   *sènko.*
     wèn-am   e-lái-ta-m-Ø   yèfene   sènko.
     tree-ERG   α.2|3NSGP-miss-IPFV.ND-REM.DLT-2|3SGA   3NSG.GEN head
     'The tree missed their heads.'

76. *Yènd*   *fèn*   *kufngongèm*   *sèkwan.*
     yènd   fèn   k-ufngo-Ø-ang-m   sèkw-an
     1.ABS   laughing   β.ØP-start-PFV.ND-INC-1NSGS_A   canoe-LOC2
     'We started laughing in the canoe.'

77. *Yènngèm*   *kafnarngèm.*
     yèn-ngèm   k-a-fènar-Ø-ang-m
     α.1NSGS_P-go.ND   β.ØP-REFL-remove-PFV.ND-INC-1NSGS_A
     'We went on and got out.'

78. *Siogham*   *tènmifongè;*   *njamke*
     siogh-am   tèn-wifo-Ø-ang-è   njamke
     hunger-ERG   β.1NSGP-finish-PFV.ND-INC-2|3SGA   food
     enegham.
     e-negh-ta-m
     α.2|3NSGP-cook-IPFV-1NSGA
     'We got hungry; we cooked food.'

79. *Sèkw*   *tèyangèm*   *njamke*   *enegham.*
     sèkw   t-y-Ø-ang-m   njamke   e-negh-ta-m
     canoe   β.3SGP-beach-PFV.ND-INC-1NSGA   food   α.2|3NSGP-cook-IPFV-1NSGA
     'We beached the canoe and cooked food.'

80. *Enetam*   *wèikor*   *kuwanongèm.*
     e-ne-ta-m   wèikor   k-uwano-Ø-ang-m
     α.2|3NSGP-eat-IPFV-1NSGA   again   β.ØP-set.off-PFV.ND-INC-1NSGS_A
     'We ate and set off again.'

81. *Yènyote*   *yènyote* *yènyote* *yènyote.*
     yèn-yo-ta-e
     α.1NSGP-travel-IPFV.ND-2|3SGA
     'We travel and travel and travel.'

334 —— Appendix D: Sample text

82. *Yènd wèikor ámb mèngotuan kákmangèm.*
yènd wèikor ámb mèngotu-an k-ákèmè-Ø-ang-m
1.ABS again another village-LOC2 β.ØP-lie.down-PFV.ND-INC-1NSGS$_A$
'Again we slept in another village.'

83. *Yènkèmang wèikor yènamè*
yèn-kèmè-ang wèikor yènamè
α.1NSGS$_P$-be.lying.down-INC again from.here
kuwanongèm.
k-uwano-Ø-ang-m
β.ØP-set.off-PFV.ND-INC-1NSGS$_A$
'Having been sleeping, we set off again from here.'

84. *Yènyote yènyote eeeeeee yabun bout*
yèn-yo-ta-e eeeeeee yabun bout
α.1NSGP-travel-IPFV.ND-2|3SGA PROL big boat
tinjongèm.
t-winjo-Ø-ang-m
β.3SGP-see-PFV.ND-INC-1NSGS$_A$
'We travel and travel and then saw a big boat.'

85. *Yabun bout tènmorwèn.*
yabun bout t-nèmo-èrwèn
big boat β.3SGS$_P$-come.ND-REM.DUR.PA
'A big boat was coming.'

86. *Korayangèm:* "*Bout yabun*
k-oray-Ø-ang-m "Bout yabun
β.ØP-say-PFV.ND-INC-1NSGS$_A$ boat big
snufane; *ndenae*
s=n-ufar-ta-e ndernae
PROX2=α.ØP-arrive-IPFV.ND-2|3SGS$_A$ how
táfáto *sèkw?".*
t-wáfáto-Ø-Ø sèkw
β.3SGP-hide-PFV.ND-2|3SGA canoe
'We said: "A big boat is coming; how to get the canoe out of the way?"'

Appendix D: Sample text — **335**

87. *Wèkeye   yèta   káráfnawèm;*
Wèkeye   yèta   k-áráfár-ta-w-m
quick   ATT   β.ØP-steer-IPFV.ND-REM.DUR-1NSGS$_A$
'Quickly we tried to steer;

88. *yèta   káráfnawèm*   *witma   ndenae   fiya*
yèta   k-áráfár-ta-w-m   witma   ndenae   fiya
ATT   β.ØP-steer-IPFV.ND-REM.DUR-1NSGS$_A$ side   this.way   OBL1
yènnmormèn.   *yásemnamèm*
yèn-nèmo-èrmèn   y-wáseman-ta-m-m
α.1NSGS$_P$-come.ND-REM.DLT.PA   α.3SGP-change.course-IPFV.ND-REM.DLT-1NSGA
sèkw.
sèkw
canoe
'we tried to steer to the side where we had to come and changed the course of
the canoe.'

89. *Mámághèrae   tè   wèi   nèlawetamèm*
mámághèrae   tè   wèi   n-èlawè-ta-m-m
disorderly   PRF   now   α.ØP-go.inside-IPFV.ND-REM.DLT-1NSGS$_A$
wènjwènjot.
wènjwènj-ot
bushy.riverbank-ALL
'Haphazardly now we went into a bushy riverbank.'

90. *Yèta   kawerangèm;*   *fèn   kufngongèm.*
yèta   k-awerè-Ø-ang-m   fèn   k-ufngo-Ø-ang-m
ATT   β.ØP-hold.on-PFV.ND-INC-1NSGS$_A$ laughing   β.ØP-start-PFV.ND-INC-1NSGS$_A$
'We tried to hold on; we started laughing.'

91. *Sèkw   nawainde;*   *sóf   yau   sèkwaene.*
sèkw   n-awaing-ta-e   sóf   yau   sèkw-e-ne
canoe   α.ØP-pass-IPFV.ND-2|3SGS$_A$ wave   NEG   canoe-DAT-GEN
'The boat passed without making a wave.'[58]

---

58 The word *sèkw* 'canoe' is used here to refer the larger boat.

336 — Appendix D: Sample text

92. *Ngawaindangè*   *sèkw*   *yènamè.*
ng-awaing-tang-è   sèkw   yènamè
AND-pass-INC.IPFV-2|3SGS$_A$ canoe   from.here
'The boat passed on from here.'

93. *Ámb*   *mèngotuan*   *kètè*   *kákmangèm.*
ámb   mèngotu-an   kètè   k-ákèmè-Ø-ang-m
another   village-LOC2   there   β.ØP-lie.down-PFV.ND-INC-1NSGS$_A$
'We slept there in another village.'

94. *Yènè*   *akwan*   *wèi*   *kuwanongèm.*
yènè   akw-an   wèi   k-uwano-Ø-ang-m
DEM   morning-LOC2   now   β.ØP-set.off-PFV.ND-INC-1NSGS$_A$
'In the morning now we set off.'

95. *Yènyote*   *yènyote*   *tènfìrtangè*
yèn-yo-ta-e   tèn-fìr-tang-è
α.1NSGP-travel-IPFV.ND-2|3SGA   β.1NSGP-get.dark-INC.IPFV-2|3SGA
ámb   *mèngotuan*   *wèi*   *wanje*   *tèndon*   *yènè*   *mèngotu*
ámb   mèngotu-an   wèi   wanje   tèndo-n   yènè   mèngotu
another   village-LOC2   now   river   side-LOC1   DEM   village
yèm.
y-m
α.3SGS$_P$-COP.ND
'We travel and travel and it got dark on us at another village now, a village on
the river side.'

96. *Mawai*   *korayangè:*   "*Yosi*   *yèna*   *so*
Mawai   k-oray-Ø-ang-è   Yosi   yèna   so
Mawai   β.ØP-say-PFV.ND-INC-2|3SGS$_A$ Yoshie   here   FUT
kákmangèm   *kèmègh".*
k-ákèmè-Ø-ang-m   kèmègh".
β.ØP-lie.down-PFV.ND-INC-1NSGS$_A$ sleeping
'Mawai said: "Yoshie, we'll sleep here."'

97. "*Yau*   *yèna*   *so*   *yènkèm;*   *sèmbár*   *aligh*   *mè.*"
yau   yèna   so   yèn-kèmè   sèmbár   aligh   mè
NEG   here   FUT   α.1NSGS$_P$-be.lying.down   night   walking   CONT
'"No, we won't be sleeping here; we'll keep going in the night."'

Appendix D: Sample text — **337**

98. *Yèndfem*   *Mawai*   *yèngwifárèm*   *Dafifè.*
Yèndfem   Mawai   y-ng-wifár-Ø-m   Dafi-fè.
1NSG.ERG   Mawai   α.3SGP-AND-chase-IPFV.DU-1NSGA   Duffy-COM
'With Duffy, we two chased Mawai away.'

99. *Fá*   *ngangongè*   *ámbro*   *ghakèr*
fá   ng-ango-Ø-ang-è   ámb-ro   ghakèr
3.ABS   AND-return-PFV.ND-INC-2|3SGSA   other-RSTR   young.man
tawaufrongè:
ta-waufro-Ø-ang-è
β.2|3NSG-tell-PFV.ND-INC-2|3SGA
'He went back and told the other men:'

100. "*O* *yau,* *mènde*   *mèrès* *syèm*
o   yau, mènde   mèrès s=y-m
oh   NEG wanting   female PROX2=α.3SGSP-COP.ND
kuwanongè   *yau* *sái* *yènkèm,*
k-uwano-Ø-ang-è   yau sái yèn-kèmè
β.ØP-set.off-PFV.ND-INC-2|3SGSA   NEG INT α.1NSGSP-be.lying.down
fá   *siya* *sèmbár* *so* *kuwanongè."*
fá   siya sèmbár so k-uwano-Ø-ang-è
3.ABS   INT night FUT β.ØP-set.off-PFV.ND-INC-2|3SGSA
'"Oh no, this woman here wants to set off; she doesn't intend to sleep; she intends to set off in the night."'

101. *Wèikor* *kènangongè;*   "*Yosi*   *so*
wèikor k-n-ango-Ø-ang-è   Yosi   so
again   β.ØP-VEN-return-PFV.ND-INC-2|3SGSA Yoshie FUT
yènkèm."
yèn-kèmè
α.1NSGSP-be.lying.down
'He came back again; "Yoshie, we'll be sleeping."'

102. "*Yau* *so*   *kuwanongèm."*
yau so   k-uwano-Ø-ang-m
NEG FUT β.ØP-set.off-PFV.ND-INC-1NSGSA
'"No, we'll set off."'

338 —— Appendix D: Sample text

103. *Yènd   kutnangèn,*   *"Ei   mèinyotyo   ghakèr-ghakèr*
    yènd   k-utár-tang-èn   ei   mèinyotyo   ghakèr~ghakèr
    1.ABS   β.ØP-shout-INC.IPFV-1SGS_A   hey   all   young.man~PL
    kuflitangi!"
    k-ufèli-tang-i
    β.ØP-get.in-INC.IPFV-3PLS_A
    'I shouted: "Hey, all you men, get in!"'

104. *Ghakèr-ghakram   fès   tè   yètáyamènd*
    Ghakèr-ghakèr-am   fès   tè   y-táya-ta-m-nd
    young.man~PL-ERG   fire   PRF   α.3SGP-make.fire-IPFV.ND-REM.DLT-2|3NSGA
    kafalawèt.
    k-a-fala-ta-w-t
    β.ØP-REFL-warm-IPFV.ND-REM.DUR-2|3NSGS_A
    'The men had made a fire and were warming themselves.'

105. *Nuflitat*   *ghakèr-ghakèr;   sèmbár*
    n-ufèli-ta-t   ghakèr-ghakèr;   sèmbár
    α.ØP-get.in-IPFV.ND-2|3NSGS_A   young.man~PL   night
    kuwanongèm.
    k-uwano-Ø-ang-m
    β.ØP-set.off-PFV.ND-INC-1NSGS_A
    'The men got in (the canoe) and we set off in the night.'

106. *Yènyote*   *yènyote   yènyote   yènyote   yènyote*
    yèn-yo-ta-e
    α.1NSGP-travel-IPFV.ND-2|3SGA
    sèmbáran.
    sèmbár-an
    night-LOC2
    'We travel and travel and travel in the night.'

107. *Yènyote*   *sèkw   kètè*
    yèn-yo-ta-e   sèkw   kètè
    α.1NSGP-travel-IPFV.ND-2|3SGA   canoe   there
    kángsongè.
    k-ángso-Ø-ang-è.
    β.ØP-get.stuck-PFV.ND-INC-2|3SGS_A
    'We travelled and the canoe got stuck there.'

Appendix D: Sample text — **339**

108. *Yèta sèmbár yèlimndamèm*   *sèkw tèfene.*
yèta sèmbár y-limán-ta-m-m   sèkw tèfene.
ATT night α.3SGP-pull-IPFV.ND-REM.DLT-1NSGA canoe 1NSG.GEN
'We tried to pull out our canoe in the night.'

109. *Tosin stamorwèn*   *kwèfitaro.*
tosin s=tá-mo-èrwèn   kwèfitè-ro
torch PROX2=β.2|3NSGSP-COP.ND-REM.DUR.PA dim-RSTR
'Our torches were dim.'

110. *Witma wèi tafandawèm*   *bout*
witma wèi ta-fanda-ta-w-m   bout
side now β.2|3NSGP-look.at-IPFV.ND-REM.DUR-1NSGA boat
tánmorwèn   *yèfenjo diyáriafayo.*
tá-nèmo-èrwèn   yèfenjo diyári-afè-yo
β.2|3NSGSP-come.ND-REM.DUR.PA 3NSG.own light-COM-EXCL
'On the other side now, we saw boats coming with their own lights.'

111. *Núrtotamèm*   *túnat.*
n-úrto-ta-m-m   túnè-t
α.ØP-get.out-IPFV.ND-REM.DLT-1NSGSA bank-ALL
'We got out [of the water] onto the bank.'

112. *Ámb ndenè korayawèt:*   *"Yau njam*
ámb ndenè k-oray-ta-w-t   "Yau njam
some like.this β.ØP-say-IPFV.ND-REM.DUR-2|3NSGSA NEG when
kúrtongi   *sófam so*
k-úrto-Ø-ang-i   sóf-am so
β.ØP-get.out-PFV.ND-INC-2|3PLSA wave-ERG FUT
ewalitote."
e-walito-ta-e.
β.2|3NSGP-sink-IPFV.ND-2|3SGA
'Some (people) said: "If you don't get out [of the water], the wave will sink you."'

113. *Núrtotamèm*   *sèkw tambèn.*
n-úrto-ta-m-m   sèkw tambèn.
α.P-get.out-IPFV.ND-REM.DLT-1NSGSA canoe from
'We got out of the canoe.'

Appendix D: Sample text

114. *Kátáyawèm*   *fès,*
     k-átáya-ta-w-m   fès,
     β.ØP-make.fire-IPFV.ND-REM.DUR-1NSGS_A   fire
     nafalam.
     n-a-fala-ta-m
     α.ØP-REFL-warm-IPFV.ND-1NSGS_A
     'We were making a fire and warming ourselves.'

115. *Sèkw   yabun   nawaindat.*
     sèkw   yabun   n-awaing-ta-t
     canoe big   α.ØP-pass-IPFV.ND-2|3NSGS_A
     'The big boats passed by.'

116. *Yènd   kuflangèm*   *tèfenjo   sèkwèn.*
     yènd   k-ufèli-Ø-ang-m   tèfe-njo sèkw-n
     1.ABS β.ØP-got.in-PFV.ND-INC-1NSGS_A   1NSG-own canoe-LOC1
     'We got in our own canoe.'

117. *Yènyote*   *yènyote yènyote yènyote yènyote yènyote*
     yèn-yo-ta-e
     α.1NSGP-travel-IPFV.ND-2|3SGA
     'We travel and travel and travel and travel.'

118. *Kètè   tènngèfáuèrngè*   *Kiunga   kaka.*
     kètè   tèn-ng-fáuèr-Ø-ang-è   Kiunga   kaka
     there β.1NSGP-dawn.on-PFV.ND-INC-2|3SGA   Kiunga   near
     'The day dawned on us there near Kiunga.'

119. *Kètè   fès   yètáyam*   *yènè   tèndon*
     kètè   fès   y-táya-ta-m   yènè   tèndo-n
     there fire α.3SGP-make.fire-IPFV.ND-1NSGA   DEM   side-LOC1
     nafalam.
     n-a-fala-ta-m
     α.ØP-REFL-warm-IPFV.ND-1NSGS_A
     'There we make a fire on the riverside and warm ourselves.'

Appendix D: Sample text — **341**

120. *Wèi yabun sèkw nuwanongè.*
wèi yabun sèkw n-uwano-Ø-ang-è
now big canoe α.ØP-set.off-PFV.ND-INC-2|3SGS_A
'Now a big boat set off.'

121. Yèndo Mawai tèmndan: "Yabun mènè yènè
Yèndo Mawai t-mèndè-Ø-èn yabun mènè yènè
1SG.ERG Mawai β.3SGP-tell-PFV.ND-1SGA big whachamacallit DEM
terang mènat so nu
t-werè-Ø-ang-Ø mènat so nu
β.3SGP-hold-PFV.ND-INC-2|3SGA so.that FUT fuel
tènmaramangi sèkw tèfnáram."
tèn-wa-ramè-Ø-ang-i sèkw tèfnár-am
β.1NSGP-APP-give-PFV.ND-INC-2|3PLA canoe practitioner-ERG
'I told Mawai: "Hold up the big whachamacallit [fuel container] so that the boat crew will give us fuel."'

122. *Yèta yerete tukmè yèta yá*
yèta y-werè-ta-e tuk-mè yèta yá
ATT α.3SGP-hold-IPFV.ND-2|3SGA upwards-PERL ATT 3SG.DAT
nutne yèta yau.
n-utár-ta-e yèta yau
α.ØP-shout-IPFV.ND-2|3SGS_A ATT NEG
'He tried to hold it up and tried to shout to him, but to no avail.'

123. *Sèkw árèm tènèmndangi: "Yau so*
sèkw ár-m tèn-mèndè-Ø-ang-i: yau so
canoe man-ERG β.1NSGP-tell-PFV.ND-INC-2|3PLA no FUT
tawaramangèm nu fèfeyot."
ta-wa-ramè-Ø-ang-m nu fèfeyot."
β.2|3NSGP-APP-give-PFV.ND-INC-1NSGA fuel 2NSG.PURP
'The men on the boat told us: "No, we won't give you fuel."'

124. *Nuflitam,* *kuwanongèm*
 n-ufèli-ta-m k-uwano-Ø-ang-m
 α.ØP-get.in-IPFV.ND-1NSGS_A β.ØP-set.off-PFV.ND-INC-1NSGS_A
 yènyote *ee* *Kiunga*
 yèn-yo-ta-e ee Kiunga
 α.1NSGP-travel-IPFV.ND-2|3SGA PROL Kiunga
 kufarngèm, *mer* *merro.*
 k-ufar-Ø-ang-m mer mer-ro
 β.ØP-arrive-PFV.ND-INC-1NSGS_A good good-RSTR
 'We get in, set off and travel until we arrived at Kiunga – good, really good.'

125. *Tane* *si* *yènaf* *tèyáto.*
 tane si yèna=f t-yáto
 1SG.GEN story here=PROX1 β.3SGS_P-be.finished
 'Here my story is just finished.'

References

Ayres, Mary C. 1983. *This side, that side: Locality and exogamous group definition in Morehead area, Southwestern Papua.* PhD dissertation, University of Chicago.

Bybee, Joan, Revere Perkins & William Pagliuca. 1994. *The Evolution of Grammar: Tense, Aspect and Modality in the Languages of the World.* Chicago/London: University of Chicago Press.

Carroll, Alice, Nicholas Evans, Darja Hoenigman & Lila San Roque. 2009. The family problems picture task. Designed for use by the Social Cognition and Language Project. A collaboration of The Australian National University, Griffith University, University of Melbourne and the Max Planck Institute for Psycholinguistics.

Carroll, Matthew J. 2016. *The Ngkolmpu language (with special reference to distributed exponence).* PhD thesis, Australian National University.

Carroll, Matthew J. 2020. The morphology of Yam languages. Reprinted from the *Oxford Research Encyclopedia, Linguistics* (oxfordre.com/linguistics). Oxford University Press, USA. Available online at https://openresearch-repository.anu.edu.au/bitstream/1885/220610/1/01_Carroll_The_morphology_of_Yam_2020.pdf, accessed 19 August 2021.

Comrie, Bernard. 1985. *Tense.* Cambridge: Cambridge University Press.

Comrie, Bernard. 1998. Rethinking the typology of relative clauses. *Language Design: Journal of Theoretical and Experimental Linguistics* 1: 59–85.

Comrie, Bernard. 2006. Syntactic typology: Just how exotic ARE European-type relative clauses? In Ricardo Mairal & Juana Gil (eds.), *Linguistic Universals*, 130–154. Cambridge: Cambridge University Press.

Comrie, Bernard & Tania Kuteva. 2013a. Relativization on subjects. In Matthew S. Dryer & Martin Haspelmath (eds.), *The World Atlas of Language Structures Online*. Leipzig: Max Planck Institute for Evolutionary Anthropology. (Available online at http://wals.info/chapter/122, accessed 6 April 2018).

Comrie, Bernard & Tania Kuteva. 2013b. Relativization on obliques. In Matthew S. Dryer & Martin Haspelmath (eds.), *The World Atlas of Language Structures Online*. Leipzig: Max Planck Institute for Evolutionary Anthropology. (Available online at http://wals.info/chapter/123, accessed 6 April 2018).

Corbett, Greville G. 2000. *Number.* Cambridge: Cambridge University Press.

Creissels, Denis. 2008. Remarks on split intransitivity and fluid intransitivity. In Oliver Bonami & Patricia Cabredo Hofherr (eds.), *Empirical Issues in Syntax and Semantics 7*, 139–168. Paris: CSSP (Colloque de Syntaxe et Sémantique à Paris).

Croft, William. 2003. *Typology and Universals* (second edition). Cambridge: Cambridge University Press.

Croft, William. 2004. Typology and universals: From implicational universals to multidimensional scaling. Tutorial given at Northwestern University. (Downloaded from http://free-tutorial-for.me/tutorial-for-northwestern/page6.html on 17 May 2012).

Crowley, Terry, John Lynch, Jeff Siegel & Julie Piau. 1995. *The Design of Language: An Introduction to Descriptive Linguistics.* Auckland: Longman Paul.

Dahl, Östen. 1985. *Tense and Aspect Systems.* Oxford: Blackwell.

Dahl, Östen & Viveka Velupillai. 2013. Perfective/imperfective aspect. In Matthew S. Dryer & Martin Haspelmath (eds.), *The World Atlas of Language Structures Online*. Leipzig: Max Planck Institute for Evolutionary Anthropology. (Available online at http://wals.info/chapter/65, accessed on 20 August 2021).

Dixon, R. M. W. 1979. Ergativity. *Language* 55. 59–138.

https://doi.org/10.1515/9783111077017-013

344 —— References

Döhler, Christian. 2018. *A Grammar of Komnzo*. (Studies in Diversity Linguistics 22). Berlin: Language Science Press.

Donohue, Mark. 2008. Complexities with restricted numeral systems. *Linguistic Typology* 12. 423–429.

Donohue, Mark & Søren Wichmann (eds.). 2008. *The Typology of Semantic Alignment*. Oxford: Oxford University Press.

ELAN (Version 6.2) [Computer software]. (2021). Nijmegen: Max Planck Institute for Psycholinguistics, The Language Archive. Retrieved from https://archive.mpi.nl/tla/elan, accessed on 14 September 2021.

Evans, Nicholas. 2009. Two pus one makes thirteen: Senary numerals in the Morehead- Maro region. *Linguistic Typology* 13. 321–335.

Evans, Nicholas. 2012a. Even more diverse than we had thought: The multiplicity of Trans-Fly languages. In Nicholas Evans & Marian Klamer (eds.), *Melanesian Languages on the Edge of Asia: Challenges for the 21st Century* (Language Documentation & Conservation Special Publication No. 5), 109–149. Manoa: University of Hawai'i Press.

Evans, Nicholas. 2012b. Nen assentives and the problem of dyadic parallelisms. In Andrea C. Schalley (ed.), *Practical Theories and Empirical Practice: Facets of a Complex Interaction*, 159–183. Amsterdam; Philadelphia: John Benjamins.

Evans, Nicholas. 2014. Positional verbs in Nen. *Oceanic Linguistics* 53. 225–255.

Evans, Nicholas. 2015a. Inflection in Nen. In Matthew Baerman (ed.), *The Oxford Handbook of Inflection*, 543–575. Oxford: Oxford University Press.

Evans, Nicholas. 2015b. Valency in Nen. In Andrej L. Malchukov & Bernard Comrie (eds.), *Valency Classes in the World's Languages: Volume 2: Case Studies from Austronesia, the Pacific, the Americas, and Theoretical Outlook*, 1049–1096. Berlin: New York: Walter de Gruyter.

Evans, Nicholas. 2017. Quantification in Nen. In Denis Paperno & Edward Keenan (eds.), *Handbook of Quantification in Natural Language, Volume II*, 571–607. New York: Springer.

Evans, Nicholas. 2019a. *Nen dictionary*. Dictionaria 8. 1–5005. https://dictionaria.clld.org/contributions/nen, accessed on 15 September 2021.

Evans, Nicholas. 2019b. Waiting for the word: Distributed deponency and the semantic interpretation of number in the Nen verb. In Matthew Baerman, Oliver Bond & Andrew Hippisley (eds), *Morphological Perspectives: Papers in Honour of Greville G. Corbett*, 100–123. Edinburgh: Edinburgh University Press.

Evans, Nicholas, Wayan Arka, Matthew Carroll, Yun Jung Choi, Christian Döhler, Volker Gast, Eri Kashima, Emil Mittag, Bruno Olsson, Kyla Quinn, Dineke Schokkin, Philip Tama, Charlotte van Tongeren & Jeff Siegel. 2017. The languages of Southern New Guinea. In Bill Palmer (ed.), *The Languages and Linguistics of New Guinea: A Comprehensive Guide*, 641–774. Berlin: Mouton de Gruyter.

Evans, Nicholas & Stephen C. Levinson. 2009. The myth of language universals: Language diversity and its importance for cognitive science. *Behavioral and Brain Sciences* 32. 429–492

Evans, Nicholas & Julia C. Miller. 2016. Nen. *Journal of the International Phonetic Association* 46. 331–349.

Greenberg, Joseph H. 1963. Some universals of grammar with particular reference to the order of meaningful elements. In Joseph Greenberg (ed.), *Universals of Language*, 73–113. London: MIT Press.

Greenberg, Joseph H. 1969. Language universals: A research frontier. *Science* 166. 473–478.

Hammarström, Harald. 2009. Whence the Kanum base–6 numeral system? *Linguistic Typology* 13. 305–319.

Haspelmath, Martin. 2001. The European linguistic area: Standard Average European. In Martin Haspelmath, Wulf Oesterreicher & Wolfgang Raible (eds.), *Language Typology and Language Universals, Handbücher zur Sprach- und Kommunikationswissenschaft*, 1492–1510. Berlin: Mouton de Gruyter.

Haspelmath, Martin. 2003. The geometry of grammatical meaning: Semantic maps and cross-linguistic comparison. In Michael Tomasello (ed.), *The New Psychology of Language, Vol. II*, 211–242. Mahwah, NJ: Lawrence Erlbaum.

Janic, Katarzyna. 2013. The Slavonic languages and the development of the antipassive marker. In Irina Kor Chahine (ed.), *Current Studies in Slavic Linguistics*, 61–74. Amsterdam: John Benjamins.

Janic, Katarzyna. 2016. On the reflexive-antipassive polysemy: Typological convergence from unrelated languages. In Nicholas Rolle, Jeremy Steffman & John Sylak-Glassman (eds.), *Proceedings of the Thirty-sixth Annual Meeting of the Berkeley Linguistics Society, February 6–7, 2010*, 158–173. Berkeley: Berkeley Linguistics Society. (Available online at https://escholarship.org/uc/item/3174s5w2, accessed 19 August 2021).

Kashima, Eri. 2020. *Language in my mouth: Linguistic variation in the Nmbo speech community of southern New Guinea.* PhD thesis. Australian National University.

Kemmer, Suzanne. 1993. *The Middle Voice*. Amsterdam: John Benjamins.

Kunze, Rui. 2011. Naming playfully: Chinese netizen's carnivalesque practice of language. (English translation of Karnevaleske Sprachpraxis chinesischer Netzbürger. In Christian Soffel, Daniel Leese & Marc Nürnberger (eds.), *Sprache und Wirklichkeit in China*, 289–302. Wiesbaden: Harrassowitz Verlag. (Available online at https://www.academia.edu/1157838/Naming_Playfully_Chinese_Netizens_Carnivalesque_Practice_of_Language, accessed on 11 September 2021.)

Lichtenberk, Frantisek. 2000. Inclusory pronominals. *Oceanic Linguistics* 39(1). 1–32.

Malchukov, Andrej. 2015. Valency classes and alternations: Parameters of variation. In Andrej Malchukov & Bernard Comrie (eds.), *Valency Classes in the World's Languages, Volume 1: Introducing the Framework, and Case Studies from Africa and Eurasia*, 73–130. Berlin: de Gruyter.

Merlan, Francesca. 1985. Split intransitivity: Functional oppositions in intransitive inflection. In Johanna Nichols & Anthony C. Woodbury (eds.), *Grammar Inside and Outside the Clause: Approaches to Theory from the Field*, 324–362. Cambridge: Cambridge University Press.

Miller, Jim. 2006. Bulgarian. In Ken Brown (ed.), *Encyclopedia of Language and Linguistics* (second edition), 149–151. Oxford: Elsevier.

Nedjalkov, Vladimir. 2007. *Reciprocal Constructions* [Typological Studies in Language 71]. Amsterdam: John Benjamins.

Parry, Mair. 1998. The reinterpretation of the reflexive in Piedmontese: 'Impersonal' SE constructions. *Transactions of the Philological Society* 96. 63–116.

Pawley, Andrew, Simon Peter Gi, Ian Saem Majnep & John Kias. 2000. Hunger acts on me: The grammar and semantics of bodily and mental process expressions in Kalam. In Videa P. De Guzman & Byron Bender (eds.), *Grammatical Analysis: Morphology, Syntax, and Semantics: Studies in honor of Stanley Starosta*, 153–185. (*Oceanic Linguistics Special Publications* 29.) Honolulu: University of Hawai'i Press.

Plank, Frans. 2009. Senary summary so far. *Linguistic Typology* 13. 337–345.

Polinsky, Maria. 2013. Antipassive constructions. In Matthew S. Dryer & Martin Haspelmath (eds.), *The World Atlas of Language Structures Online*. Leipzig: Max Planck Institute for Evolutionary Anthropology. (Available online at http://wals.info/chapter/108, accessed 19 August 2021).

Price, Mavis. 2000. Alphabet development workshop for sociolinguistic orthography [Nama]. ms. Ukarumpa: Summer Institute of Linguistics.

San Roque, Lila, Alan Rumsey, Lauren Gawne, Stef Spronck, Darja Hoenigman, Alice Carroll, Julia Miller & Nicholas Evans. 2012. Getting the story straight: language fieldwork using a narrative problem-solving task. *Language Documentation and Conservation* 6. 134–173.

Siegel, Jeff. 1996. *Vernacular Education in the South Pacific* (International Development Issues No.45). Canberra: Australian Agency for International Development.

Siegel, Jeff. 1997. Using a pidgin language in formal education: Help or hindrance? *Applied Linguistics* 18(1). 86–100.

Siegel, Jeff. 2014. The morphology of tense and aspect in Nama, a Papuan language of southern New Guinea. *Open Linguistics* 1. 211–231.

Siegel, Jeff. 2017. Transitive and intransitive verbs in Nama, a Papuan language of southern New Guinea. *Oceanic Linguistics* 56. 123–142.

Siegel, Jeff. 2019. The Relative Pronoun strategy: New data from southern New Guinea. *Studies in Language* 43(4). 997–1014.

Smith, Anne-Marie. 1978. *The Papua New Guinea Dialect of English*. Port Moresby: University of Papua New Guinea (ERU Research Report no. 25).

Smith, Anne-Marie. 1988. The use of aspect in Papua New Guinea English. In *Asian Pacific Papers 10*, 109–134. Melbourne: Applied Linguistics Association of Australia.

Tapari, Budai. 1995. Development or deterioration: Socio-economic change in the Morehead District. Ok-Fly Social Monitoring Report no.12. (Available online at https://crawford.anu.edu.au/rmap/archive/Ok-Fly_social_monitoring/Ofsmp12-Tapari1995-development-in-the-Morehead-District.pdf, accessed 19 August 2021).

Williams, F. E. 1936. *Papuans of the Trans-Fly*. Oxford: Clarendon.

Wonderly, William L., Lorna F. Gibson & Paul L. Kirk. 1954. Number in Kiowa: Nouns, demonstratives, and adjectives. *International Journal of American Linguistics* 20(1). 1–7.

Youngdu Church. 2018. *Mer Si Yesuenemèn Makam Yèfarotam* (Good talking about Jesus that Mark wrote) in Nama [nmx] language of Papua New Guinea.

Index

abilitative (quasi-modal) 203–204
ablative (case) 71–73, 86
absolutive (case) 32, 56–57, 147
action nominals 77, 87, 95, 175–176
adjectival functions 85–87, 171–174, 175, 205, 207, 208, 211
adverbial clauses:
– concessive 273–275
– conditional 277–279
– of place 272–273
– of purpose 275–277
– of reason 279–280
– of time 267–272
adverbial functions 60, 64, 67, 70, 97, 176–177, 182, 187–189, 207
adverbial subordinating conjunctions 98, 180
adverbs 97, 180, 204, 205, 206, 207, 208, 210–214, 218–219, 232–234, 256, 286, 289, 290, 293, 295, 296
affricates 35–36
agentless transitive verbs 223–224, 264
alignment 30, 31
allative (case) 74–77, 78–79, 86, 116, 119, 276
andative prefix 112–115, 119, 155–156, 245–246, 249–250
anticausative 90, 157–159, 161, 162, 163
antipassive 156–157, 162, 163
applicative prefix 57, 110, 116–120, 121, 154–155, 247, 248
apprehensive constructions 198, 280
aspect (verbal). See imperfective, perfective, durative, delimited, irrealis, inceptive imperfective
aspect markers (periphrastic) 192–196, 200–201, 202–203
associative (case) 71, 77, 79–82, 86
attemptive (modal) 201–202, 274–275
attributive (case) 85–87
attributive (nominal) 95, 99, 281, 286
autobenefactive imperatives 250
autobenefactive prefix 120–121, 250
autocausative 160–161, 163

benefactive imperatives 247–250
beneficiary argument 57, 107, 116, 117, 119, 199, 120, 248

case suffixes 29, 54–88, 97–98, 105–106
clause chaining 266–267
comitative (case) 63–65, 86, 92, 174, 282–283
comparative constructions 256
complement clauses 287–295
complementiser 287–290
complex sentences 267–298
compound sentences 265–267
conjunctive adverbs 98, 259–260
consonant clusters 45–46
consonants 34–39, 45–46
continuative (aspect marker) 195–196
contrastive adverbial clause 273–275
coordinating conjunctions 224–225, 265–266
copula 151–156, 217, 246–247, 263, 264, 295–296, 302
core argument indexing affixes 29–31, 107–111
counterfactual clauses 191, 278–279
cultural practices 7–12
current tense:
– imperfective 29, 30, 107, 108, 109, 122, 127–133, 146–147, 300
– imperfective inceptive 144, 147
– perfective (inceptive) 135–138, 144, 148, 149, 151, 154

data sources 25–28
dative (case) 53, 56, 57–59, 60, 72, 73, 75, 79, 80, 81, 82, 84, 86, 99, 100, 102, 116
declarative sentences 226
deictic prefixes 110, 112–115, 119, 121, 245–246, 249–250
delimited (aspect) 109, 133–134, 150–151, 152–154, 169
demonstrative 179–180, 212
demonstrative pronoun 180
deponents 162, 163
derivational nominal affixes 98–100

https://doi.org/10.1515/9783111077017-014

348 —— Index

dialects 2, 309
diminutive 103, 263
diphthongs 42–43
discourse markers 205, 213, 254–257, 258–262, 271
discourse particles 229, 254, 256–257, 260–262
distributed exponence 109, 122
ditransitive 57, 107, 116, 120
dual 79, 122–123, 126, 127, 129–132, 134, 135, 139–140, 141, 144, 145, 148, 151, 153, 154, 155–156, 168, 225, 234–236, 238, 239–241, 242, 248–249, 251, 261, 262–264, 295, 301–304, 305
dual argument suffix 127, 129–130, 131, 132, 240, 264
dual imperative suffix 240
dual suffix (with P-aligned intransitives) 148, 151, 168, 262
dubitative (modal) 198–199
durative (aspect) 109, 134–135, 150–151, 152–154, 169, 235

emphasis 205–206, 209, 258
epenthesis 47–48
ergative (case) 30, 32, 51, 54–57, 86, 120, 178, 190, 229, 283–284
evaluative responses 254
evidential (modal) 201
exclamative sentences 257–258
exclusive suffix 89–93, 99, 184, 189
experiencer object constructions. See P-focussed constructions

fieldwork 23–25
focus marking clauses 295–298
food 8–9
fricatives 35–36
future imperatives 241–244
future tense marker 129, 140, 190–191, 253fn

geminate consonants 39
gender roles 7
genitive (case) 60–63, 64, 82, 84, 86
glides 38
grammatical number 301–304
greater plural 304–305, 308
greetings 252

hortative 207

immediacy adverb 212–213, 217, 256–257, 295–296
immediate imperatives 140, 234–241
imperative sentences 133, 140, 144, 234–251
imperfective 30, 49, 107, 108, 109, 121–123, 123–127, 128–135, 136, 138, 146–147, 236–239, 240–243, 247–248, 301–302
implicational hierarchy 302, 303
inceptive 123, 135–140, 142–143, 144, 148, 149, 150, 152, 154, 235, 263, 301
inceptive imperfective 142, 144–146, 146–147, 236, 239, 248, 250, 283, 301–302
inchoate (aspect marker) 192–195
inchoative construction 106, 115, 119, 175–176
inclusory construction 225
indefinite pronouns 182–183, 231–232, 294–295
independent pronouns 56–57
indirect speech 288, 290–291
instrumental (case) 60, 210
intentional (modal) 199–200, 277
interjections 252–254, 257–258
internally headed relative clauses 286
interrogative adverbs 232–234, 286, 293
interrogative pronouns 229–232, 291–295
interrogative sentences 227–234
intransitive verbs:
– A-aligned intransitives 30, 105, 107–111, 112–147, 156–164, 166–167, 220
– P-aligned intransitives 30, 31, 105, 107–111, 147–156, 164–169, 220
investigative (quasi-modal) 207–208
irrealis 132, 146, 198, 201, 204, 206, 207, 209, 236, 242, 278

Komnzo 3, 4, 17, 37fn, 41, 44, 127, 225, 253fn, 301, 304, 309fn

landscape 4–7
language use 20–21
large plural 262, 264, 304–305, 308
ligature 59
liquids 38
locational adverbs 211
locational proximal clitic 45, 216–217, 296
locative (case) 67–71, 86, 210, 268

marriage 7–8
mental predicates 287, 289
minor sentences 251–257
mirative proclitic 217–18, 264
modals (modality markers) 196–203
Morehead-Upper Maro family. See Yam languages
morphological discord 262–264, 304–308
morphological processes 29–30, 51
morphological subclasses (of verbs) 112
morphophonemic changes (with verbs) 30, 49–50, 106, 109–111, 114, 116, 123–127, 139, 142
multi-verb constructions 225–226

Namat 2, 44, 86, 199fn
Nambo 3, 302, 300, 304
nasals 36–37
negative imperatives 244–245
negative sentences 226–227
Nen 3, 42, 108, 122, 302, 304, 305, 308
Ngkolmpu 3, 41, 300
nominal morphology 29, 51–104
nominal phrase structure 190
nominal subclasses 168–189
nominalised verbs 75, 86, 87, 95, 105–107, 175, 268, 276, 280
nominalising suffix 105–107, 175
nondual 122–123, 123—125, 127, 128, 133, 134, 137, 139, 141, 142, 143, 144, 145, 146, 148, 150, 151–152, 156–157, 217, 238, 264, 295, 302–304
noun adjunct 171–174
nouns 168–178
number marking suffixes 51–53
numbers 13, 16
numerals 186–187

obligative (quasi-modal) 204–207
obligative continuative (modal) 200–201
orientational nouns 73, 91, 176–178
originative (case) 82–84, 86, 280
orthography 1, 7fn, 28, 44

P-focussed constructions 221–225
paucal 52–53, 57, 62, 72, 75, 82
perfect (aspect marker) 192

perfective 121–123, 135–143, 146–147, 148, 149, 150, 151, 153, 235–237, 238, 239, 240, 241–242, 262–263, 301–302
perlative (case) 73–74, 86, 282–283
permissive (modal) 200
personal names 55, 64, 66, 72, 79, 80, 86, 102, 175
phonology 34–50
phonotactics 45–50
phrasal coordination 224–225
picture task 27, 89fn, 253, 256
PNG English 21
Police Motu 20–21
politeness expressions 252
possessive prefixing 100–102
possibility (modal) 196–197
potential (modal) 197–198
privative (case) 65–67, 86
pronouns 32, 56–57, 58, 59, 62–63, 65, 67, 71, 73, 77, 79, 82, 84, 91, 99–100, 100–102, 122–123, 147, 178, 180, 182–183, 190, 229—232, 280–286, 291–295, 299–300, 301
proximal clitics 45, 194, 215–217, 296
punctual 121, 135, 136, 141–142, 146–147, 150, 235, 236, 248, 301
purposive (case) 76, 77–79, 86, 282–283

quantifiers 181–187, 190
quasi-modals 203–209
question marker 261
question word questions 229–234
quotatives/quotations 289, 290–291

recent tense:
- imperfective 29, 107, 108, 109, 122, 127–133, 146–147, 236
- inceptive imperfective 144, 146–147, 236
- perfective (inceptive) 135–138, 144, 146–147, 148, 149, 151, 154, 235
recipient argument 57, 107, 116, 117, 119, 199, 120, 248
reciprocal 107, 112, 115–116, 159–160, 162, 163–164
recordings 22, 26–28, 311, 312
reduplication 29, 44, 46, 102–104
reflexive 90, 107, 112, 115–116, 120, 121, 159–160, 161, 162, 163–164

350 —— Index

relationship nouns 100, 101, 102, 175
relative adverbs 286–287
relative clauses 280–287, 299–300
relative pronouns 280–286, 291– 295, 299–300
religion 10, 12
remote tense (P-aligned intransitive
    verbs) 150–151, 152–154, 169
remote tense (transitive and A-aligned
    intransitive verbs):
– imperfective delimited 109, 133–134, 146–147
– imperfective durative 109, 134–135, 146–147
– inceptive imperfective 146–146, 146–147
– perfective inceptive 142–143, 146–147
– perfective punctual 141–142, 146–147
restrictive suffix 93–95, 99, 180, 183, 184,
    187, 189

sections 7
semblative (case) 85
senary counting system 13, 16
sentences. See simple sentences, compound
    sentence, complex sentences
simple sentences 220–264
sister exchange 7–8
spatial adverbs 179, 212
split ergativity 32
split intransitivity 31, 147
stative nominals 87, 175–176
stative verbs 147, 164–166, 304
stops 35
stress 46–47
subordinating conjunctions 267–280, 289
superlative constructions 266fn
syllable structure 45
syncope 47–50

TAM markers 190–209
temporal (case) 87–88
temporal adverbs 213–214
temporal nominals 187–189, 272
temporal nouns 91, 176–178
temporal proximal clitic 45, 194, 215–216
tense (verbal). See current tense, recent tense,
    remote tense
tense marker (periphrastic) 190–191, 202–203
third person imperatives 250–251
Tok Pisin 21, 25, 100, 214
transitive verbs 30, 107–111, 112–147, 156–169,
    220, 221, 223–224
transitiviser/transitive prefix 119–120, 120–121
tribes 7
typological implications 299–308

valency reduction:
– A-aligned intransitive 156–164
– P-aligned intransitive 164–169
validative (quasi-modal) 208–209
venitive prefix 112–115, 119, 155, 245–246,
    249–250
verb phrase structure 218–219
verbal morphology 29, 105–169, 215–218
verbs. See transitive verbs, intransitive verbs
vowels 39–43, 47–50

word formation 92, 96, 97–98
word order 31, 220–221, 222

Yam languages 2, 3, 13, 41–42, 47, 109, 212fn,
    289, 299, 300
yams 8, 12–17
yes/no questions 227–229